The Jahn–Teller Effect and Vibronic Interactions in Modern Chemistry

MODERN INORGANIC CHEMISTRY

Series Editor: John P. Fackler, Jr.
Texas A&M University

METAL INTERACTIONS WITH BORON CLUSTERS
Edited by Russell N. Grimes

HOMOGENEOUS CATALYSIS
WITH METAL PHOSPHINE COMPLEXES
Edited by Louis H. Pignolet

THE JAHN–TELLER EFFECT AND
VIBRONIC INTERACTIONS IN MODERN CHEMISTRY
I. B. Bersuker

The Jahn–Teller Effect and Vibronic Interactions in Modern Chemistry

I. B. Bersuker
Institute of Chemistry
Moldavian Academy of Sciences
Kishinev, USSR

Plenum Press • New York and London

Library of Congress Cataloging in Publication Data

Bersuker, I. B. (Isaak Borisovich)
The Jahn–Teller effect and vibronic interactions in modern chemistry.

(Modern inorganic chemistry)
Bibliography: p.
Includes index.
1. Jahn–Teller effect. 2. Chemical reactions. I. Title. II. Series.
QD461.B46 1983 541′.22 83-16070

DOI: 10.1007/978-1-4613-2653-3

MyCopy version of the original edition 1984

A Division of Plenum Publishing Corporation
233 Spring Street, New York, N.Y. 10013

Preface

The first half of the title of this book may delude the uninitiated reader. The term "Jahn–Teller effect," taken literally, refers to a special *effect* inherent in particular molecular systems. Actually, this term implies a new *approach* to the general problem of correlations between the structure and properties of any molecular polyatomic system, including solids. Just such a new approach, or concept (in some sense, a new outlook or even a new way of thinking), which leads not to one special effect but to a series of different effects and laws, is embodied in the many ($\sim$4000) studies devoted to the investigation and application of the Jahn–Teller effect. The term "vibronic interactions" seems to be most appropriate to the new concept, and this explains the origin of the second half of the title.

The primary objective of this book is to present a systematic development of the concept of vibronic interactions and its applications, and to illustrate its possibilities and significance in modern chemistry. In the first three chapters (covering about one-third of the book) the theoretical background of the vibronic concept and Jahn–Teller effect is given. The basic ideas are illustrated fully, although a comprehensive presentation of the theory with all related mathematical deductions is beyond the scope of this book. In the last three chapters the applications of theory to spectroscopy, stereochemistry and crystal chemistry, reactivity, and catalysis, are illustrated by a series of effects and laws. Greatest consideration is given to those applications which initiate new trends in appropriate areas of investigation.

Both the content and structure of this book are in principle different from those of monographs in chemistry devoted to a definite class of chemical compounds or a certain method of research. The concept of vibronic interactions, aimed at improving the bridge between the structure and properties of substances, is to a certain degree concerned with all kinds of molecular systems and solids, and affects all methods of investigation employed in chemistry. This universality of aims and methods is a characteristic feature of the vibronic approach to the problems in modern chemistry demonstrated in this book. Joined by a common concept, the various phenomena arising in different systems (and observable by different methods of study) may serve as a source of new ideas: the reader may find quite unexpected similarities between very different systems united by the concept of vibronic interactions.

The author has tried to present the material in a way suitable for a large number of chemists and physicists (such as scientific workers, university professors and students, and high school teachers), as well as engineers engaged in investigating the structure and properties of matter. For this purpose most attention is paid to the physical meaning of the results. A limited number of mathematical expressions is analyzed briefly (mainly in the first three chapters) in order to maintain the quantitative level of the theory. More detailed deductions can be found easily in the original papers cited in the References. On the other hand, applications of the theory are given in such a form that they can be used without a detailed study of the theory. Nevertheless, for a full understanding of this book the reader is expected to know the basic ideas of quantum mechanics and the theory of symmetry.

I should like to express my thanks to my co-workers and colleagues, especially to S. A. Borshch, S. S. Budnikov, A. S. Dimoglo, M. D. Kaplan, I. Ya. Ogurtsov, Yu. E. Perlin, V. Z. Polinger, Yu. B. Rosenfeld, B. S. Tsukerblat, and B. G. Vekhter, with whom research on, and discussion of, vibronic phenomena (with most of them during a long period of more than twenty years) made possible the general presentation of the problem treated in this book. I am grateful to A. Abragam, J. Ammeter, M. Bacci, C. J. Ballhausen, F. Basolo, C. A. Bates, H. Bill, L. A. Boatner, J. Brickmann, L. S. Cederbaum, R. E. Coffman, T. M. Dunn, J. Duran, R. Englman, J. Gazo, G. L. Hofacker, B. Hoffman, O. Kahn, H. Köppel, S. Kirschner, D. I. Khomskii, R. Lacroix, M. Lambert, S. Leach, A. D. Liehr, K. A. Müller, M. C. M. O'Brien, R. Pearson, O. E. Polansky, A. Ranfagni, M. Ratner, D. Reinen, J. S. Slonczewski, M. Thomas, M. Wagner, and their co-workers for cooperation and discussions. Many thanks are due to Professor J. P. Fackler, Jr., who read the manuscript thoroughly and made important suggestions. Also I apologize to those readers whose contribution to works on vibronic problems are not cited here. A comprehensive list of references can be found in the specially prepared Bibliographic Review.*

I. B. Bersuker

Kishinev

* *The Jahn–Teller Effect: A Bibliographic Review,* I. B. Bersuker, IFI/Plenum, New York, 1984.

Contents

Mathematical Notation

Symbols used more than once, but with another meaning, have local concern only in the corresponding chapter or section indicated in parentheses

A, B	— rotational constants
A	— hyperfine constants (§4.1)
C_{ij}	— LCAO coefficients
$C^{(i)}$	— anharmonicity correction coefficients (§6.3)
D	— barrier height
d	— barrier width
E	— doubly degenerate representation (term)
E	— energy
E_{JT}	— JT stabilization energy
e	— strain components
e^*	— effective charge (§5.1)
$\mathscr{E}$	— electric field intensity
F	— linear vibronic constant
$F(\Omega)$	— form function for spectroscopic transition band shapes (§4.3)
f	— linear orbital vibronic constant
f	— angular-dependence function in ESR spectra (§4.1)
$f_{\theta,\varepsilon}$	— molecular field intensity (§5.2)
G	— quadratic vibronic constant
G_{ij}	— cross-section of light scattering (§4.2)
g	— quadratic orbital vibronic constant (§1.3)
g	— EPR g-factor (§4.1)
g^I	— statistical weight (§4.2)
H	— Hamiltonian
$\mathscr{H}$	— magnetic field intensity
$\mathscr{H}_n$	— Hermite polynomials (§1.2)

I — nuclear spin
J — quantum number
$K_{\bar{\Gamma}}^{\Gamma}$ — force (elastic) constant
$K_{\Gamma}(\bar{\Gamma})$ — vibronic reduction factor
$K(\Omega)$ — coefficient of light absorption (Ch. 4)
k_{Γ}^{i} — orbital force constant
$k^{(i)}$ — force constant coefficients (§§1.4, 6.3)
L — sound propagation factor (§4.1)
l, m, n — quantum numbers of vibronic energy levels (Ch. 3,4)
l, m, n — direction cosines (§4.1)
M_{12} — transition moment
M — nucleus mass (Ch. 1)
m — electron mass (Ch. 1)
n — vibrational quantum number
P — quadrupole interaction constant (§4.1)
P_a — amplification coefficient (§5.1)
P — electron–strain interaction
p — dipole moment
$p = K_E(A_2)$ — vibronic reduction factor in $E-e$ problem
p — ratio of θ and ε components of vibration frequency (§3.2)
Q — normal (symmetrized) nuclear coordinates
$q = K_E(E)$ — vibronic reduction factor in $E-e$ problem
q_i — MO occupation numbers
$q_{1,2}$ — transformed nuclear coordinates (§2.4)
R — nuclear coordinates, interatomic distance
r_i — electron coordinates
S_{ij} — overlap integral
S — electron spin
$T_{1,2}$ — triply degenerate representation
T — temperature
T — period of vibration (§5.1)
U_x — nondiagonal matrix element of Hamiltonian
V — electron–nuclear plus nuclear–nuclear interaction
v — sound velocity
W — vibronic interaction term
W — energy of external and relaxational perturbations (Ch. 4)
x, y, z — Cartesian coordinates
Z — statistical sum
Z_{α} — effective charge (§§1.5, 4.2)
α — magnetic field direction
β — Bohr magneton
β — anharmonicity correction (§6.3)
Γ — irreducible representation (term)

3Γ	— tunneling splitting (§4.1)
γ	— line of degenerate irreducible representations
γ	— correlation parameter in CJTE (§5.2)
γ	— anharmonicity coefficient (§6.3)
Δ	— electron energy gap
$\bar{\Delta}$	— mean value of random strain splitting
ΔX	— change in X
δ	— tunneling splitting
ε	— electronic energy
θ, ε	— two components of doubly degenerate E representation
ξ, η, ζ	— three components of triply degenerate T_2 representation
η	— asymmetry parameter (§5.3)
ζ	— Coriolis constant (§4.2)
κ	— vibrational quantum number
λ_Γ	— dimensionless vibronic constant
λ	— spin–orbit interaction constant (§4.1)
λ_j	— power of atomic function exponent (§1.5)
μ	— correlation parameter in CJTE (§5.2)
ν^I	— quantum number of hyperfine interaction
Ω	— incident irradiation frequency, electron transition frequency
ω	— vibrational frequency
Ψ	— total wave function
ψ	— electronic MO wave function
ρ, ϕ	— nuclear polar coordinates
ρ	— light depolarization (§4.2)
σ	— ordering parameter (§5.2)
τ	— lifetime, relaxation time
Φ	— angular wave function
$\phi_{i\kappa}$	— approximate wave function for local state at AP minimum (§3.2)
φ	— one-electron atomic or MO wave function
φ	— random strain direction (§4.1)
φ_{JKM}	— spherical top molecule rotational wave function

Abbreviations Commonly Used in Text

JT	— Jahn–Teller
JTE	— Jahn–Teller effect
PJTE	— pseudo-Jahn–Teller effect
CJTE	— cooperative Jahn–Teller effect
CPJTE	— cooperative pseudo-Jahn–Teller effect

AP — adiabatic potential(s)
VI — vibronic interaction(s)
VC — vibronic constant(s)
OVC — orbital vibronic constant(s)
MO — molecular orbital(s)
HOMO — highest occupied molecular orbital(s)
LUMO — lowest unoccupied molecular orbital(s)
IR — infrared
UV — ultraviolet

Introduction — The Concept of Vibronic Interaction

The detailed treatment of the influence of electrons on molecular nuclear configuration and dynamics, carried out by means of nonadiabatic vibronic mixing of electronic states, forms a new trend in modern chemistry.

The structure and properties of a molecular system are determined by the motion of its electrons and nuclei and by their interaction. The main laws governing these motions were established in the early thirties, just after the discovery of quantum mechanics. However, because of mathematical difficulties, the quantum-mechanical treatment of the molecular structure in most cases can be carried out only if some simplifying (but physically justified) approximations are introduced. The most general of them is the adiabatic approximation.[1,2]

According to the adiabatic approximation, the large difference between the masses of the electron and nucleus causes a significant difference in their velocities of motion, and therefore a stationary electronic state is attained for every instantaneous nuclear configuration. In other words, the electrons adiabatically and noninertially follow the motions of the nuclei, whereas the latter are moving in the averaged field produced by the electrons. It is clear that in this approximation the role of the electrons in the nuclear dynamics is mainly classical — they are the source of the averaged electrostatic field. It will be shown that some important quantum features of the electronic structure, in particular its nonadiabatic variation under nuclear displacement, are lost in this averaged field.

In the thirties, the most striking deviations from the adiabatic approximation were understood. *In 1934, L. D. Landau* in a discussion with E. Teller[3] *first formulated the idea of instability and spontaneous distortion of the nuclear configuration of a molecule in an orbitally degenerate electronic term* formed by two or more orbital states with the same energy. This idea was later verified by Jahn and Teller and shown to be true for all nonlinear molecular systems, including those with spin degeneracy (except the spin twofold degeneracy of Kramers type). It resulted in the Jahn–Teller theorem published in 1937,[4] which is now widely used in research on the structure and properties of polyatomic systems.[5–14]

In the presence of the Jahn–Teller effect, the electrons do not adiabatically follow the motions of the nuclei, and the nuclear states are determined not only by the averaged field of the electrons, but also by the details of the electronic structure and their changes under nuclear displacements. The resulting special coupling between the electronic and nuclear motions is the first essential deviation from the adiabatic approximation. Note that electronic degeneracy is present in all molecular systems, especially if excited states and oxidized, reduced, radical, or activated (under the influence of another system) forms are taken into account.

In 1957, Öpik and Pryce[15] first noted that effects, similar to that of Jahn and Teller, may be inherent in systems with near (quasi-degenerate or pseudodegenerate) electronic energy levels. The criterion of this pseudo-Jahn–Teller effect includes a relationship among three parameters of the system, and hence the notion of near energy levels is relative (see below). Therefore the PJTE covers, in principle, all molecular systems without exception. The question now becomes: "when is the effect so small that it may be neglected?" The PJTE is concerned with virtually all intermediate cases between the presence and absence of electronic degeneracy the latter being understood in the sense that the energy gap between the electronic states is so large that these states do not mix strongly under nuclear displacements. However, there are cases when the PJTE leads in principle to new phenomena impossible in the JTE (e.g., dipole instability for systems with an inversion center). The origin of analogous phenomena in linear molecules — the Renner effect (RE) — is quite similar.

From a theoretical point of view the JTE, PJTE, and RE result from a new approach in which, unlike the adiabatic approximation, it is assumed that the electronic states depend strongly on nuclear coordinates. If stationary electronic states (ground, first excited, etc.) are obtained as solutions of the Schrödinger equation for fixed nuclei, an accounting of the vibronic interaction (VI) terms in the Hamiltonian (interaction of electrons with nuclear displacements) mixes these electronic states, and this mixing is especially strong in the cases of electronic degeneracy or pseudodegeneracy. *Vibronic mixing of electronic states is an important starting point in the* systematic *analysis of the origin of molecular properties*, revealing a series of new phenomena and laws.

In the early sixties, the development of the theory of vibronic interactions and its applications increased dramatically, resulting in a new trend in the physics and chemistry of molecules and crystals known under the general title of JTE. The applications cover spectroscopy in all frequency ranges, stereochemistry, and crystal chemistry, including structural phase transitions and ferroelectricity.

In the last few years, *application of* VI *theory to chemical reactivity has been initiated.* This includes an examination of the origin and mechanism of chemical reactions (orbital symmetry rules[16,17]) and of the mutual influence of molecular groups in chemical interactions. In particular, this involves chemical activation of a given molecule under the influence of another, e.g., in catalysis,[18–21] and the estimation of absolute rates of reactions. These developments introduce quite new aspects in the VI outlook, expanding it toward molecular transformations. In particular, the idea of vibronic chemical activation essentially is based on new parameters of the molecular structure — orbital vibronic constants (OVC). In addition to the well-known molecular obital (MO) description of the electronic structure of molecules (the static picture), *OVC allow us to take into account dynamic properties of the molecule and their possible changes under perturbation.* The OVC characterize quantitatively the contribution of a single MO electron to the formation (deformation) of the nuclear configuration, its force constants, anharmonicity, etc.

The dynamic structure of the molecule introduced by OVC enables prediction of the behavior of the nuclear configuration under cartain (not very large) electronic rearrangements. These electronic rearrangements cover excitation, ionization, oxidation, reduction, and activation by chemical binding with another molecule.

If vibronic effects due to electronic rearrangement are taken into account, one can distinguish *three main areas wherein vibronic effects are evident.* For simplicity, they are illustrated in Figure 1 in the one-coordinate approximation (Q is the coordinate of the nuclear configuration). The first case is the proper JTE when, for the maximum symmetric nuclear configuration at the point $Q = 0$, two (or several) branches of the adiabatic potential (AP) (the energy as a function of nuclear coordinates) coincide resulting in electron degeneracy (see Figure 1a, twofold). The JT theorem states that at the point of electronic degeneracy the AP has no minimum, and hence at this point the system under consideration is unstable. If in the electronic state under consideration the system is in principle stable, then the minima of the AP are situated at the points $Q = \pm Q' \neq 0$, at which the system has a lower energy and the nuclear configuration has a lower symmetry. The AP as a whole becomes very complicated and leads to a series of new properties of the system.

The second case (Figure 1b) illustrates the PJTE, when at $Q = 0$ the energy levels do not cross but are closely spaced (the crossing is removed by some interaction and only a "pseudocrossing" remains). If the effect is small the structure of the molecule in the ground state is softened but still not unstable. In the strong limit (the criterion of weak and strong PJTE is

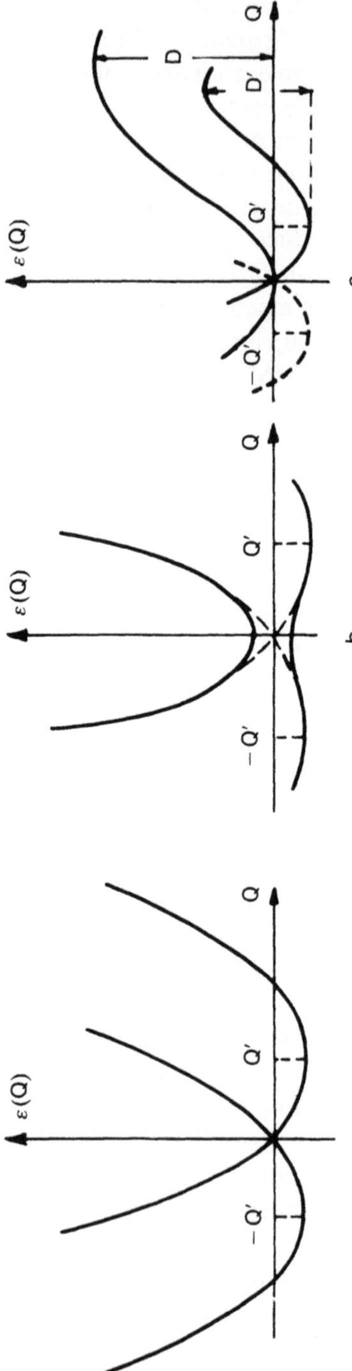

Figure 1. Schematic illustration of the three main types of specific adiabatic potential behavior due to vibronic interactions: (a) electronic degeneracy — the Jahn–Teller effect; (b) electronic pseudodegeneracy — the pseudo-Jahn–Teller effect; (c) electronic rearrangement effect.

given below) the system is unstable at $Q = 0$ and has two minima of the AP at $Q = \pm Q' \neq 0$, quite similar to the JTE.

The third case (Figure 1c) illustrates the variation in the AP at long distances due to the JTE or electronic rearrangement. In the general case, these changes include the shift of the nuclear configuration equilibrium (minimum) position, the change of the force constants, and anharmonicity changes. As a result the reactivity and the activation energy for certain chemical reactions are also changed.

It is evident that these three important main cases of *vibronic effects influence all significant areas of modern chemistry*. The diversity of the effects and the laws revealed by means of the vibronic ideas demonstrated in this book shows that there are not only some new trends in this area of scientific activity but there is *a new concept, the concept of vibronic interactions*, which enables us to analyze all molecular properties, including molecular transformations, from a common point of view. *The theoretical background for this concept lies in the successive account of* VI (*vibronic mixing of electronic states by nuclear displacements*). From the standpoint of a better general understanding of chemical and physical phenomena, *the new concept takes into account the role of the details of the electronic structure in the formation of the nuclear configuration and its dynamics*, and the influence of the latter on spectroscopy, stereochemistry, crystal chemistry, and reactivity.

In some sense the introduction of this concept develops *a new and more adequate way of thinking in chemistry*. Indeed, in books on molecular structure the treatment of the adiabatic approximation (and of molecular properties related to the respective adiabatic electronic state) is covered comprehensively, but with few exceptions nothing is said about the nonvalidity of this approach for many problems in modern chemistry. In this way a stereotype thinking is developed, in which important features of the electronic structure and its influence on nuclear dynamics are lost. As a consequence a proper understanding of the origin of the changes in the nuclear configuration due to changes in the electronic structure, and hence the origin of molecular transformations, is also lost. The concept of VI is, in principle, devoid of this fault. It directs thinking toward a deeper understanding of the origin of the structure and properties of polyatomic systems, including chemical transformations.

In addition to this conceptual meaning *the vibronic approach is of indispensable practical importance*. Vibronic effects directly influence the structure and properties of molecules and crystals, and should be taken into account in any investigation of the origin of their properties and the prediction of new properties. It should be particularly emphasized that *vibronic effects are especially important for unstable molecular species* — ions, radicals, and excited and intermediate states. These play a primary role in chemical transformations and in biological processes, i.e., in some of the most important areas of modern chemistry. Information about the origin of the structure and properties of these systems, as supplied by the vibronic approach, is even more important than for stable molecules, due to the difficulties associated with experimental study of unstable systems.

The above vibronic effects are displayed in two circumstances:

(1) In the resonance interaction of the system with external perturbations. This includes spectroscopy in the whole range of frequencies from radio waves (radio spectroscopy, ESR, NMR, NQR) through vibrational (IR) and electronic (UV and visible) to X-ray and γ-ray spectra.

(2) In nonresonance interactions. This covers magnetic and electric polarizabilities, stereochemistry, and crystal chemistry, including crystal structure formation, isomers, structural phase transitions, and ferroelectricity, as well as chemical reactivity. The investigation of all these effects is far from being exhausted; some studies are only in an initial stage. This book only illustrates some of the applications, mainly those which develop a new trend or have a tendency toward forming such a trend in the appropriate area of research. *At present any extensive investigation of molecular properties without taking into account vibronic effects is impossible*, unless it is shown that these effects are negligible.*

* A more complete treatment of the theory of vibronic interactions is given in a recent book with detailed rubrication and short abstracts.[13] A list of references on the Jahn–Teller effect and related phenomena accompanied by section bibliographic reviews, and author and subject index, can be found in the Bibliographic Review.[14]

1

Theoretical Background

The theory of vibronic effects is based on the consistent account of the interactions of the electrons with the nuclei, especially in the areas where their motions cannot be separated in the framework of the adiabatic approximation.

1.1. The Adiabatic Approximation

One of the most important simplifications of the Schrödinger equation for a molecular system is separation of the motions of electrons and nuclei in the adiabatic approximation. Otherwise, even the formulation of the problem of the molecular structure and properties becomes difficult. The adiabatic approximation is based on the fundamental inequality of the masses and velocities of electrons and nuclei. Since the nuclear mass is about 2000 times that of the electron, the velocity of the latter is much greater than that of the former. Therefore, it can be assumed that every instantaneous (fixed) position of the nuclei corresponds to a stationary electronic state, and the motions of the nuclei are governed by the averaged field of the electrons.

This assumption enables us to solve the problem in two stages: (1) to ignore nuclear motions when solving the electronic part of the problem, and (2) to use the mean electronic energy as the potential for the nuclear motions. The second part of this procedure, ignoring the role of the detailed electronic structure and its nonadiabatic change under nuclear displacements, is most unsatisfactory in this treatment.

Let us consider a more rigorous approach. Divide the total Hamiltonian of the Schrödinger equation into three components:

$$H = H_r + H_Q + V(r, Q) \tag{1.1}$$

where H_r is the electronic component including the kinetic energy of the electrons and the interelectronic electrostatic interaction, H_Q is the kinetic energy of the nuclei, and $V(r, Q)$ is the energy due to interaction of the electrons with the nuclei and internuclear repulsion (r and Q denote the whole set of coordinates of the electrons r_i, $i = 1, 2, \ldots, n$, and nuclei Q_α, $\alpha = 1, 2, \ldots, N$, respectively).

The operator $V(r, Q)$ can be expanded as a series of small displacements of the nuclei about the point $Q_\alpha = Q_{\alpha 0} = 0$ (chosen as origin):

$$V(r,Q) = V(r,0) + \sum_\alpha \left(\frac{\partial V}{\partial Q_\alpha}\right)_0 Q_\alpha + \frac{1}{2}\sum_{\alpha,\beta}\left(\frac{\partial^2 V}{\partial Q_\alpha \partial Q_\beta}\right)_0 Q_\alpha Q_\beta + \cdots \tag{1.2}$$

If the first term of this expansion is regarded as the potential energy of the electrons in the field of fixed nuclei, one can solve the electronic part of the Schrödinger equation

$$[H_r + V(r,0) - \varepsilon'_k]\varphi_k(r) = 0 \tag{1.3}$$

and obtain a set of energies ε'_k and wave functions $\varphi_k(r)$ for the given nuclear configuration corresponding to the point $Q_{\alpha 0}$.

In order to see how these solutions vary under nuclear displacements, the full Schrödinger equation

$$(H - E)\Psi(r, Q) = 0 \tag{1.4}$$

must be solved. In this approach the total wave function $\Psi(r, Q)$ is expanded in terms of electronic functions $\varphi_k(r)$,

$$\Psi(r, Q) = \sum_k \chi_k(Q)\varphi_k(r) \tag{1.5}$$

where the expansion coefficients $\chi_k(Q)$ are functions of the nuclear coordinates. Substituting equation (1.5) into equation (1.4), one obtains after some simple transformations the following system of coupled equations for the functions $\chi_k(Q)$:

$$[H_Q + \varepsilon_k(Q) - E]\chi_k(Q) + {\sum_{m \neq k}}' W_{km}(Q)\chi_m(Q) = 0 \tag{1.6}$$

where $W_{km}(Q)$ denotes the electronic matrix element of vibronic interactions (VI), i.e., that part of the electron–nuclear interaction $V(r, Q)$ which depends on Q,

$$\begin{aligned} W(r,Q) &= V(r,Q) - V(r,0) \\ &= \sum_\alpha \left(\frac{\partial V}{\partial Q_\alpha}\right)_0 Q_\alpha + \frac{1}{2}\sum_{\alpha,\beta}\left(\frac{\partial^2 V}{\partial Q_\alpha \partial Q_\beta}\right)_0 Q_\alpha Q_\beta + \cdots \end{aligned} \tag{1.7}$$

and

$$\varepsilon_k(Q) = \varepsilon'_k + W_{kk}(Q) \tag{1.8}$$

is the potential energy of the nuclei in the mean field of the electrons in state $\varphi_k(r)$. (In the absence of electronic degeneracy or pseudodegeneracy $\varepsilon(Q)$ is the adiabatic potential for this state; see below.)

It is seen from the coupled system of equations (1.6) that if vibronic mixing of different electronic states can be ignored ($W_{km}(Q) = 0$ for $k \neq m$), coupling between these states vanishes, and the system of equations decomposes into a set of simple equations:

$$[H_Q + \varepsilon_k(Q) - E]\chi_k(Q) = 0 \tag{1.9}$$

each of which, for given k, represents the Schrödinger equation for the nuclei moving in the mean field of the electrons in state $\varphi_k(r)$.

In other words, in the case under consideration the motions of the nuclei and electrons are separated and the problem as a whole can be solved in two stages. In the first stage, the electronic states $\varphi_k(r)$ are determined as solutions of equation (1.3) and used to calculate the potential energy of the nuclei $\varepsilon_k(Q)$ by equation (1.8). In the second stage, the wave functions $\chi_k(Q)$ and energies E of the nuclei are determined by equation (1.9), the total wave function being $\Psi(r, Q) = \varphi_k(r)\chi_k(Q)$. This is the simple adiabatic approximation, or the Born–Oppenheimer approximation.[1,2]

Thus the simple adiabatic approximation is valid if and only if the terms of the vibronic mixing of different electronic states in equation (1.6) can be ignored. It can be shown[22] that the perturbation of the total wave function by vibronic interactions is sufficiently small if

$$\hbar\omega \ll |\varepsilon'_m - \varepsilon'_k| \tag{1.10}$$

where $\hbar\omega$ is the energy quantum of vibrations in the electronic state under consideration (k or m), and ε'_m and ε'_k are the energy levels mentioned above. Equation (1.10) may be considered as the criterion governing the adiabatic approximation. Its deduction implies that the electronic state under consideration is stable and generates localized vibrational states (otherwise, the theory is much more complicated).

If criterion (1.10) is satisfied, the error introduced by the simple adiabatic approximation is of the order of $(m/M)^{1/2}$, where m and M are the electron and nuclear masses, respectively (details are given elsewhere[13,22]). Here, $(m/M)^{1/2} \sim 2.3 \cdot 10^{-2}$, which is sufficiently small.

In a more comprehensive treatment of the adiabatic approximation (see, for instance, Bersuker and Polinger[13]) the vibronic interaction terms

$W(r, Q)$ are introduced fully into electronic equation (1.3) in which the Q coordinates are regarded as parameters. In this case, the resulting energies $\varepsilon_k(Q)$ and wave functions $\varphi_k(r, Q)$ depend on the nuclear coordinates Q, and $W_{km}(Q)$ in the system of coupled equations (1.6) are replaced by matrix elements of the operator of nonadiabaticity. The latter contains combinations of derivatives of $\varphi_k(r, Q)$ with respect to Q, their values being determined by the same operator $W(r, Q)$ of vibronic interactions. It follows that the separation of the motions of the electrons and nuclei is ultimately determined by the same reasoning — ignoring vibronic mixing of different electronic states.

However, this approach, called the adiabatic approximation, is somewhat different in detail from the Born–Oppenheimer (or simple adiabatic) approximation referred to earlier. The adiabatic approximation, if applicable (namely when criterion (1.10) is satisfied), gives more exact results than the simple approach. The error introduced by the adiabatic approximation is of the order of $(m/M)^{3/4}$, instead of $(m/M)^{1/2}$ in the Born–Oppenheimer approximation. But if criterion (1.10) is not satisfied, equation (1.6) is a more suitable starting point from which to consider the nonadiabatic effects (the alternative vibronic approach). Attempts to solve numerically the Schrödinger equation without adiabatic separation of the electron and nuclear coordinates were undertaken by Buenker.[22a]

1.2. Vibronic Interactions. Linear Vibronic Constants

Consider a molecular system in its nuclear configuration where the electronic state is degenerate, or there are near-lying electronic states (pseudodegeneracy). Criterion (1.10) is not satisfied for these electronic states and vibronic mixing plays a significant role in determining new molecular properties. It is seen from equation (1.6) that the mixing is put into effect by the matrix elements $W_{km}(Q)$ of the vibronic interaction operator, which is given by equation (1.7) and contains linear, quadratic, cubic, etc., terms. For most cases it is enough to take into account the linear and quadratic terms in order to reveal the vibronic effects, provided the initial configuration has been chosen correctly. In the quadratic approximation the use of normal coordinates essentially simplifies the investigation due to the use of symmetry considerations and the results of group theory.

Let us determine first the initial nuclear configuration of the molecule. In a great number of cases this structure can be taken from experiment (for example, by means of diffraction methods). However, the structure is not always determined exactly. In general there is no evidence that the observed configuration is the true one, that is, it is not an averaged configuration resulting from complicated nuclear dynamics. Besides, if for a free molecule the observed configuration is not the structure of highest symmetry, it is not clear whether it can be chosen as a starting point for the vibronic problem.

For instance, if $PtCl_4^{2-}$ is square planar, the question arises why it is not tetrahedral, as expected in isotropic space in the absence of external (and internal) low symmetry forces. If the quadratic configuration results from the vibronic instability of the tetrahedral configuration (see Chapter 5), the latter has to be chosen as a starting nuclear configuration in the vibronic interaction problem. In the simple example under consideration both configurations may be investigated.

In general, *the initial nuclear configuration must be chosen at the point where the electronic state is degenerate or nearly degenerate.* Since the electronic (nonaccidental) degeneracy is strongly related to the symmetry of the molecular system (the higher the symmetry, the higher the degeneracy), this point is often just the one of highest symmetry. But there are cases when electronic degeneracy remains for lower symmetries (a more detailed theoretical discussion of this question is given elsewhere[13]).

It is well known that normal coordinates can be related to normal vibrations. The latter have a real meaning only for stable configurations. In the adiabatic approximation for nondegenerate electronic states, normal vibrations can be determined by equation (1.9), in which the adiabatic potential $\varepsilon_k(Q) = \varepsilon'_k + W_{kk}(Q)$ is taken in the aforementioned quadratic (harmonic) approximation. The quadratic form of $W_{kk}(Q)$ expressed as

$$W_{kk}(Q) = \sum_{\alpha} \left(\frac{\partial V_{kk}}{\partial Q_\alpha}\right)_0 Q_\alpha + \frac{1}{2} \sum_{\alpha,\beta} \left(\frac{\partial^2 V_{kk}}{\partial Q_\alpha \partial Q_\beta}\right)_0 Q_\alpha Q_\beta \tag{1.11}$$

can be reduced by means of normal coordinates to canonical (diagonal) shape, in which the cross-terms containing $Q_\alpha Q_\beta$ disappear (and only terms of type $\Sigma_\alpha c_\alpha Q_\alpha^2$ remain), while the kinetic energy operator maintains its additive form, $H_Q = -(\hbar^2/2)\Sigma_\alpha M_\alpha^{-1}(\partial/\partial Q_\alpha)^2$.

The normal coordinates can be determined by means of symmetrized displacements. By symmetrized displacements are meant collective (concerted) nuclear displacements which, under the symmetry operation of the molecular point group, transform according to one of its irreducible representations; they can be easily found by the methods of group theory[22,23] and are available in tabular form. If a molecule has N atoms, then the number of vibrational degrees of freedom and hence the number of symmetrized displacements is $3N-6$ (or $3N-5$ for linear molecules). Examples of their classification by irreducible representations are given in Table 1.1; the appropriate atomic displacements are illustrated in Figure 1.1. Expressions for symmetrized displacements in terms of Cartesian ones (Figure 1.2a, b) are given in Table 1.2.

Table 1.1. Classification of Symmetrized Displacements for Several Types of Molecules

Number of atoms N (number of normal vibrations $3N-6$)	Symmetry	Example, shape	Types of symmetrized displacements
4 (6)	C_{3v}	NH_3, pyramid	A'_1, A''_1, E', E''
5 (9)	T_d	MnO_4, tetrahedron	A_1, E, T'_2, T''_2
7 (15)	O_h	CrF_6^{3}, octahedron	A_{1g}, E_g, T_{2g}, T_{2u}, T'_{1u}, T''_{1u}
7 (15)	D_{4h}	MA_4B_2, tetragonally distorted octahedron	A'_{1g}, A''_{1g}, A'_{2u}, A''_{2u}, B_{1g}, B_{2g}, B_{2u}, E_g, E_u, E'_u, E''_u
9 (21)	O_h	CsF_8, cube	A_{1g}, A_{2u}, E_g, E_u, T_{2u}, T'_{1u}, T''_{1u}, T'_{2g}, T''_{2g}

The f-fold degenerate representations (symmetry types) Γ have f lines γ. For instance, the twofold degenerate representation $\Gamma = E$ has two lines, $\gamma = \theta, \varepsilon$, and for the threefold one $\Gamma = T$, $\gamma = \xi, \eta, \zeta$. As usual, for degenerate representations the division into components corresponding to the γ lines is rather conventional. The displacement types shown in Figure 1.1 are commonly used. Below, the notation introduced by Mulliken will be used wherever possible: A, B for nondegenerate terms, E for twofold and T for threefold representations.

The symmetrized displacements, with the type of symmetry which occurs only once in the group-theoretical classification for the system under consideration (Table 1.1), are automatically normal coordinates. For repeated types of symmetrized displacements (e.g., T'_2 and T''_2 in tetrahedral systems, E' and E'' in C_{3v} symmetry, etc.) the normal coordinates are linear combinations of the symmetrized ones and can be obtained by means of an additional (sometimes involved) procedure.

With normal coordinates, equation (1.9) for nuclear motions separates into $3N-6$ (or $3N-5$) equations of harmonic oscillators:

$$-\frac{\hbar^2}{2M_\alpha}\frac{\partial^2\chi_{n_\alpha}}{\partial Q_\alpha^2} + \frac{1}{2}\omega_\alpha^2 Q_\alpha^2 \chi_{n_\alpha} = E_{n_\alpha}\chi_{n_\alpha}, \qquad \alpha = 1,2,\ldots,3N-6 \tag{1.12}$$

where M_α is the reduced mass of the αth normal vibration, ω_α being its frequency.

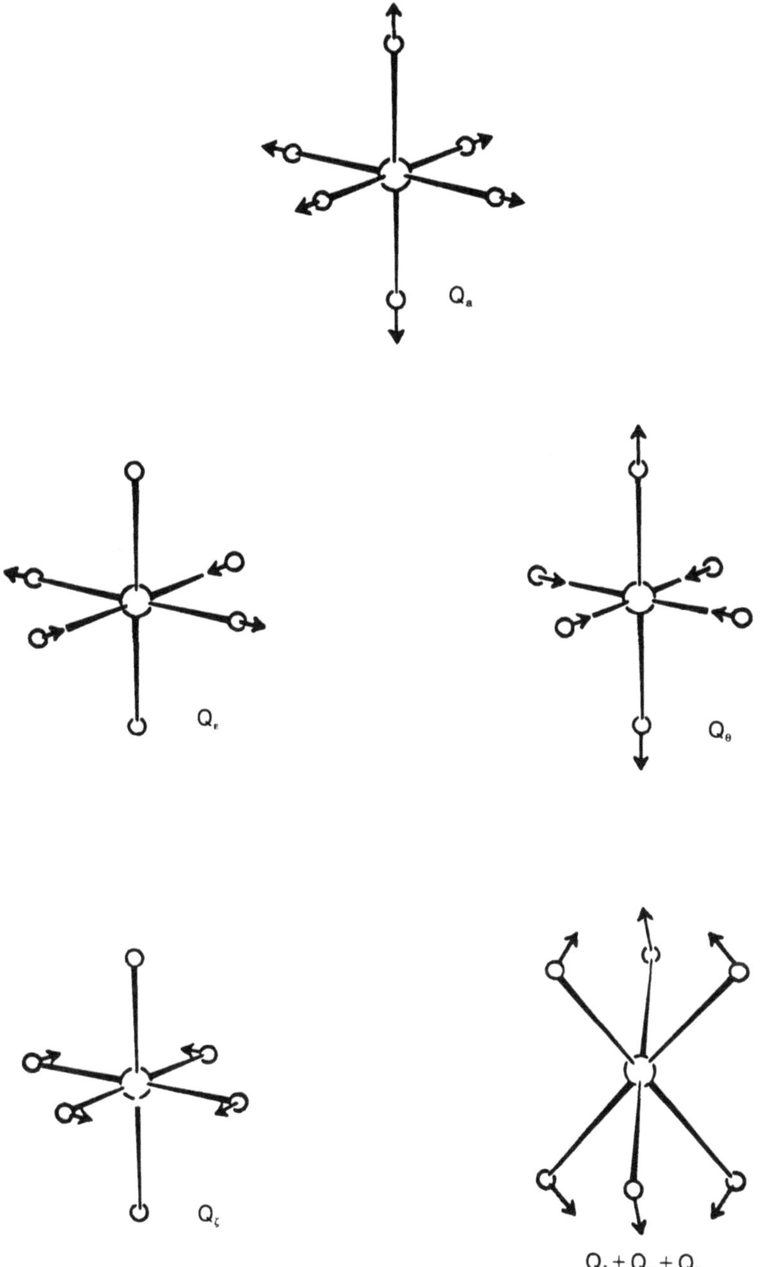

Figure 1.1a. Shape of symmetrized (normal) displacements in case of octahedral ML_6 molecules.

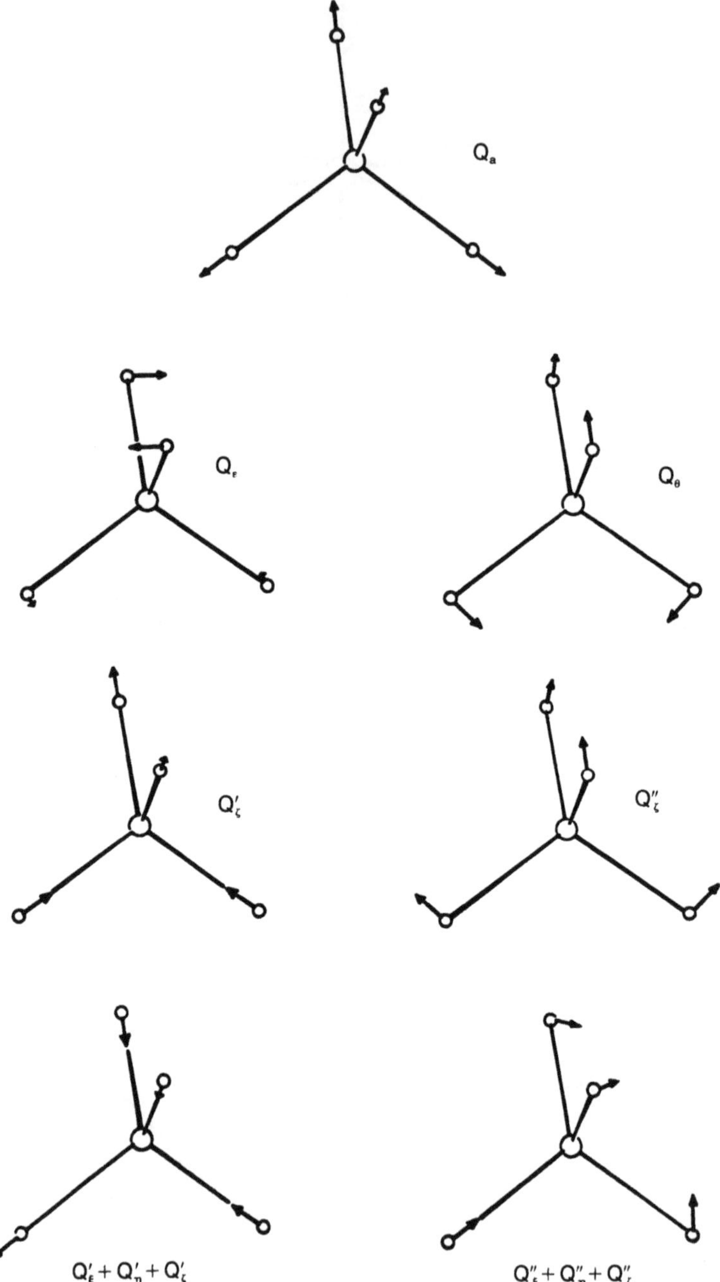

Figure 1.1b. Shape of symmetrized (normal) displacements in case of tetrahedral ML_4 molecules.

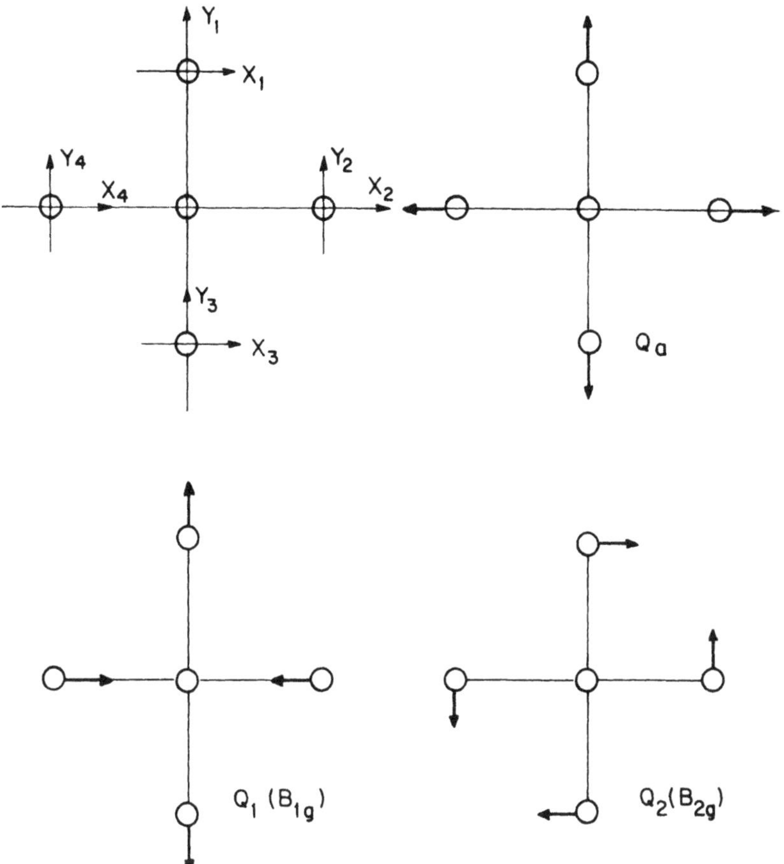

Figure 1.1c. Shape of symmetrized (normal) displacements in case of quadratic X_4 molecules.

The solutions of equation (1.12) are well known[22]:

$$E_{n_\alpha} = \hbar\omega_\alpha\left(n_\alpha + \frac{1}{2}\right), \qquad n_\alpha = 1, 2, \ldots$$

$$\chi_{n_\alpha}(Q_\alpha) = (n!2^n\sqrt{\pi})^{-1/2}\mathcal{H}_n(Q_\alpha)\exp\left(-\frac{M_\alpha\omega_\alpha Q_\alpha^2}{2\hbar}\right) \tag{1.13}$$

where $\mathcal{H}_n(Q_\alpha)$ are Hermite polynomials.

The frequencies ω_α of the normal vibrations are just those which occur in the experimental vibrational spectra, and this illustrates the significance of the normal coordinates and normal vibration approach in polyatomic problems.

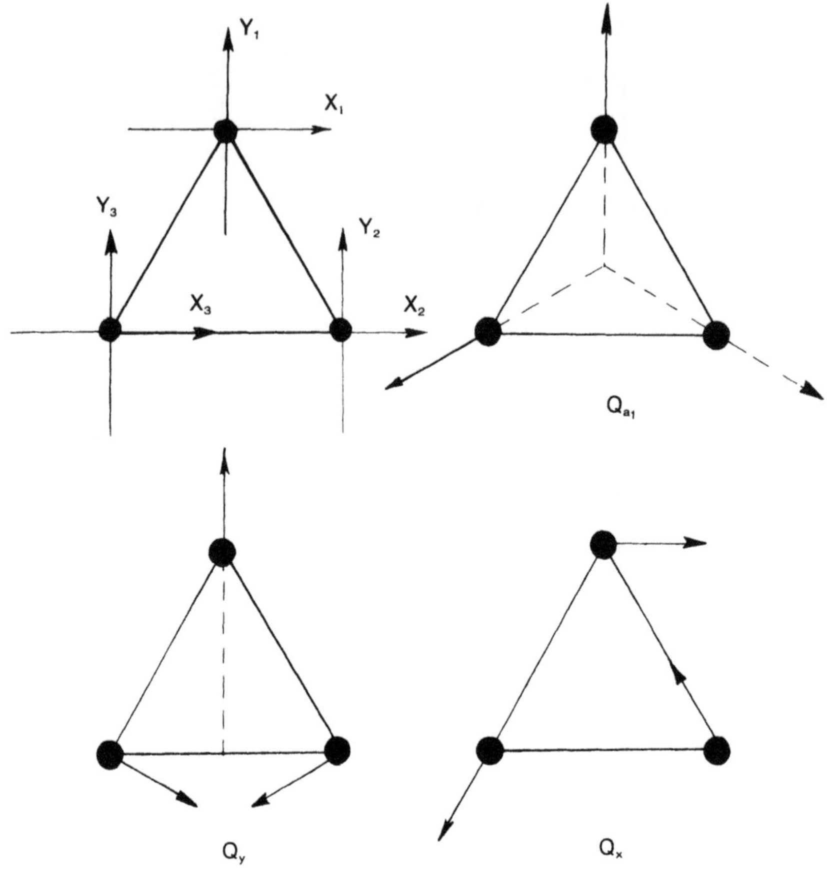

Figure 1.1d. Shape of symmetrized (normal) displacements in case of triangular X_3 molecules.

The operator of vibronic interactions in normal coordinates $Q_{\Gamma\gamma}$ may be written in the form

$$W(r, Q) = \sum_{\Gamma\gamma} \left(\frac{\partial V}{\partial Q_{\Gamma\gamma}}\right)_0 Q_{\Gamma\gamma} + \frac{1}{2} \sum_{\Gamma_1\gamma_1} \sum_{\Gamma_2\gamma_2} \left(\frac{\partial^2 V}{\partial Q_{\Gamma_1\gamma_1} \partial Q_{\Gamma_2\gamma_2}}\right)_0 Q_{\Gamma_1\gamma_1} Q_{\Gamma_2\gamma_2} \quad (1.14)$$

Consider now the coefficients of this expansion which are derivatives of the electron–nuclear interaction. The matrix elements of these coefficients are the *constants of vibronic coupling* or *vibronic constants.* They are very important in the analysis of vibronic interaction effects: *vibronic constants characterize the measure of coupling between the electronic structure and nuclear displacements,* i.e., *the measure of influence of the nuclear displacements on the electron distribution* and, conversely, *the effect of the changes in the electronic structure upon nuclear dynamics.*

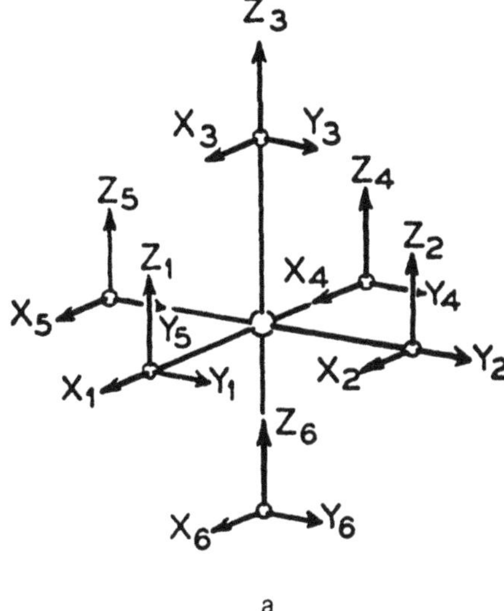

a

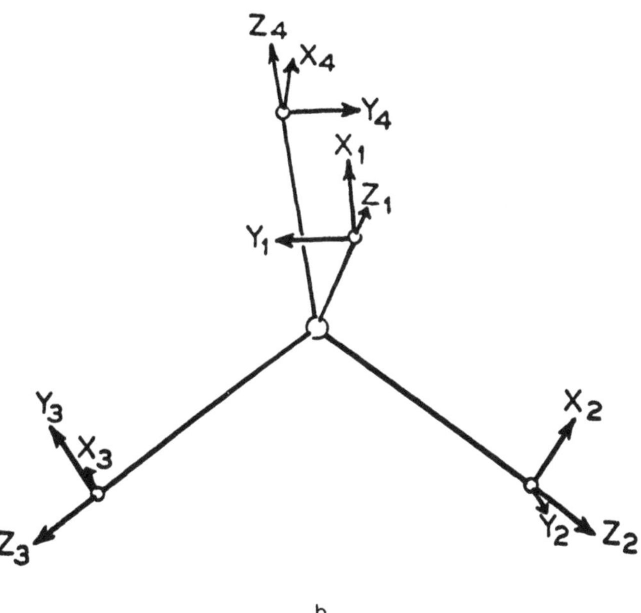

b

Figure 1.2. General and local ligand coordinate systems for: (a) octahedral and (b) tetrahedral molecules.

Table 1.2. Symmetrized Displacements (Normal Coordinates) Expressed by Cartesian Coordinates (Figure 1.2) for Some Trigonal, Tetragonal, Tetrahedral and Octahedral Systems

Symmetrized coordinates	Symmetry type	Transformation properties	Expressions in terms of Cartesian coordinates
Trigonal systems X_3. Symmetry D_{3h}			
$Q_x^{(t)}$	E'	x	$(X_1 + X_2 + X_3)/\sqrt{3}$
$Q_y^{(t)}$		y	$(Y_1 + Y_2 + Y_3)/\sqrt{3}$
$Q_{a_2}^{(r)}$	A_2'	axial vector	$(2X_1 - X_2 - \sqrt{3}Y_2 - X_3 + \sqrt{3}Y_3)/\sqrt{12}$
Q_{a_1}	A_1'	$x^2 + y^2$	$(2Y_1 + \sqrt{3}X_2 - Y_2 - \sqrt{3}X_3 - Y_3)/\sqrt{12}$
Q_x	E'	$2xy$	$(2X_1 - X_2 + \sqrt{3}Y_2 - X_3 - \sqrt{3}Y_3)/\sqrt{12}$
Q_y		$x^2 - y^2$	$(2Y_1 - \sqrt{3}X_2 - Y_2 + \sqrt{3}X_3 - Y_3)/\sqrt{12}$
Tetragonal systems ML_4. Symmetry D_{4h}			
Q_a	A_{1g}	$x^2 + y^2$	$\frac{1}{2}(Y_1 + X_2 - Y_3 - X_4)$
A_1	B_{1g}	$x^2 - y^2$	$\frac{1}{2}(Y_1 - X_2 - Y_3 + X_4)$
Q_2	B_{2g}	xy	$\frac{1}{2}(X_1 + Y_2 - X_3 - Y_4)$
Tetrahedral systems[a] ML_4. Symmetry T_d			
Q_a	A	$x^2 + y^2 + z^2$	$\frac{1}{2}(Z_1 + Z_2 + Z_3 + Z_4)$
Q_θ	E	$2z^2 - x^2 - y^2$	$\frac{1}{2}(X_1 - X_2 - X_3 + X_4)$
Q_ϵ		$3(x^2 - y^2)$	$\frac{1}{2}(Y_1 - Y_2 - Y_3 + Y_4)$
Q_ξ'	T_2'	yz	$\frac{1}{2}(Z_1 - Z_2 + Z_3 - Z_4)$
Q_η'		xz	$\frac{1}{2}(Z_1 + Z_2 - Z_3 - Z_4)$
Q_ζ'		xy	$\frac{1}{2}(Z_1 - Z_2 - Z_3 + Z_4)$
Q_ξ''	T_2''	yz	$\frac{1}{4}(-X_1 + X_2 - X_3 + X_4) + \sqrt{3}(-Y_1 + Y_2 - Y_3 + Y_4)/4$
Q_η''		xz	$\frac{1}{4}(-X_1 - X_2 + X_3 + X_4) + \sqrt{3}(Y_1 + Y_2 - Y_3 - Y_4)/4$
Q_ζ''		xy	$\frac{1}{2}(X_1 + X_2 + X_3 + X_4)$
Octahedral systems[a] ML_6. Symmetry O_h			
Q_a	A_{1g}	$x^2 + y^2 + z^2$	$(X_2 - X_5 + Y_3 - Y_6 + Z_1 - Z_4)/\sqrt{6}$
Q_θ	E_g	$2z^2 - x^2 - y^2$	$(2Z_3 - 2Z_6 - X_1 + X_4 - Y_2 + Y_5)/\sqrt{3}$
Q_ϵ		$x^2 - y^2$	$\frac{1}{2}(X_1 - X_4 - Y_2 + Y_5)$
Q_ξ	T_{2g}	yz	$\frac{1}{2}(Z_2 - Z_5 + Y_3 - Y_6)$
Q_η		xz	$\frac{1}{2}(X_3 - X_6 + Z_1 - Z_4)$
Q_ζ		xy	$\frac{1}{2}(Y_1 - Y_4 + X_2 - X_5)$
Q_x'	T_{1u}'	x	$\frac{1}{2}(X_2 + X_3 + X_5 + X_6)$
Q_y'		y	$\frac{1}{2}(Y_1 + Y_3 + Y_4 + Y_6)$
Q_z'		z	$\frac{1}{2}(Z_1 + Z_2 + Z_4 + X_5)$

Table 1.2 — *continued*

Symmetrized coordinates	Symmetry type	Transformation properties	Expressions in terms of Cartesian coordinates
Q''_x	T''_{1u}	x	$(X_1+X_4)/\sqrt{2}$
Q''_y		y	$(Y_2+Y_5)/\sqrt{2}$
Q''_z		z	$(Z_3+Z_6)/\sqrt{2}$
$\tilde{Q}_\xi$	T_{2u}	$x(y^2-z^2)$	$\frac{1}{2}(X_2+X_5-X_3-X_6)$
$\tilde{Q}_\eta$		$y(z^2-x^2)$	$\frac{1}{2}(Y_3+Y_6-Y_1-Y_4)$
$\tilde{Q}_\zeta$		$z(x^2-y^2)$	$\frac{1}{2}(Z_1+Z_4-Z_2-Z_5)$

[a] The expressions for T'_{1u} and T''_{1u} coordinates of octahedral systems and T_{2u} coordinates of tetrahedral systems are given for L_6 and L_4 molecules respectively, i.e., without the displacements of the central atom M.

We shall denote the electronic states by the appropriate irreducible representations $\Gamma, \Gamma', \ldots,$ of the symmetry group of the molecular system, and assume first that the states Γ and Γ' are not degenerate. The matrix element

$$F_{\bar{\Gamma}}^{(\Gamma\Gamma')} = \left\langle \Gamma \middle| \left(\frac{\partial V}{\partial Q_{\bar{\Gamma}}}\right)_0 \middle| \Gamma' \right\rangle \tag{1.15}$$

is called the linear vibronic constant. Following the rules of group theory, $F_{\bar{\Gamma}}^{(\Gamma\Gamma')}$ is nonzero if and only if $\Gamma \times \Gamma' = \bar{\Gamma}$. If Γ or Γ' or both are degenerate (in this case $\bar{\Gamma}$ may also be degenerate), a set of linear vibronic constants appropriate to all the lines γ and γ' of the two representations Γ and Γ' and their combinations $F_{\bar{\Gamma}\bar{\gamma}}^{(\Gamma\gamma\Gamma'\gamma')}$ must be introduced instead of one vibronic constant (1.13). This can be easily done if one takes into account the fact that the matrix elements within a degenerate term differ solely in numerical coefficients, their values being known.

Indeed, according to the group theory theorem of Wigner–Eckart[24,25] for any physical operator $X_{\bar{\Gamma}\bar{\gamma}}$, which transforms according to line $\bar{\gamma}$ of representation $\bar{\Gamma}$, we have

$$\langle \Gamma\gamma | X_{\bar{\Gamma}\bar{\gamma}} | \Gamma'\gamma' \rangle = \langle \Gamma \| X_{\bar{\Gamma}} \| \Gamma' \rangle \langle \bar{\Gamma}\bar{\gamma}\Gamma'\gamma' | \Gamma\gamma \rangle \tag{1.16}$$

where $\langle \Gamma \| X_{\bar{\Gamma}} \| \Gamma' \rangle$ is the reduced matrix element (which does not depend on $\bar{\gamma}$, γ, and γ') and $\langle \bar{\Gamma}\bar{\gamma}\Gamma'\gamma' | \Gamma\gamma \rangle$ are the Clebsch–Gordan coefficients available from tabular data.

Using formula (1.16) for the operator $X_{\bar{\Gamma}\bar{\gamma}} = (\partial V/\partial Q_{\bar{\Gamma}\bar{\gamma}})_0$ we obtain the following relation between different components of the linear vibronic constants:

$$F_{\bar{\Gamma}\bar{\gamma}}^{(\Gamma\gamma\Gamma'\gamma')} = F_{\bar{\Gamma}}^{(\Gamma\Gamma')}\langle\bar{\Gamma}\bar{\gamma}\bar{\Gamma}'\gamma' \mid \Gamma\gamma\rangle \tag{1.16'}$$

It follows that if one knows at least one vibronic constant $F_{\Gamma\gamma}^{(\Gamma\gamma\Gamma'\gamma')}$, all the others can be easily calculated by (1.16′) using tabular data on the Clebsch–Gordan coefficients.

Some of the linear vibronic constants have a clear physical meaning. *The diagonal constant of the linear coupling* $F_{\bar{\Gamma}\bar{\gamma}}^{(\Gamma\gamma\Gamma\gamma)} \equiv F_{\bar{\Gamma}\bar{\gamma}}^{\Gamma\gamma}$ *has the sense of the force with which the electrons in state* $\Gamma\gamma$ *affect the nuclei in the direction of symmetrized displacements* $Q_{\bar{\Gamma}\bar{\gamma}}$. For instance, $F_{E\theta}^{E\varepsilon}$ denotes the force with which the electrons in the $E\varepsilon$ state distort the nuclear configuration in the direction of the $E\theta$ displacements, illustrated in Figure 1.1.

For degenerate states Γ, according to the group-theoretical condition, the diagonal matrix element $F_{\bar{\Gamma}}^{\Gamma}$ is nonzero if the symmetrical product $[\Gamma \times \Gamma]$ contains $\bar{\Gamma}$, $\bar{\Gamma} \in [\Gamma \times \Gamma]$ (compare this with the condition for nondiagonal elements $\bar{\Gamma} \in \Gamma \times \Gamma'$). For nondegenerate states $[\Gamma \times \Gamma] = \Gamma \times \Gamma = A_1$, where A_1 is the totally symmetric representation. It follows that in nondegenerate states $\bar{\Gamma} = A_1$, and the electrons can distort the nuclear configuration only in the direction of totally symmetric displacements, for which the symmetry of the system does not change. If the electronic state Γ is degenerate, the symmetric product $[\Gamma \times \Gamma]$ contains representations not totally symmetric along with the symmetric one. Indeed, for cubic symmetry systems $[E \times E] \rightarrow A_1 + E$, $[T \times T] \rightarrow A_1 + E + T_1 + T_2$; for a D_{4h} symmetry system $[E \times E] \rightarrow A_1 + B_1 + B_2$, etc. In these cases $\bar{\Gamma}$ may not be totally symmetric (degenerate E, T_1, T_2, or nondegenerate B_1, B_2, etc.). Thus under the influence of the electrons the nuclear configuration undergoes appropriate distortions which are not totally symmetric. It is just these distortions that are predicted by the Jahn–Teller theorem (see Section 1.6).

1.3. Force Constants and Quadratic Vibronic Constants

The quadratic (or second-order) vibronic constants can be introduced in principle in a fashion similar to the linear ones. However, complications arise here because the diagonal matrix elements of some of the second derivatives $(\partial^2 V/\partial Q_{\Gamma_1\gamma_1}\partial Q_{\Gamma_2\gamma_2})_0$ in equation (1.14) in appropriate combinations form the curvature of the adiabatic potential or the force constants (in the equilibrium position). The remaining terms and the nondiagonal matrix elements contain the quadratic vibronic constants, which must be distinguished from the force constants.

In order to investigate this question we subject the second (quadratic) term in equation (1.14) to further group-theoretical transformations. We introduce the tensor convolution $\{(\partial^2 V/\partial Q_{\Gamma_1}\partial Q_{\Gamma_2})_0\}_{\Gamma\gamma}$, which means that the linear combination of second derivatives with respect to the $Q_{\Gamma_1\gamma_1}$ and $Q_{\Gamma_2\gamma_2}$ coordinates transform according to line γ of representation $\Gamma \in \Gamma_1 \times \Gamma_2$. This combination can be easily found by means of group theory and the Wigner–Eckart theorem (1.16). Similarly, the appropriate tensor convolution for the coordinates $\{Q_{\Gamma_1} \times Q_{\Gamma_2}\}_{\Gamma_\gamma}$ may be introduced. It can be shown[13] that in this notation the vibronic interaction operator (1.14) takes the form

$$W(r,Q) = \sum_{\Gamma\gamma} \left(\frac{\partial V}{\partial Q_{\Gamma\gamma}}\right)_0 Q_{\Gamma\gamma} + \frac{1}{2}\sum_{\Gamma\gamma}\sum_{\Gamma_1\Gamma_2}\left\{\left(\frac{\partial^2 V}{\partial Q_{\Gamma_1}\partial Q_{\Gamma_2}}\right)_0\right\}_{\Gamma\gamma}\{Q_{\Gamma_1}\times Q_{\Gamma_2}\}_{\Gamma\gamma} \quad (1.14')$$

In this expression all the terms are grouped such that the terms in each group transform according to a specific representation of the symmetry group of the system.

Now we introduce matrix elements of the operators — quadratic term coefficients in equation (1.14′):

$$K_{0\bar{\Gamma}\bar{\gamma}}^{(\Gamma\Gamma')}(\Gamma_1 \times \Gamma_2) = \frac{1}{2}\left\langle \Gamma \left| \left\{\left(\frac{\partial^2 V}{\partial Q_{\Gamma_1}\partial Q_{\Gamma_2}}\right)_0\right\}_{\bar{\Gamma}\bar{\gamma}} \right| \Gamma' \right\rangle \quad (1.17)$$

Similar to the linear case considered above, in the diagonal matrix element $K_{0\bar{\Gamma}}^{(\Gamma\Gamma)} \equiv K_{0\bar{\Gamma}}^{\Gamma}$ the representation $\bar{\Gamma} \in \Gamma_1 \times \Gamma_2$ for nondegenerate states can only be totally symmetric, $\bar{\Gamma} = A_1$, while for degenerate states $\bar{\Gamma}$ can be both totally symmetric and nontotally symmetric. It is clear that since $\bar{\Gamma} \in \Gamma_1 \times \Gamma_2$, for the totally symmetric part $\bar{\Gamma} = A_1$, $\Gamma_1 = \Gamma_2 = \bar{\Gamma}$ (the product of only equal irreducible representations contains A_1). In this case the totally symmetric combination of second derivatives in equation (1.17) is as follows:

$$\left\{\left(\frac{\partial^2 V}{\partial Q_{\bar{\Gamma}}^2}\right)_0\right\}_{A_1} = \sum_{\bar{\gamma}} \left(\frac{\partial^2 V}{\partial Q_{\bar{\Gamma}\bar{\gamma}}^2}\right)_0$$

For instance,

$$\left\{\left(\frac{\partial^2 V}{\partial Q_E^2}\right)_0\right\}_{A_1} = \left(\frac{\partial^2 V}{\partial Q_{E\theta}^2}\right)_0 + \left(\frac{\partial^2 V}{\partial Q_{E\epsilon}^2}\right)_0$$

Similarly $\{Q_{\bar{\Gamma}}^2\}_{A_1} = \Sigma_{\bar{\gamma}} Q_{\bar{\Gamma}\bar{\gamma}}^2$; in particular, $\{Q_E^2\}_{A_1} = Q_{E\theta}^2 + Q_{E\epsilon}^2$.

The totally symmetric part of the diagonal matrix element (1.17), $K_{0\bar{\Gamma}}^{\Gamma}$ has a clear physical meaning (see also Section 1.5): together with the nondiagonal linear vibronic constant $F_{\bar{\Gamma}}^{(\Gamma\Gamma')}$, it forms the curvature, $K_{\bar{\Gamma}}^{\Gamma}$, of the adiabatic potential in the direction (section) $Q_{\bar{\Gamma}}$ (in case of degeneracy the curvature is the same in all directions $Q_{\bar{\Gamma}\bar{\gamma}}$ with different $\bar{\gamma}$),

$$K_{\bar{\Gamma}}^{\Gamma} = K_{0\bar{\Gamma}}^{\Gamma} - \sum_{\Gamma'}{}' |F_{\bar{\Gamma}}^{(\Gamma\Gamma')}|^2/\Delta_{\Gamma'\Gamma} \tag{1.18}$$

where $\Delta_{\Gamma'\Gamma} = \frac{1}{2}(\varepsilon_{\Gamma'} - \varepsilon_{\Gamma})$ is the appropriate energy semidifference between the states Γ' and Γ.

This value of $K_{\bar{\Gamma}}^{\Gamma}$ has been obtained as the coefficient of $\{Q_{\bar{\Gamma}}^2\}_{A_1}$ in the expression for the energy as a function of nuclear coordinates, calculated by means of perturbation theory (by considering the vibronic interaction terms as small perturbations). At the points of equilibrium (stable) configurations where $F_{\bar{\Gamma}}^{\Gamma} = 0$, $K_{\bar{\Gamma}}^{\Gamma}$ equals the force constant.

As mentioned above for the degenerate term Γ, there are nontotally symmetric contributions to the diagonal matrix element $K_{0\bar{\Gamma}\bar{\gamma}}^{\Gamma}$ $(\Gamma_1 \times \Gamma_2)$ in equation (1.17) which correspond to the nontotally symmetric representations of the product $[\Gamma \times \Gamma]$. These representations, as shown below in Section 1.6, are the Jahn–Teller active representations. Using the Wigner–Eckart theorem (1.16) only one reduced constant for every representation $\bar{\Gamma} \in \Gamma_1 \times \Gamma_2$ can be introduced. We denote this constant as $G_{\bar{\Gamma}}^{\Gamma}(\Gamma_1 \times \Gamma_2)$ and call it the quadratic (or second-order) vibronic constant.

Thus *the totally symmetric part of the diagonal matrix element* (1.17) *is the essential component of the adiabatic potential curvature or force constant, while the nontotally symmetric part forms the quadratic vibronic constant.* The nondiagonal matrix elements (1.17) are the nondiagonal quadratic vibronic constants $G_{\bar{\Gamma}}^{(\Gamma\Gamma')}(\Gamma_1 \times \Gamma_2)$.

One useful feature of the vibronic constants to be mentioned here is the additivity of the contributions of the core and valence electrons to the vibronic constant values. This property is based on the additivity of the electron–nuclear interaction operator V in equation (1.1) which can be divided into two parts, describing the interactions within the core V^c and the electron–core interaction V^v, $V = V^c + V^v$, where

$$V^v = -\sum_i \sum_\alpha \frac{e^2 Z_\alpha}{|\mathbf{r}_i - \mathbf{R}_\alpha|}; \qquad V^c = \frac{1}{2} \sum_{\substack{\alpha,\beta \\ \alpha \neq \beta}}{}' \frac{e^2 Z_\alpha Z_\beta}{|\mathbf{R}_\alpha - \mathbf{R}_\beta|} \tag{1.19}$$

Here Z_α is the effective charge of the α atomic core (or the α nucleus), and $\mathbf{R}_\alpha$ is its coordinate vector. Consequently, the vibronic constants can be presented in the form of appropriate sums (the force constants (1.18) cannot be presented as such simple sums):

$$\begin{aligned} F_\Gamma &= F_\Gamma^c + F_\Gamma^v \\ G_\Gamma &= G_\Gamma^c + G_\Gamma^v \end{aligned} \tag{1.19'}$$

Note that *the division of the system into core and valence electrons is not trivial.* Indeed, this division implies that the inner electrons of the core completely follow the nuclear displacements and form together with the nucleus a rigid ion (otherwise, the Z_α values are not constants). This assumption is approximately valid only for electrons strongly coupled with the nucleus; the error is larger, the less bounded the electrons included in the core.

The above consideration implies that the introduction of vibronic constants is not based on any assumptions of core–electron division. The latter becomes necessary for specific quantitative calculations.

1.4. Orbital Vibronic Constants as Parameters of Molecular Dynamic Structure

The introduction of orbital vibronic constants (OVC)[18-21] *is a first step forward in the development of a theory of vibronic interactions which combines the idea of vibronic coupling with the molecular orbital (MO) approach* to the investigation of molecular structure and properties. Taken together, the concept of electronic structure (defined by means of one-electron MO) and the concept of vibronic interaction result in vibronic molecular orbitals (characterized by OVC). The latter present a more refined picture of molecular structure, a description which includes nuclear dynamic parameters.

We denote the one-electron MO energies by ε_i and their wave functions by $\varphi_i(r) = |i\rangle$. Taking into account the additivity of the electron–nuclear interaction operator $V(r, Q)$ according to equation (1.1) (see also (1.19)), we have

$$V = \sum_{i=1}^{n} V_i(r_i); \qquad V_i(r_i) = -\sum_{\alpha} \frac{e^2 Z_\alpha}{|\boldsymbol{r}_i - \boldsymbol{R}_\alpha|} + (1/n)V^c \tag{1.20}$$

where n is the total number of electrons. Based on this notation the orbital vibronic constants can be introduced similarly to the vibronic constants (VC) for the system as a whole, equations (1.15)–(1.17). In contradistinction to the OVC, the VC may be called integral vibronic constants.

For the linear OVC we have

$$f_{\bar{\Gamma}}^{(ij)} = \left\langle \varphi_i(r_i) \left| \left(\frac{\partial V_i}{\partial Q_\Gamma}\right)_0 \right| \varphi_j(r_i) \right\rangle \tag{1.21}$$

Similar to equation (1.17), the orbital matrix elements of the quadratic vibronic interactions (1.14′) are the following:

$$k_{0\bar{\Gamma}\bar{\gamma}}^{(ij)}(\Gamma_1 \times \Gamma_2) = \frac{1}{2} \left\langle \varphi_i(r_i) \left| \left\{ \left(\frac{\partial^2 V_i}{\partial Q_{\Gamma_1} \partial Q_{\Gamma_2}}\right)_0 \right\}_{\bar{\Gamma}\bar{\gamma}} \right| \varphi_j(r_i) \right\rangle \tag{1.22}$$

where the brace has the same meaning as in equation (1.17). The totally symmetric part ($\bar{\Gamma} = A_1$) of the diagonal element $k_{0\bar{\Gamma}}^{(ii)} \equiv k_{0\bar{\Gamma}}^{i}$ ($\bar{\Gamma} = \Gamma_1 = \Gamma_2$) contributes to the orbital force constant $k_{\bar{\Gamma}}^{i}$ (see below), whereas the nontotally symmetric parts, being nonzero only for the degenerate MO, are the appropriate diagonal OVC $g_{\bar{\Gamma}}^{i}(\Gamma_1 \times \Gamma_2)$. The nondiagonal matrix elements (1.22) are the nondiagonal OVC $g_{\bar{\Gamma}}^{(ij)}(\Gamma_1 \times \Gamma_2)$.

The physical meaning of the OVC may be clarified by means of the addition theorem: the linear diagonal integral vibronic constants equal the sum of linear diagonal OVC multiplied by the appropriate MO population numbers q_i^{Γ}:

$$F_{\bar{\Gamma}\bar{\gamma}}^{\Gamma} = \sum_i q_i^{\Gamma} f_{\bar{\Gamma}\bar{\gamma}}^{i} \tag{1.23}$$

The proof of this theorem is based on the additivity properties of the VC mentioned above. If the total wave function of the system in equation (1.15) for $F_{\bar{\Gamma}}^{\Gamma}$ is expressed by the determinant (or a linear combination of determinants) of the one-electron MO functions $\varphi_i(r_i)$, and expressions (1.20) for V and (1.21) for $f_{\bar{\Gamma}}^{i}$ are taken into account, we can easily derive formula (1.23).

We thus obtain from equation (1.23) that the distorting influence of the electrons on the nuclear framework with a force $F_{\bar{\Gamma}\bar{\gamma}}^{\Gamma}$ is formed additively by all the appropriate MO single electron effects, $f_{\bar{\Gamma}\bar{\gamma}}^{i}$. A clear physical meaning for *the linear diagonal OVC* follows immediately: $f_{\bar{\Gamma}\bar{\gamma}}^{i}$ *equals the force with which the electron of the ith MO distorts the nuclear configuration* in the direction of the symmetrized displacements $Q_{\bar{\Gamma}\bar{\gamma}}$ minus the appropriate part of the internuclear repulsion in this direction (see equation (1.20)).

For quadratic VC, quite similarly to equation (1.23), we obtain the following addition formula:

$$G_{\bar{\Gamma}\bar{\gamma}}^{\Gamma}(\Gamma_1 \times \Gamma_2) = \sum_i q_i^{\Gamma} g_{\bar{\Gamma}\bar{\gamma}}^{i}(\Gamma_1 \times \Gamma_2) \tag{1.24}$$

The deduction of similar expressions for the nondiagonal VC is more difficult, since they involve excited states for which the calculation of the contribution of the one-electron MO states is complicated. The analysis can be simplified by the "frozen orbital" approximation, equivalent to the approximation of the Koopmans theorem in quantum chemistry.[26] In this approximation the change of only one MO $i \to j$ is taken into account in a one-electron excitation of the system $\Gamma \to \Gamma'$, the changes of all the others due to the alteration of the interelectronic repulsion being neglected. In this case simple relations exist between integral and orbital linear VC:

$$F^{(\Gamma\Gamma')}_{\bar{\Gamma}\bar{\gamma}} = f^{(ij)}_{\bar{\Gamma}\bar{\gamma}} \tag{1.25}$$

Using this relation the adiabatic potential (AP) curvature (or the force constant) $K^{\Gamma}_{\bar{\Gamma}}$ can be presented as a sum of orbital contributions $k^{i}_{\bar{\Gamma}}$:

$$\begin{aligned} K^{\Gamma}_{\bar{\Gamma}} &= \sum_i q_i k^{i}_{\bar{\Gamma}} \\ k^{i}_{\bar{\Gamma}} &= k^{i}_{0\bar{\Gamma}} - \sum_j{}' |f^{(ij)}|^2/\Delta_{ji} \end{aligned} \tag{1.26}$$

where $\Delta_{ij} = \frac{1}{2}(\varepsilon_i - \varepsilon_j)$ (cf. equation (1.26) with its other form (6.19)). The proof of these relations is quite similar to that of equation (1.23).

As with the case of the linear and quadratic VC, cubic and higher-order VC can be introduced.

In the MO approach the electronic structure of a molecule with fixed nuclei is represented by an electron charge distribution for the one-electron MO, and by the energies of the latter in the field of fixed nuclei. This picture of the electronic structure is static; it does not characterize sufficiently the back influence of the electrons on the nuclear positions. In general, it does not indicate whether the chosen nuclear configuration will be stable for the electronic structure under consideration. It should also be noted that this static MO picture does not allow us to predict how the nuclear configuration and its dynamics will vary with changes in the electronic structure.

This deficiency of the MO approach can be overcome by means of the vibronic constants, since they characterize the forces (and force constants, anharmonicities, etc.) which influence the nuclei. The orbital vibronic constants are of special interest: if the OVC and the orbital contributions to the force constants are known, the behavior of the nuclear configuration (changes of its stable configuration, force constants, anharmonicities) with small changes of the electronic structure (changes of electronic MO population numbers) can be predicted. These results can be used to predict changes of the reactivity of molecules under electronic rearrangements (see Chapter 6). Thus *the OVC complement the static picture of the electronic structure with dynamic parameters.*

In Figure 1.3 the new parametrization of the molecular structure is illustrated for the highest-occupied MO (HOMO) 5σ and lowest-unoccupied MO (LUMO) 2π of two molecules N_2 and CO, taken as examples (the numerical values are obtained in Section 6.3). It is seen that in addition to the MO energy value and wave function (shown schematically in Figure 1.3 by its symmetry), the electron of the HOMO 5σ in the N_2 molecule tightens the nuclei with force $f^{5\sigma}_R(N_2) = 3.51 \cdot 10^{-4}$ dyn (R is the distance between the nuclei), while in the CO molecule the electron of the

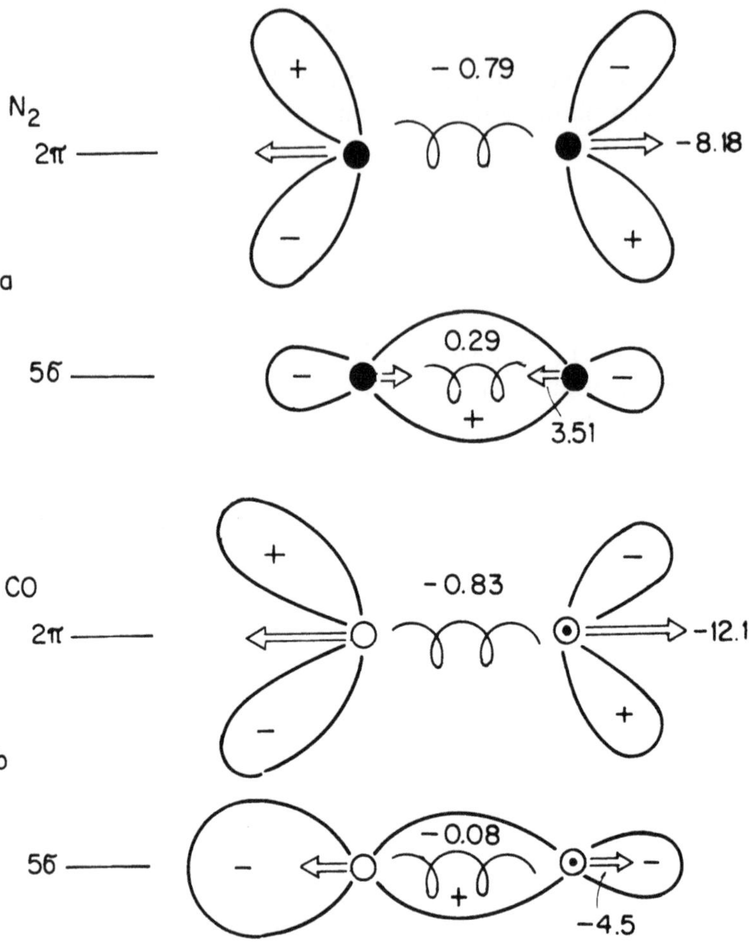

Figure 1.3. Illustrations of the vibronic structure of molecules. In addition to the MO energies and wave function, the OVC characterize the contribution of the orbital electron to the distorting force (shown by arrows; values given in 10^{-4} dyn) and force constant (shown by springs; values given in 10^6 dyn/cm): (a) HOMO and LUMO of the nitrogen molecule — the HOMO 5σ is bonding whereas the LUMO 2π is antibonding; (b) HOMO and LUMO of carbon monoxide — the HOMO is weakly antibonding whereas the LUMO is strongly antibonding.

analogous MO pushes them away with force $f_R^{5\sigma}(CO) = -4.5 \cdot 10^{-4}$ dyn. (More precisely, the binding of the nuclei by the electron is less than their interrepulsion by this amount.) The electron of the LUMO 2π pushes away the nuclei in both the N_2 and CO molecules with forces $f_R^{2\pi}(N_2) = -8.18 \cdot 10^{-4}$ dyn and $f_R^{2\pi}(CO) = -12.1 \cdot 10^{-4}$ dyn, respectively. For the appropriate one-electron MO contribution to the force constant k^i shown in Figure 1.3 by a spring (see equation (6.28)), we have: $k_R^{5\sigma}(N_2) = 0.29 \cdot 10^{-6}$

dyn/cm, $k_R^{2\pi}(N_2) = -0.79 \cdot 10^{-6}$ dyn/cm, $k_R^{5\sigma}(CO) = -0.08 \cdot 10^{-6}$ dyn/cm, and $k_R^{2\pi}(CO) = -0.83 \cdot 10^{-6}$ dyn/cm.

It follows from these data that the orbital 5σ is bonding in N_2 and antibonding in CO, whereas the 2π MO is antibonding in both cases. Note that when there are several orbitals of the same type, as in the cases under consideration ($5\sigma, 4\sigma, 3\sigma, \ldots$), it is very difficult to reveal the bonding nature of each of them, even qualitatively, without *the vibronic approach.* The latter *also provides quantitative information about the degree of the bonding or antibonding nature of the appropriate MO, and its contribution to the distorting force, force constants, as well as anharmonicity constants* (see Chapter 6). As mentioned above, this information is of special importance for analysis of the influence of electronic rearrangements on the nuclear configuration behavior.

The diatomics N_2 and CO, taken as examples, are the simplest. In the case of polyatomic molecules, the OVC and the orbital contributions to the force constants contain much more information than the measure of the bonding or antibonding nature of the MO. Indeed, from the group theory rules,[23] the linear OVC $f_{\bar{\Gamma}}^{(ij)} = \langle i | (\partial V / \partial Q_{\bar{\Gamma}})_0 | j \rangle$ is nonzero, if the direct product of the irreducible representations Γ_i and Γ_j of the i and j MO contain the $\bar{\Gamma}$ representation of the symmetrized displacements $Q_{\bar{\Gamma}}$. For different Γ_i and Γ_j, $\bar{\Gamma}$ may be of any type possible for the symmetry group of the system under consideration. For the diagonal OVC $f_{\bar{\Gamma}}^i$, $\bar{\Gamma}$ must be the component of the symmetrized product $[\Gamma_i \times \Gamma_i]$. Therefore (quite similar to the VC case considered above), if the Γ_i is nondegenerate, $\bar{\Gamma}$ is totally symmetric. In other words, the electrons of nondegenerate MO distort the nuclear configuration along Q_{A_1} while leaving its symmetry unchanged. Depending on the sign of $f_{\bar{\Gamma}}^i$, these MO are either bonding ($f_{\bar{\Gamma}}^i > 0$) or antibonding ($f_{\bar{\Gamma}}^i < 0$).

For degenerate MO the product $[\Gamma_i \times \Gamma_i]$ contains (in addition to the A_1 representation) nontotally symmetric representations. It follows that the electrons of degenerate MO distort the nuclear framework, changing its symmetry in accordance with (and in directions determined by) the Jahn–Teller effect (Section 1.6). In addition to this distortion (in limited directions), the orbital electron softens, or hardens, the nuclear framework according to equation (1.26). The latter contains the nondiagonal OVC $f_{\bar{\Gamma}}^{(ij)}$, for which $\bar{\Gamma}$ may be of either type, and therefore the softening or hardening may be of any direction, depending on the mixing orbitals.

1.5. Anharmonicity and Instability of Molecular Systems

Let us consider now the most essential features of the behavior of stable molecular systems which occur due to the vibronic mixing of the ground state with the excited states. Assume that the system is in a nondegenerate electronic state which, neglecting vibronic mixing, within the harmonic approximation generates small harmonic vibrations near the minimum of the AP in accordance with equation (1.12) and its solutions (1.13). As pointed out in Section 1.1, use of the simple adiabatic approximation in those cases when criterion (1.10) is satisfied introduces an error of the order of $(m/M)^{1/2} \cong 0.02$. On the other hand, use of the harmonic approximation (i.e., expansion of the AP up to and including quadratic terms) implies that the nuclear displacements from the stable configurations are small, as compared with the interatomic distances.

Using the quasiclassical approximation to the problem of harmonic vibrations, it can be shown[13] that the root-mean-square displacements in the ground state are $(\overline{\Delta Q^2})^{1/2} \sim d(m/M)^{1/4}$, where d is the dimension of the system; hence the harmonic approximation introduces an error of $(\overline{\Delta Q^2})^{1/2}/d \sim (m/M)^{1/4} \cong 0.15$. It follows that *the error of the harmonic approximation is much greater than that of the adiabatic approximation.* In other words, in order to improve the solutions obtained in the adiabatic and harmonic approximations, the anharmonicity has to be taken into account first, and then, in the next order of the $(m/M)^{1/4}$ magnitude, the nonadiabaticity of the simple adiabatic approximation must be included.

Consider the AP in the harmonic approximation having the form of a parabola, $\varepsilon(Q) = \frac{1}{2}KQ^2$ (Figure 1.4). Its distinguishing feature is a constant curvature $K = \partial^2\varepsilon/\partial Q^2$ at all points Q. If the cubic and higher-order terms in Q are taken into account, the AP becomes anharmonic and the curvature becomes dependent on Q (in Figure 1.4 the curvature decreases with positive, and increases with negative Q, which is characteristic for cubic anharmonicity).

What is the origin of the anharmonicity in the AP? An obvious source of anharmonicity is proved by higher-order terms in the expansion (1.2) of the electron–nuclear and nuclear–nuclear interactions $V(r, Q)$. If the diagonal matrix elements of the expansion coefficients of higher-than-quadratic terms in equation (1.2) (the anharmonicity constants and anharmonic VC) are nonzero, appropriate anharmonic terms occur in the AP (proper anharmonicity). However, *the proper anharmonicity is in general small* and manifests itself in large distance (from the stable configuration) effects only.

Another source of anharmonicity of particular interest is the vibronic mixing of the ground state with the excited states (*vibronic anharmonicity*). As opposed to proper anharmonicity, vibronic anharmonicity occurs already in linear and quadratic approximations of the Hamiltonian (in the $V(r, Q)$ operator), if the appropriate nondiagonal VC mixing the electronic states are taken into account.

Consider the general expression (1.18) for the curvature of the AP

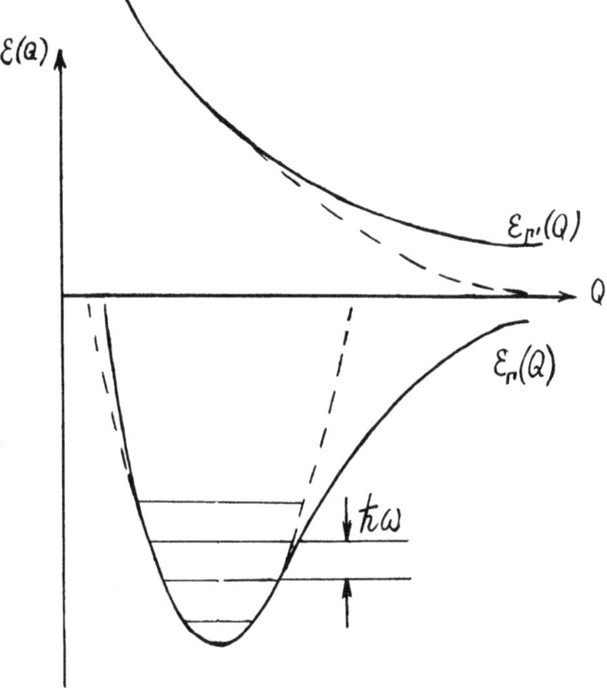

Figure 1.4. Anharmonicity of the AP due to vibronic mixing of the ground state with the near-lying excited state (vibronic anharmonicity). AP of noninteracting states are shown by dashed lines.

surface of electronic state Γ in direction $Q_{\bar{\Gamma}}$, expressed in the more convenient form (index $\bar{\Gamma}$ omitted)

$$K^{\Gamma} = \left\langle \Gamma \left| \left(\frac{\partial^2 V}{\partial Q^2} \right)_0 \right| \Gamma \right\rangle - 2 \sum_{\Gamma'}{}' \frac{|\langle \Gamma | (\partial V / \partial Q)_0 | \Gamma' \rangle|^2}{\varepsilon_{\Gamma'} - \varepsilon_{\Gamma}} \tag{1.27}$$

The first term of this expression, K_0^{Γ}, represents the contribution to K^{Γ} from the nuclear displacement in a fixed (rigid) electronic distribution that does not follow the nuclei. The second term, K_v^{Γ}, accounts for the contribution of changes in electronic states due to nuclear displacements (the "relaxation" term[27]). It is always negative due to the fact that when the electrons follow the nuclear displacements, the resistance to these displacements (the force constants) decreases.

It is evident that the second term, K_v^{Γ}, in equation (1.27) is nonzero only due to vibronic mixing of the ground state with the excited states. Therefore, the two terms K_0^{Γ} and K_v^{Γ} may be called the proper curvature and the vibronic contribution, respectively. In the Born–Oppenheimer (simple adiabatic) approximation the proper curvature K_0^Γ is independent of Q, whereas the vibronic contribution K_v^Γ is strongly dependent on Q and increases with the convergence of the AP $\varepsilon_\Gamma(Q)$ and $\varepsilon_{\Gamma'}(Q)$, thus decreasing the total curvature K^Γ. As a result the AP of the ground state becomes anharmonic (Figure 1.4). It follows that vibronic mixing causes anharmonicity, and this vibronic anharmonicity is determined by the magnitude of the nondiagonal linear VC $F^{(\Gamma\Gamma')}$ and by the convergence of the energy levels (AP) $\varepsilon_{\Gamma'}$ and ε_Γ.

If the negative vibronic contribution K_v^Γ is larger than the proper curvature K_0^Γ, the total curvature $K^\Gamma = K_0^\Gamma + K_v^\Gamma$ becomes negative. At the point of equilibrium where $F^\Gamma = 0$, the inequality $K^\Gamma \leqslant 0$ means that the AP has no minimum at this point (for $K^\Gamma < 0$ there is a maximum), and hence the system is dynamically unstable (as opposed to static instability for which $F^\Gamma \neq 0$). If K_0^Γ is positive, dynamic instability is completely caused by vibronic mixing. At points where $K^\Gamma > 0$ the positiveness of K_0^Γ is obvious, since $K_v^\Gamma < 0$. However, if $K^\Gamma < 0$ it remains unknown, *a priori*, whether this inequality is due only to the vibronic contribution K_v^Γ, or whether K_0^Γ is also negative.

A proof that $K_0^\Gamma > 0$ has been presented recently[28] with a self-consistent field MO LCAO approximation (a more rigorous proof was carried out recently[28a]). With atomic $1s$ functions $\exp(-\lambda R)$ (with other functions the results are similar) and with the Mulliken approximation for three-center integrals, the expression for the energy of the molecule as a function of internuclear distances and for its second derivatives (under the condition that the wave functions do not depend on the nuclear displacements) can be reduced to an additive function of pairwise interactions between the atoms. For the proper curvature K_0 of the AP for a system with closed shells we have (the index Γ is omitted)

$$K_0 = \sum_{\alpha,\beta} K_0^{\alpha\beta}$$

$$K_0^{\alpha\beta} = \tfrac{1}{3}(Z_\alpha q_\alpha \lambda_\alpha^3 + Z_\beta q_\beta \lambda_\beta^3) + Z_\alpha Z_\beta / R_0^3 - \sum_{i,j=\alpha,\beta}{}' (Z_i q_j / 2R_0^3)[1 - (1 + 2\lambda_j R_0 + 2\lambda_j^2 R_0^2 + 2\lambda_j^3 R_0^3)\exp(-2\lambda_j R_0)] \quad (1.28)$$

where Z_i is the effective charge of the core of the ith atom, q_i is its total MO electron population ($q_i = \Sigma_{m,v} 2C_{mi}^* C_{vi} S_{mv}$, where C_{mi} are the LCAO coefficients, and S_{mv} is the overlap integral) and R_0 is the distance between the α and β atoms.

It follows from this expression that for any neutral system in any state, ground or excited (specified by the values of q_i), except the uninteresting case of rigid ions of different signs (without electrons), $K_0^{\alpha\beta} > 0$ and hence

$K_0 > 0$ in either direction. This result means that any displacement of the nuclei from their initial positions, for which the electronic distribution is self-consistent with the nuclear configuration, increases the potential energy, provided the electronic distribution does not change under the nuclear displacements. But if $K_0 > 0$, the dynamic instability of the molecular system in a given electronic state is due to, and only due to, vibronic mixing with the excited states.(28)

Based on the inequalities $K_0 > 0$ and $K_v < 0$, the total curvature $K = K_0 + K_v$ equals the difference between two contributions of opposite signs, each of which may be very large. The values of K_0 and K_v depend on the number of electrons, the states of which remain unchanged when the nuclei displace. If part of the electron system is assumed to follow the nuclei exactly, i.e., if the system is divided approximately into a rigid core and valence electrons (see Section 1.3), the absolute values of K_0 and K_v decrease. For example, using ab initio electronic structure calculation data for the N_2 molecule,(29) the values of $K_0 = 6018.688$ a.u. and $K_v = -6017.216$ a.u., resulting in $K = 1.472$ a.u. $= 2.292 \cdot 10^6$ dyn/cm, have been obtained. On the other hand, the assumption that all the electrons follow their own nuclei while taking into account the polarization of the electronic distribution by the nuclear displacements results in $K_0 = 4.564$ a.u. and $K_v = -3.061$ a.u. with practically the same difference K_0 as above. For force constant calculations the second approach is obviously preferable.

Note that when disregarding electronic polarization (i.e., under the assumption that all the electrons follow the nuclei exactly), the electrons do not contribute to the force constant,(29) and according to equation (1.28) $K_0^{\alpha\beta} = Z_\alpha Z_\beta / R_0^3$. However, in a polyatomic system the electronic cloud cannot follow the displacements of all the nuclei simultaneously. If each electron exactly follows one (its own) nucleus (resulting in the model of rigid atoms or ions), then $K = 0$ for a neutral, statically stable homeopolar system ($Z_\alpha = Z_\beta = 0$), and $K < 0$ for a heteropolar system (Z_α and Z_β have opposite signs). These values of K illustrate the unacceptability of the model of electrons exactly following the nuclei (the rigid ion model).

If the instability of the given state Γ is due to vibronic mixing with mainly one excited state Γ', the latter has some interesting features. In particular, the Γ' state has been shown to be dynamically stable.(28) It suggests the conclusion that for any group of atoms unstable in the ground state, there may be stable excited states. A theorem was formulated(28) as follows (see Chapter 6): *for any nuclear configuration of a neutral group of atoms* (*which in principle can be statically stable*) *there are bonding electronic* (*ground or excited*) *states*. The electronic state is considered to be bonding if its AP surface has a well (minimum), and the nuclear configuration under consideration corresponds to any point of the well area (not necessarily to the minimum point itself). Experimental observation of the bonding state is possible only if the minimum depth is larger than the kinetic energy of the motions in it.[22]

Thus *vibronic mixing creates the main contribution to the anharmonicity of nuclear motions, and determines completely the dynamic instability of molecular systems.*

1.6. The Jahn–Teller Theorem

The Jahn–Teller (JT) theorem is based on the group-theoretical analysis of the behavior of the AP of a polyatomic system near the point of electronic degeneracy. Similar to other group-theoretical statements, the JT theorem allows one to deduce a series of qualitative results without performing specific calculations, or essentially reduces the extent of such calculations. However, as opposed to usual situations in quantum chemistry, in which the group-theoretical treatment arises to simplify the calculations, the proof of the JT theorem preceded the calculations of the AP and stimulated such calculations.

Suppose that by solving the electronic Schrödinger equation (1.3) for the nuclei fixed at the point $Q_{\Gamma\gamma} = Q^0_{\Gamma\gamma} = 0$, we obtain an f-fold degenerate electronic term, i.e., f states $\varphi_k(r)$, $k = 1, 2, \ldots, f$, with identical energies $\varepsilon'_k = \varepsilon_0$. How do these energy levels vary under nuclear displacements $Q_{\Gamma\gamma} \neq 0$? To answer this question the AP for arbitrary coordinates $Q_{\Gamma\gamma}$ near the point of degeneracy must be determined. This can be conducted by estimating the effect of the vibronic interaction (VI) terms $W(r, Q)$ after equation (1.14) on the energy level positions ε'_k.

For sufficiently small nuclear displacements $Q_{\Gamma\gamma}$, the AP $\varepsilon_k(Q)$ for an f-fold degenerate electronic term Γ can be obtained as solutions of the secular equation

$$\begin{vmatrix} W_{11} - \varepsilon & W_{12} & \cdots & W_{1f} \\ W_{21} & W_{22} - \varepsilon & \cdots & W_{2f} \\ \vdots & \vdots & & \vdots \\ W_{f1} & W_{f2} & \cdots & W_{ff} - \varepsilon \end{vmatrix} = 0 \qquad (1.29)$$

where W_{ij} are the matrix elements of the VI operator calculated with the wave functions of the degenerate term. Since the degeneracy is assumed due to the high symmetry of the system, the totally symmetric displacements Q_A, not changing the symmetry, do not remove the degeneracy and will not be considered in this section as nonessential. Again, second-order terms may also be omitted due to the assumed small values of $Q_{\Gamma\gamma}$. As a result the matrix elements of the VI contain only linear, nontotally symmetric terms $W_{ij} = \Sigma_{\Gamma\gamma} \langle i | (\partial V / \partial Q_{\Gamma\gamma}) | j \rangle Q_{\Gamma\gamma}$ or, taking into account equations (1.15) and (1.16′),

$$W_{\Gamma\gamma\Gamma\gamma'} = \sum_{\bar{\Gamma}\bar{\gamma}} F_{\bar{\Gamma}\bar{\gamma}}^{\Gamma\gamma\Gamma\gamma'} Q_{\bar{\Gamma}\bar{\gamma}} = \sum_{\bar{\Gamma}\bar{\gamma}} F_{\bar{\Gamma}}^{\Gamma} Q_{\bar{\Gamma}\bar{\gamma}} \langle \bar{\Gamma}\bar{\gamma}\Gamma\gamma' | \Gamma\gamma \rangle \tag{1.30}$$

where $F_{\bar{\Gamma}}^{\Gamma}$ are the linear vibronic constants (1.15). If at least one of them is nonzero, then at least one of the roots ε of equation (1.29) contains linear terms in the appropriate displacements $Q_{\bar{\Gamma}\bar{\gamma}}$, and hence the AP $\varepsilon_k(Q)$ has no minimum at the point $Q_{\Gamma\gamma}^{0} = 0$ with respect to these displacements.

On the other hand, the question whether the VC $F_{\bar{\Gamma}}^{\Gamma} = \langle \Gamma | (\partial V / \partial Q_{\bar{\Gamma}})_0 | \Gamma \rangle$ is zero or not may be easily answered by means of the well-known group-theoretical rule: $F_{\bar{\Gamma}}^{\Gamma}$ is nonzero if and only if the symmetric product $[\Gamma \times \Gamma]$ contains the representation $\bar{\Gamma}$ of the symmetrized displacement $Q_{\bar{\Gamma}}$. For instance, for an E term $[E \times E] = A_1 + E$ and thus, if the system under consideration has E vibrations (see the classification of vibrations given in Table 1.1), it has no minimum at the point of degeneracy with respect to the E displacements.

Jahn and Teller[4] examined all types of degenerate terms of all symmetry point groups and showed that for any orbital degenerate term of any molecular system there are nontotally symmetric displacements with respect to which the AP of the electronic term (more precisely, at least one of its branches) has no minimum; molecules with linear arrangement of atoms are exceptions (see below). It was later shown by Jahn[30] that a similar statement is valid also in case of spin degeneracy, with the exception of twofold degeneracy for systems with total spin 1/2 (Kramers degeneracy). The statement about the absence of extrema at the point of degeneracy is just *the Jahn–Teller theorem*[4] which may be formulated more rigorously as follows: *if the adiabatic potential of a nonlinear polyatomic system has several* ($f > 1$) *sheets coinciding at one point* (f*-fold degeneracy*), *at least one of them has no extremum at this point, the cases of Kramers degeneracy being exceptions.*

The variation of the AP in the simplest case of a double-degenerate electronic term in the space of only one coordinate Q is shown schematically in Figure 1 of the Introduction: other examples are considered in Chapter 2. It is seen that the two curves intersect at the point of degeneracy. Away from this point the energy term splits, and the degeneracy is removed. As a result, the energy is lowered so that the small nuclear displacements Q are of advantage. For larger values of Q the quadratic, cubic and higher-order terms become important and further distortion of the system may be inconvenient energetically (see Chapter 2).

The exclusion of linear molecules from the JT statement needs clarification. For linear molecules the nontotally symmetric displacements are of odd type with respect to reflections in the plane containing or perpendicular to the molecular axis, whereas the product of wave functions of the degenerate term is always even with respect to such reflections. This means that the appropriate VC equals zero, and hence the AP at the point of linear configuration, as opposed to nonlinear molecules, has no linear terms, i.e., it is extremal.

However, if the linear terms of the VI $W(r, Q)$ are zero, the quadratic terms are of primary importance (see equation (1.7)). It can be shown that the matrix elements of terms, second order in Q, in W in the case of linear molecules in degenerate states, are nonzero, and the solution of secular equation (1.29) results in splitting of the energy term and instability. This was shown by Renner[31] prior to the publication of the paper of Jahn and Teller.[4] Figure 1.5 illustrates this case schematically. It is seen that, although at the point of degeneracy $Q = 0$ the two branches of the AP for the double-degenerate term are extremal (there are no linear terms in the AP), in case of strong coupling (Figure 1.5b) one of them has a maximum, i.e., the system is also unstable. However, as distinct from the JT linear (static) instability, in case of linear molecules the instability is quadratic (dynamic). As with the Jahn–Teller effect, *the dynamic instability of linear molecules in electron degenerate states is called the Renner effect* (more detailed considerations are given elsewhere[32,32a,32b]).

The lack of a minimum of the AP at the point of electronic degeneracy is usually interpreted as instability of the nuclear configuration at this point. Therefore, the formulation of the JT theorem is often given as follows: a nonlinear polyatomic system in the nuclear configuration with a degenerate electronic ground state term is unstable. Also, this statement of instability is treated in the sense that the system distorts itself spontaneously so that the electronic term splits and the ground state becomes nondegenerate.

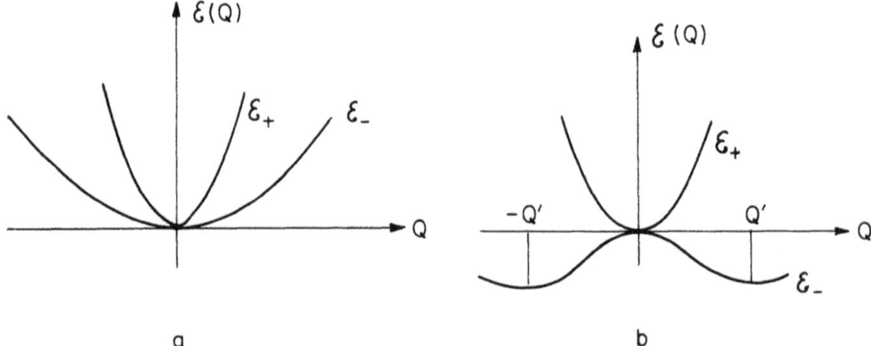

Figure 1.5. Behavior of AP in the case of the Renner effect: (a) weak coupling, term splitting without instability; (b) strong coupling, dynamic instability in lower sheet.

This formulation and interpretation of the JT theorem initiated by its authors[4] now appears in monographs and handbooks and has widespread use in the general treatment of experimental results. Meanwhile, as will be shown later in this book, the actual situation in systems with electron degeneracy is much more complicated than implied by the simple statement of instability. Moreover, taken literally, this statement is not true and may lead to misunderstanding.[33]

The conclusion about the lack of a minimum of the AP $\varepsilon(Q)$ at the point of degeneracy was reached as a consequence of the solution of the electronic part of the Schrödinger equation, and therefore it cannot be attributed to the nuclear behavior. The latter is determined by the solution of the nuclear motion equations and can be obtained only after their solution. The absence of a minimum of the function $\varepsilon(Q)$ may generally be interpreted as instability only when there is no degeneracy. Indeed, in the absence of degeneracy (or in areas far from the point of degeneracy) the electronic and nuclear motions can be separated in the adiabatic approximation (Section 1.1), so that the AP $\varepsilon(Q)$ has the meaning of potential energy of the nuclei in the mean field of the electrons, and hence the derivative $(d\varepsilon/dQ)_0$ means the force acting upon the nuclei at the point $Q^0_{\Gamma\gamma}$. Here the condition $(d\varepsilon/dQ)_0 \neq 0$ may be interpreted as a nonzero distorting force (in the Q direction), due to which the nuclear configuration becomes unstable.

However, in the presence of electronic degeneracy the AP $\varepsilon(Q)$ loses the meaning of potential energy of the nuclei in the mean field of the electrons, since the motions of the electrons and nuclei near the point of degeneracy cannot be separated. In this area the notions $\varepsilon(Q)$ (or the AP as a whole) become formal without unambiguous physical meaning. Accordingly, the reasoning given above about distorting forces and instability is, strictly speaking, invalid here. In these cases the term "instability" should be taken formally as an indication of the lack of a minimum of the AP, but not as a nuclear feature. The latter, as indicated above, must be deduced from the solutions of equations (1.6) of nuclear dynamics.

The nuclear distortions due to the Jahn–Teller effect (JTE) and the pseudo-Jahn–Teller effect (PJTE) in the absence of low symmetry perturbations are in general of a dynamic nature, for which the nuclear configuration (determined quantum-mechanically as the Ψ averaged values of the nuclear coordinate operators) is not distorted.[33] *It is not simply the nuclear configuration distortion, but special nuclear dynamics which are predicted by the JTE and PJTE in free (unperturbed) molecular systems.* The electronic degeneracy is not removed either, but it is transferred from an electronic to a vibronic degeneracy (see Chapter 3).

The specific JT behavior of the AP due to electronic degeneracy or pseudodegeneracy (as other features of electrostatic origin) can be ex-

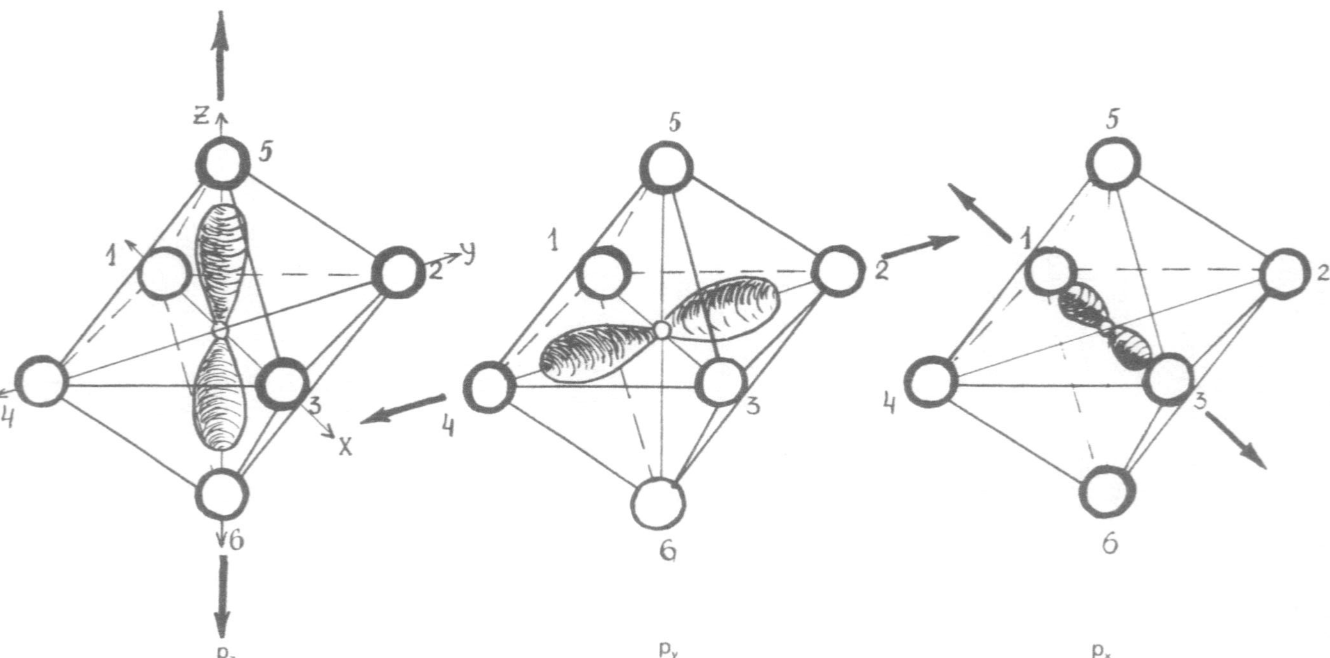

Figure 1.6. Rough illustration of the electrostatic origin of the JTE. If the electron of the central atom of an octahedral complex falls in one of the three states equivalent in energy (threefold degeneracy), it repels (or attracts) the appropriate pair of ligands resulting in tetragonal distortion of the octahedron. The three equivalent directions of distortion are shown by arrows.

Table 1.3. JT Active Symmetrized Displacements (Normal Coordinates) Determined as Components Γ_i of the Expansion $[\Gamma^2] = \Sigma_i \Gamma_i$, Except the Totally Symmetric One. For Double-Valued Representations the Expansion of the Antisymmetric Products $\{\Gamma^2\} = \Sigma_i \Gamma_i$ Must be Used (after Jahn and Teller[4])

Symmetry group[a]	Expansion of the symmetric or antisymmetric products of representations
C_∞^v	$[E_k^2] = A_1 + E_{2k} \quad (k = 1, 2, \ldots)$
C_∞^{vi}	$[E_{kg}^2] = [E_{ku}^2] = A_{1g} + E_{2k,g} \quad (k = 1, 2, \ldots)$
C_{2p+1}	$[E_k^2] = \begin{cases} A + E_{2k} & \text{for } k \leqslant p/2 \\ A + E_{2p+1-2k} & \text{for } k > p/2 \end{cases} \quad (k = 1, 2, \ldots, p)$
C_{2p}	$[E_k^2] = \begin{cases} A + E_{2k} & \text{for } k < p/2 \\ A + 2B & \text{for } k = p/2 \\ A + E_{2p-2k} & \text{for } k > p/2 \end{cases} \quad (k = 1, 2, \ldots, p-1)$
D_{2p+1} and C_{2p+1}^v	$[E_k^2] = \begin{cases} A_1 + E_{2k} & \text{for } k \leqslant p/2 \\ A_1 + E_{2p+1-2k} & \text{for } k > p/2 \end{cases} \quad (k = 1, \ldots, p)$
D_{2p} and C_{2p}^v	$[E_k^2] = \begin{cases} A_1 + E_{2k} & \text{for } k < p/2 \\ A_1 + B_1 + B_2 & \text{for } k = p/2 \\ A_1 + E_{2p-2k} & \text{for } k > p/2 \end{cases} \quad (k = 1, \ldots, p-1)$
C_{2p+1}^i	$[E_{kg}^2] = [E_{ku}^2] = \begin{cases} A_g + E_{2k,g} & \text{for } k < p/2 \\ A_g + E_{2p+1-2k,g} & \text{for } k > p/2 \end{cases} \quad (k = 1, \ldots, p)$
C_{2p}^i	$[E_{kg}^2] = [E_{ku}^2] = \begin{cases} A_g + E_{2k,g} & \text{for } k < p/2 \\ A_g + 2B_g & \text{for } k = p/2 \\ A_g + E_{2p-2k,g} & \text{for } k > p/2 \end{cases} \quad (k = 1, \ldots, p-1)$
D_{2p+1}^i	$[E_{kg}^2] = [E_{ku}^2] = \begin{cases} A_{1g} + E_{2k,g} & \text{for } k \leqslant p/2 \\ A_{1g} + E_{2p+1-2k,g} & \text{for } k > p/2 \end{cases} \quad (k = 1, \ldots, p)$
D_{2p}^i	$[E_{kg}^2] = [E_{ku}^2] = \begin{cases} A_{1g} + E_{2k,g} & \text{for } k < p/2 \\ A_{1g} + B_{1g} + B_{2g} & \text{for } k = p/2 \\ A_{1g} + E_{2p-2k,g} & \text{for } k > p/2 \end{cases} \quad (k = 1, \ldots, p-1)$
C_{2p+1}^h and D_{2p+1}^h	$[E_K^2] = [E_k^2] = \begin{cases} A' + E'_{2k} & \text{for } k \leqslant p/2 \\ A' + E'_{2p+1-2k} & \text{for } k > p/2 \end{cases} \quad (k = 1, \ldots, p)$
S_{4p}	$[E_k^2] = \begin{cases} A + E_{2k} & \text{for } k < p \\ A + 2B & \text{for } k = p \\ A + E_{4p-2k} & \text{for } k > p \end{cases} \quad (k = 1, \ldots, 2p-1)$
S_{4p}^v	$[E_k^2] = \begin{cases} A_1 + E_{2k} & \text{for } k < p \\ A_1 + B_1 + B_2 & \text{for } k = p \\ A_1 + E_{4p-2k} & \text{for } k > p \end{cases} \quad (k = 1, \ldots, 2p-1)$

continued

Table 1.3 — *continued*

Symmetry group[a]	Expansion of the symmetric or antisymmetric products of representations
T	$[E^2]=A+E$ $[T^2]=A+E+T$ $\{\Gamma_8^2\}=A+E+T$
T_d and O	$[E^2]=A_1+E$ $[T_1^2]=[T_2^2]=A_1+E+T_2$ $\{\Gamma_8^2\}=A_1+E+T_2$
T_h	$[E_g^2]=[E_u^2]=A_g+E_g$ $T_g^2=T_u^2=A_g+E_g+T_g$ $\{\Gamma_{8g}^2\}=\{\Gamma_{8u}^2\}=A_g+E_g+T_{2g}$
O_h	$[E_g^2]=[E_u^2]=A_{1g}+E_g$ $[T_{1g}^2]=[T_{1u}^2]=[T_{2g}^2]=[T_{2u}^2]=A_{1g}+E_g+T_{2g}$ $\{\Gamma_{8g}^2\}=\{\Gamma_{8u}^2\}=A_{1g}+E_g+T_{2g}$
I	$[T_1^2]=[T_2^2]=A+H$ $[G^2]=A+G+H$ $[H^2]=A+G+2H$ $\{\Gamma_8^2\}=A+H$ $\{\Gamma_{10}^2\}=A+G+2H$
I_h	$[T_{1g}^2]=[T_{1u}^2]=[T_{2g}^2]=[T_{2u}^2]=A_g+H_g$ $[G_g^2]=[G_u^2]=A_g+G_g+H_g$ $[H_g^2]=[H_u^2]=A_g+G_g+2H_g$ $\{\Gamma_{8g}^2\}=\{\Gamma_{8u}^2\}=A_g+H_g$ $\{\Gamma_{10g}^2\}=\{\Gamma_{10u}^2\}=A_g+G_g+2H_g$

[a] Superscripts *i*, *v* and *h* denote the addition (to the appropriate symmetry axis) of an inversion center (*i*), reflection plane containing the symmetry axis (*v*), and reflection plane perpendicular to the symmetry axis (*h*), respectively.

plained by simple images. Consider, for example, the case of one electron in one of three equivalent p-orbitals (p_x, p_y, and p_z) of the central atom of a hexacoordinated molecular system MX_6 (term T_{1u}), the ligands X bearing negative charges (Figure 1.6). It is clear that if the electron is on the p_x orbital, it interacts more strongly with the nearest ligands 1 and 3, and pushes them away. As a result the octahedral complex becomes tetragonally distorted along the $0x$ axis.

Similarly, the electron on the p_y-orbital pushes away ligands 2 and 4 distorting the complex equivalently to the previous case, but along the $0y$ axis, and for the electron on the p_z-orbital the distortion is along $0z$. It follows that in this case the AP of the system has three minima, corresponding to the appropriate distortions, and no minimum in the high-symmetry octahedral configuration (see Section 2.2) due to obvious electrostatic forces. The same result emerges from the MO description with the atomic p-orbital taking part in the appropriate antibonding MO (for bonding orbitals the distortions are of opposite sign).

The proof of the JT theorem by means of an examination of all types of degenerate terms in all symmetry point groups[4] as illustrated above, although rigorous, cannot be considered elegant. More general proofs have been obtained recently[34] (see also Bersuker and Polinger[13]). However, an examination of all possible degenerate terms has its advantages, one of which is to reveal the JT active modes, i.e., the nuclear displacements $Q_{\bar{\Gamma}}$, for which the VC $F_{\bar{\Gamma}}^{\Gamma}$ is nonzero and which remove the electron degeneracy of the electronic term Γ. These active modes (active vibrations) are very important basic components of the JT problem since they form the space of the nontotally symmetric nuclear displacements in which the Jahn–Teller effect is realized.

If the types of possible vibration of the system under consideration are known (Table 1.1), the JT active nuclear displacements $Q_{\bar{\Gamma}}$ may be obtained easily. As indicated above, $F_{\bar{\Gamma}}^{\Gamma}$ is nonzero if the symmetric product $[\Gamma \times \Gamma]$ contains $\bar{\Gamma}$. It follows that the $\bar{\Gamma}$ displacements can be found as the irreducible parts of the reducible representation $[\Gamma \times \Gamma]$. The nontotally symmetric components of the latter are just the representations $\bar{\Gamma}$ of the active displacements $Q_{\bar{\Gamma}}$. For instance, for an E_g term in an octahedral system, $[E_g \times E_g] = A_{1g} + E_g$, and hence the JT active displacements are of E_g type. Similarly, for a T term in a tetrahedral system $[T \times T] = A_1 + E + T_2$, and both the E- and T_2-type displacements are JT active. The JT active displacements for all important point groups are given in Table 1.3.

[illegible]

2

Adiabatic Potentials

The determination of the adiabatic potentials is an important stage in the solution of vibronic problems. Some qualitative features of the vibronic effects may be judged by the potential shapes without solving the complicated equations of nuclear motion.

2.1. General Considerations. The Orbital Doublet (E term)

It follows from the Jahn–Teller theorem that at the point of the nuclear configuration where the electronic state is degenerate, the surface of the potential energy of the nuclei in the mean field of electrons has no minimum. The question arises whether this surface possesses any minimum, and if so, where it is situated. A more general formulation of this problem is: what is the stable configuration of the nuclei, their dynamics, and energy spectra in the presence of the Jahn–Teller effect? To answer this question the shape of the adiabatic potential of the system $\varepsilon(Q)$ in the configurational space of all nuclear displacements Q must first be determined. For a nondegenerate electronic state $\varphi_k(r)$ the expression for $\varepsilon_k(Q)$ is given by equation (1.8). If the electronic state is f-fold degenerate, the adiabatic potential (AP) has f sheets $\varepsilon_k(Q)$, $k = 1, 2, \ldots, f$, which intersect at the point of degeneracy. The functions $\varepsilon_k(Q)$ are determined by secular equation (1.29).

Before proceeding to the solution of equation (1.29) for specific types of systems, some general considerations are necessary. First we separate the totally symmetric part of the diagonal matrix elements of the vibronic interactions, which give rise to the force constants K_Γ (Section 1.3). Then we choose the initial configuration of the system at the point $Q_{\Gamma\gamma} = 0$, where the AP without the vibronic interaction (VI) has a minimum, and assume that the proper anharmonicity (Section 1.5) may be neglected. Under these conditions (see Bersuker and Polinger[13] for more details) the f sheets of the AP of an f-fold degenerate electronic term are given by the following expressions (the $\tilde{\Gamma}$ subscript of the terms is omitted where possible):

$$\varepsilon_k(Q_{\Gamma\gamma}) = \tfrac{1}{2}\sum_{\Gamma\gamma} K_\Gamma Q_{\Gamma\gamma}^2 + \varepsilon_k^{\mathrm{v}}(Q_{\Gamma\gamma}), \qquad k = 1, 2, \ldots, f \tag{2.1}$$

where $\varepsilon_k^{\mathrm{v}}(Q_{\Gamma\gamma})$ are the roots of the secular equation

$$\| W_{\tilde{\gamma}\tilde{\gamma}'}^{\mathrm{v}} - \varepsilon^{\mathrm{v}} \| = 0, \qquad \tilde{\gamma}, \tilde{\gamma}' = 1, 2, \ldots, f \tag{2.2}$$

in which, unlike equation (1.29), the diagonal matrix elements $W_{\tilde{\gamma}\tilde{\gamma}'}^{\mathrm{v}}(Q_{\Gamma\gamma})$ do not contain the totally symmetric part of the quadratic terms used in the force constant formation.

With the adiabatic potential $\varepsilon_k(Q)$ known, the vibronic system of equations (1.6) which determine the nuclear energy spectrum and dynamics can be solved (in principle). However, *the determination of the AP is of special interest.* In the region of nuclear configurations far from the point of degeneracy (especially near the minima of the AP in the case of strong vibronic coupling), the energy gap between different sheets of the AP can be sufficiently large. In this case *the nuclei may be approximately treated semiclassically as moving along the AP surface.* Knowledge of the shape of this surface makes possible a qualitative analysis of the nuclei behavior, and the information thus obtained may sometimes be as important as that deduced from a numerical solution of the vibronic equations (1.6).

A molecular system in a doubly degenerate orbital electronic state (the E term case) is the simplest and most widespread JT system. The tables of irreducible representations of symmetry point groups show that E terms are possible for molecules which possess at least one axis of not less than order three. The triangular molecules of type X_3, and triangular-pyramidal, quadratic, quadratic-pyramidal, tetrahedral, octahedral, cubic, pentagonal, and hexagonal molecular systems are those having twofold degenerate E terms in the ground or excited states.

To solve secular equation (2.2) for the functions $\varepsilon_k^{\mathrm{v}}(Q)$, *the matrix elements* $W_{\tilde{\gamma}\tilde{\gamma}'}^{\mathrm{v}}$ *must be calculated first.* According to the JT theorem (Section 1.6) the nonzero parts of $W_{\tilde{\gamma}\tilde{\gamma}'}^{\mathrm{v}}(Q_{\Gamma\gamma})$ for an E term contain only type E or type B_1 and B_2 nontotally symmetric displacements $Q_{\Gamma\gamma}$ (see Table 1.3). The type B_1 and B_2 JT active modes are specific to systems possessing symmetry axes of even-order multiples of four (C_4, C_8, etc.; for these cases there are E terms for which $[E \times E] = A_1 + B_1 + B_2$), whereas the type E terms are inherent in all the other cases containing symmetry axes of both odd and even orders or only of odd order (here $[E \times E] = A_1 + E$). The role of totally symmetric displacements, mentioned briefly in Section 1.6, will be taken into account below. The two types of Jahn–Teller

E-term problems are specified as the E–e and E–(b_1+b_2) problems, respectively.

Consider first the E–e problem. The two electronic wave functions of the E term may be denoted by $|\theta\rangle$ and $|\varepsilon\rangle$ with symmetry properties of the well-known functions $\theta \sim 3z^2 - r^2$ and $\varepsilon \sim x^2 - y^2$ (or $\theta \sim d_{z^2}$ and $\varepsilon \sim d_{x^2-y^2}$ in the transition metal d function nomenclature). The two components of the normal E-type (tetragonal) displacements Q_θ and Q_ε are illustrated in Figure 1.1, while their expressions in Cartesian coordinates of the nuclei are given in Table 1.2. Accordingly the matrix elements $W^{\text{v}}_{\bar{\gamma}\bar{\gamma}'}$ and hence $\varepsilon^{\text{v}}_k(Q_{\Gamma\gamma})$ in equation (2.2) depend only on these two coordinates. Consequently, the AP in all the other coordinates, in line with equation (2.1), retains a simple parabolic form

$$\varepsilon_k(Q_{\Gamma\gamma}) = \tfrac{1}{2}\sum_{\Gamma\gamma}{}' K_\Gamma Q^2_{\Gamma\gamma} \qquad (\Gamma \neq E)$$

The remaining nonzero matrix elements of the VI assume a simple form, if the vibronic constants introduced above are used. Denote

$$F_E = \left\langle \theta \left| \left(\frac{\partial V}{\partial Q_\theta}\right)_0 \right| \theta \right\rangle, \qquad G_E = \frac{1}{2}\left\langle \theta \left| \left(\frac{\partial^2 V}{\partial Q_\theta \partial Q_\varepsilon}\right)_0 \right| \varepsilon \right\rangle$$

Then, retaining only the linear and second-order VI terms, the explicit form of equation (2.2) for the E–e problem is given by[35–37]

$$\begin{vmatrix} F_E Q_\theta + G_E(Q_\theta^2 - Q_\varepsilon^2) - \varepsilon^{\text{v}} & -F_E Q_\varepsilon + 2G_E Q_\theta Q_\varepsilon \\ -F_E + 2G_E Q_\theta Q_\varepsilon & -F_E Q_\theta - G_E(Q_\theta^2 - Q_\varepsilon^2) - \varepsilon^{\text{v}} \end{vmatrix} = 0 \quad (2.3)$$

In polar coordinates

$$Q_\theta = \rho\cos\phi, \qquad Q_\varepsilon = \rho\sin\phi$$

the solution to equation (2.3) is

$$\varepsilon^{\text{v}}_\pm(\rho,\phi) = \pm\rho[F_E^2 + G_E^2\rho^2 + 2F_E G_E \rho\cos 3\phi]^{1/2} \qquad (2.4)$$

Substitution of these values into equation (2.1) yields the following expression for the AP in the space of Q_θ and Q_ε coordinates:

$$\varepsilon_\pm(\rho,\phi) = \tfrac{1}{2}K_E\rho^2 \pm \rho[F_E^2 + G_E^2\rho^2 + 2F_E G_E \rho\cos 3\phi]^{1/2} \qquad (2.5)$$

In particular, when the quadratic VI may be neglected ($G_E = 0$) this surface simplifies (the linear case has been studied previously[38]) to

$$\varepsilon_{\pm}(\rho, \phi) = \tfrac{1}{2}K_E\rho^2 \pm |F_E|\rho \tag{2.6}$$

In this linear approximation the AP is independent of the angle ϕ and has the form of a rotation surface, often called the "*mexican hat*" (Figure 2.1). The radius ρ_0 of the circle at the bottom of the trough and its depth determined from the degeneracy point at $\rho = 0$ (the JT stabilization energy E_{JT}) are given by the relationships

$$\rho_0 = |F_E|/K_E, \qquad E_{JT} = F_E^2/2K_E \tag{2.7}$$

Taking into account the quadratic terms of VI this surface warps, and along the bottom of the trough of the "mexican hat" three wells occur, alternating regularly with three humps (Figures 2.2 and 2.3). The extremal points of the surface (ρ_0, ϕ_0) are

$$\rho_0 = \frac{\pm F_E}{K_E \mp (-1)^n 2G_E}, \qquad \phi_0 = n\pi/3, \quad n = 0, 1, \ldots, 5 \tag{2.8}$$

the upper and lower signs corresponding to cases $F_E > 0$ and $F_E < 0$,

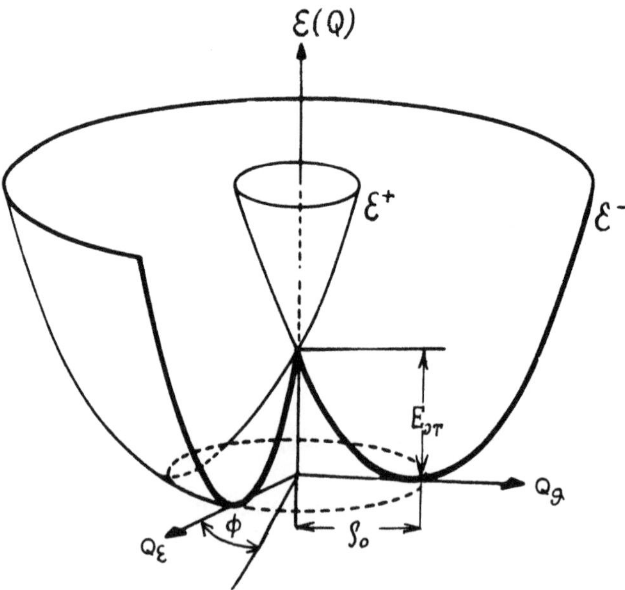

Figure 2.1. "Mexican hat"-type AP shape for a twofold degenerate E term interacting linearly with the twofold degenerate E type vibrations, described by Q_θ and Q_ϵ coordinates (linear E–e problem).

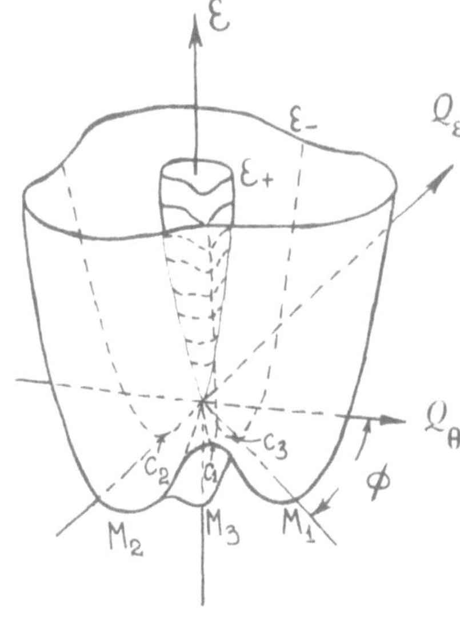

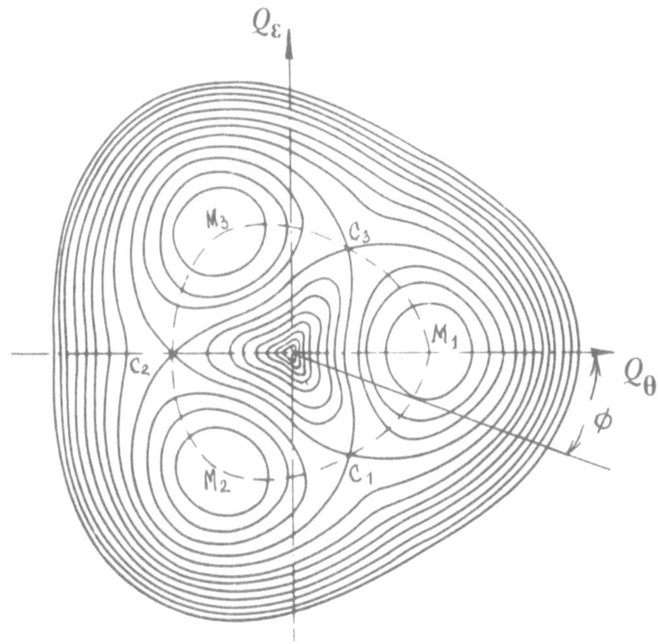

Figure 2.2. Three minima (M_1, M_2, and M_3) and three saddle points (C_1, C_2, and C_3) of the AP of the E–e problem taking into account quadratic terms of VI: (a) in the space of Q_θ and Q_ε coordinates; (b) in the equipotential cross-section of the lower sheet of the surface (the dashed line shows the steep slope from the saddle points to the minima).

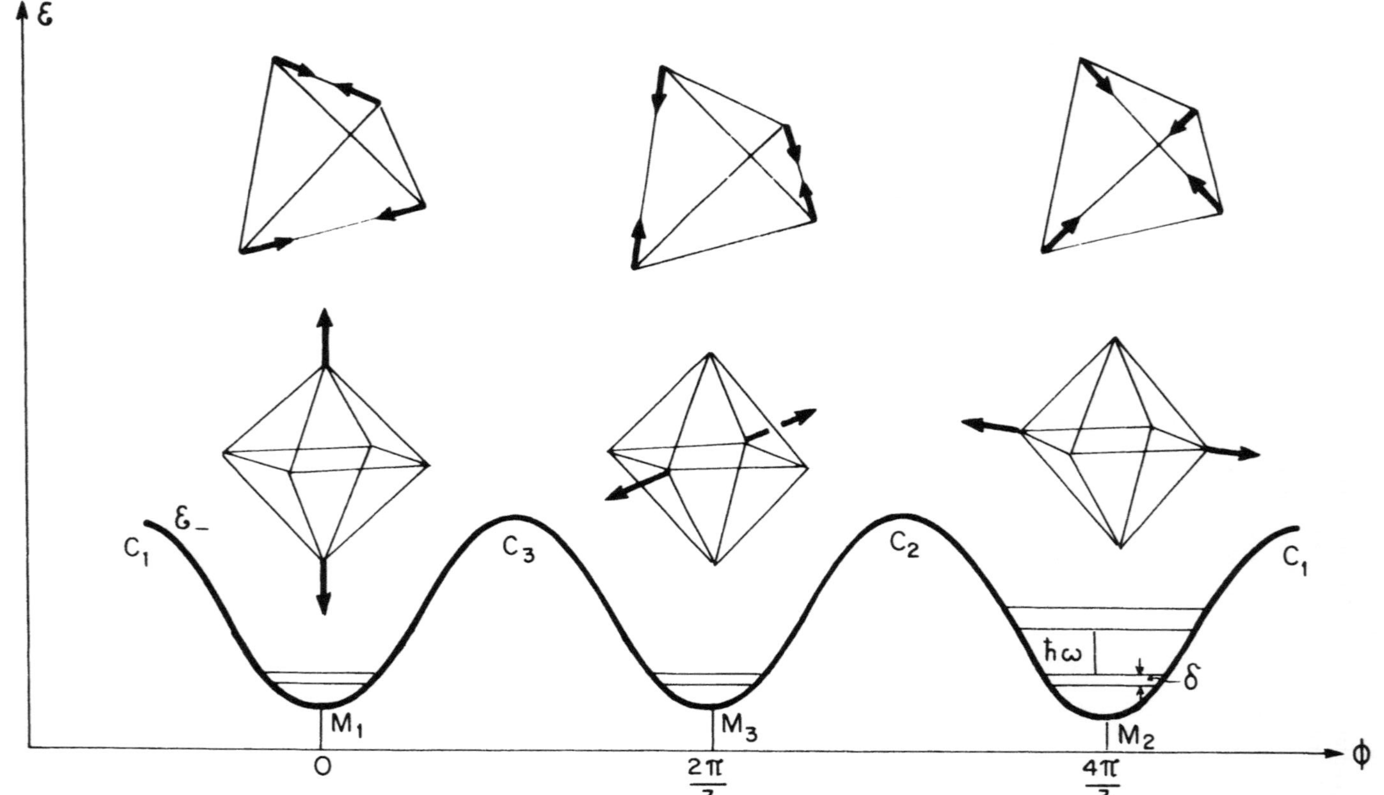

Figure 2.3. Section of the lowest sheet of the AP of the quadratic $E - e$ problem along the steep slope (the dashed line of Figure 2.2b) illustrating distortions of an octahedral or tetrahedral system at the minima.

respectively. If $(F_E \times G_E) > 0$ the points at which $n = 0, 2, 4$ are minima, and those at which $n = 1, 3, 5$ are saddle points, whereas for $(F_e \times G_E) < 0$ these two types of extremal points interchange. For the JT stabilization energy E

$$E_{JT} = F_E^2/2(K_E - 2|G_E|) \tag{2.9}$$

and the (minimum) barrier height Δ between the minima is

$$\Delta = 4E_{JT}|G_E|/(K_E + 2|G_E|) \tag{2.10}$$

In the linear approximation the curvature K of the surface is the same at all points in the bottom of the trough. Along the trough $K_\phi = 0$, and in the perpendicular (radial) direction $K_\rho = K_E$. (Note that in the absence of VI, $K_\phi = K_\rho = K_E$.) Taking into account the quadratic terms one can obtain[13]

$$K_\rho = K_E - 2|G_E|, \qquad K_\phi = 9|G_E|(K_E - 2|G_E|)/(K_E - |G_E|) \tag{2.11}$$

It follows that if $2|G_E| \geqslant K_E$, the system under consideration has no minima at the point ρ_0 and decomposes, provided higher-order terms in Q of the VI neglected above do not make the system stable at larger distances. At the minimal points the curvature equals the force constant, which is proportional to the square of the frequency of the corresponding vibrations.

The two wave functions $\Psi_\pm$, which are the solutions of the perturbation problem corresponding to the two sheets of equation (2.3), are

$$\begin{aligned} \Psi_- &= \cos(\Omega/2)|\theta\rangle - \sin(\Omega/2)|\varepsilon\rangle \\ \Psi_+ &= \sin(\Omega/2)|\theta\rangle + \cos(\Omega/2)|\varepsilon\rangle \end{aligned} \tag{2.12}$$

where

$$\tan\Omega = \frac{F_E \sin\phi - |G_E|\rho\sin 2\phi}{F_E\cos\phi + |G_E|\rho\cos 2\phi}$$

It is often assumed that $\Omega \equiv \phi$, which is true only in the absence of quadratic VI when $G_E = 0$.

The known shapes of the symmetrized displacements Q_θ and Q_ε and their values at the minima can be used to determine the appropriate JT distortions of different types of molecules. Some distortions for tetrahedral and octahedral systems are illustrated in Figure 2.3.

The motions of the nuclei under the AP obtained above can be investigated properly only by solving equations (1.6) (see Chapter 3).

However, some qualitative features of the nuclear behavior, as indicated at the beginning of this section, can be clarified by considering the nuclei moving along the AP. This treatment has some physical basis when the energy gap between different AP sheets is sufficiently large, i.e., for strong vibronic coupling and for nuclear configurations near the minima of the lowest (ground) sheet (where, in the case of the E term, the gap equals $4E_{JT}$).

If only linear terms are considered, i.e., when the lowest sheet of the AP has the shape of a "mexican hat", *the nuclear configuration performs free rotations in the space of Q_θ and Q_ε coordinates along the circle of minima in the trough.* In this case,[39] each atom of, say, a triangle molecule X_3 describes a circle of radius $\rho_0\sqrt{3}$. The variation of the Cartesian coordinates of the atoms x_i, y_i $(i = 1, 2, 3)$ during the motion of the system along the trough is given by

$$x_1 = (1/\sqrt{3})\rho \cos\phi \qquad y_1 = (1/\sqrt{3})\rho \sin\phi$$

$$x_2 = (1/\sqrt{3})\rho \cos[\phi - (2\pi/3)] \qquad y_2 = (1/\sqrt{3})\rho \sin[\phi - (2\pi/3)]$$

$$x_3 = (1/\sqrt{3})\rho \cos[\phi - (4\pi/3)] \qquad y_3 = (1/\sqrt{3})\rho \sin[\phi - (4\pi/3)]$$

It is seen that the circular motions of these atoms are correlated: the vectors of their displacements are shifted in phase through an angle of $2\pi/3$ (Figure 2.4). At any instant of time the equilateral triangle X_3 is distorted

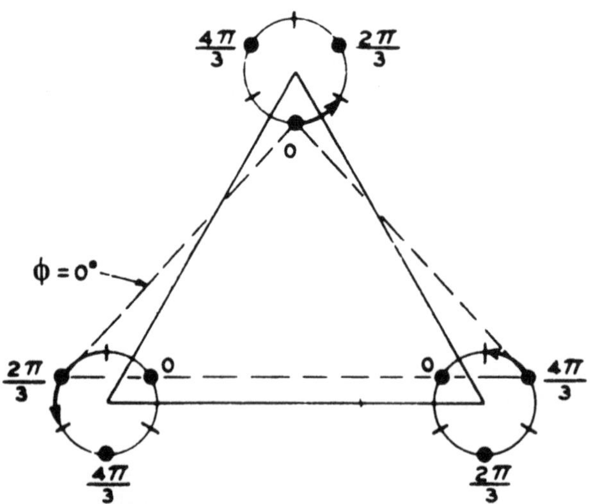

Figure 2.4. Distortions of a triangle X_3 molecule at the points in the bottom of the trough of the AP in the linear E–e problem. Each atom moves along a circle, their phases being concerted. The dashed triangle corresponds to the point $\phi = 0$ in Figure 2.3 (only compressed triangles are shown, although elongated ones are also possible).

into an isosceles triangle, and *this distortion travels as a wave around the triangle's geometric center* (performing a specific internal rotation). In the case of an octahedral molecule, which in a trigonal projection looks like two equilateral triangles, the two deformation waves traveling around each of them are opposite in phase. As a result the octahedron becomes elongated (or compressed) alternatively along one of the three fourfold axes and simultaneously compressed (or elongated) along the remaining two axes (Figure 2.5).

When the quadratic terms of the VI are taken into account, the lowest sheet of the AP has three minima at each of which the octahedron is elongated (or compressed) along one of the three axes of order four (Figure 2.3). *The nuclear motions along the AP surface* when allowing for quantum effects *are likewise hindered rotations and tuneling transitions between the*

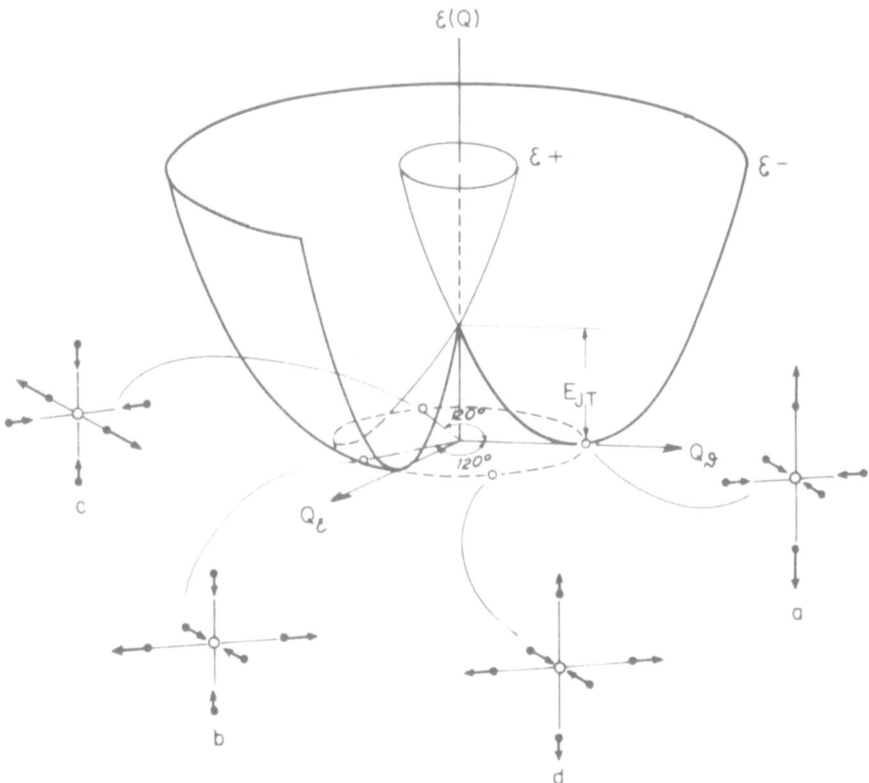

Figure 2.5. Distortions of an octahedral system ML_6 at the points along the bottom of the trough of the "mexican hat" in the linear E–e problem. At points $\phi = 0$, $2\pi/3$, $4\pi/3$ the octahedron is tetragonally distorted along the three fourfold axes, respectively (a, b, c). In between these points the system has D_{2h} symmetry and varies continuously from one tetragonal configuration to another.

minima; these may be presented by way of illustration as "pulse" motions which are discussed in Chapter 3.

As indicated earlier, since $[E \times E] = A_1 + E$, the totally symmetric displacements A_1 are also JT active in the twofold degenerate E state (as in all other cases): strictly speaking, the E–$(e + a_1)$ problem must be solved instead of the E–e problem considered above. However, totally symmetric displacements do not change the molecular symmetry; they change proportionally only the interatomic distances. Therefore, considering vibronic effects, it may be assumed that an origin is taken at a new minimum position with regard to the A_1 coordinates, so that the interaction with the A_1 displacements, as with all other JT inactive modes, becomes unimportant. This cannot be done when one has to compare vibronic effects in different systems or in a series of systems (see Chapter 5) for which the A_1 displacement contributions may be different. For this and other reasons the more rigorous expressions for the AP which include interaction with all the active modes may be useful.

Let us denote the linear VC for totally symmetric displacements ρ_A by F_A and the appropriate force constant by K_A. Then[35]

$$\varepsilon(\rho_A, \rho, \phi) = \tfrac{1}{2}K_A\rho_A^2 - F_A\rho_A + \tfrac{1}{2}K_E\rho^2$$
$$\pm \rho[F_E^2 + G_E^2\rho^2 + 2F_EG_E(\rho\cos 3\phi - 2\sqrt{2}\rho_A)$$
$$+ 4G_E^2\rho_A(2\rho_A - \sqrt{2}\rho\cos 3\phi)]^{1/2} \tag{2.13}$$

The extremal points for this surface are (cf. equation (2.8))

$$\left.\begin{aligned} &\rho_0 = \frac{\pm(F_E + 2\sqrt{2}\beta F_A)}{K_E \mp (-1)^n 2G_E - 4\beta G_E}, \quad \phi_0 = \frac{n\pi}{3}, \qquad n = 0, 1, 2, \ldots, 5 \\ &\rho_{A0} = F_A/K_A \pm 2\sqrt{2}\beta\rho_0 = \rho_{A_0}^0 \pm 2\sqrt{2}\beta\rho_0 \end{aligned}\right\} \tag{2.14}$$

where $\beta = G_E/K_A$ characterizes the role of the quadratic terms of VI with respect to the "homogeneous hardness" of the system K_A.

In the E–$(b_1 + b_2)$ problem, corresponding to the two irreducible representations B_1 and B_2 of the two JT active modes $Q_{b_1} \equiv Q_1$ and $Q_{b_2} \equiv Q_2$, two linear, $F_{b_1} \equiv F_1$ and $F_{b_2} \equiv F_2$ (and two quadratic) VC have to be introduced. Denoting also $K_{b_1} \equiv K_1$ and $K_{b_2} \equiv K_2$, one can easily derive the following (linear approximation) solutions of secular equation (2.2) for the AP of the E–$(b_1 + b_2)$ problem (Figure 2.6):

$$\varepsilon_\pm(Q_1, Q_2) = \tfrac{1}{2}(K_1Q_1^2 + K_2Q_2^2) \pm [F_1^2Q_1^2 + F_2^2Q_2^2]^{1/2} \tag{2.15}$$

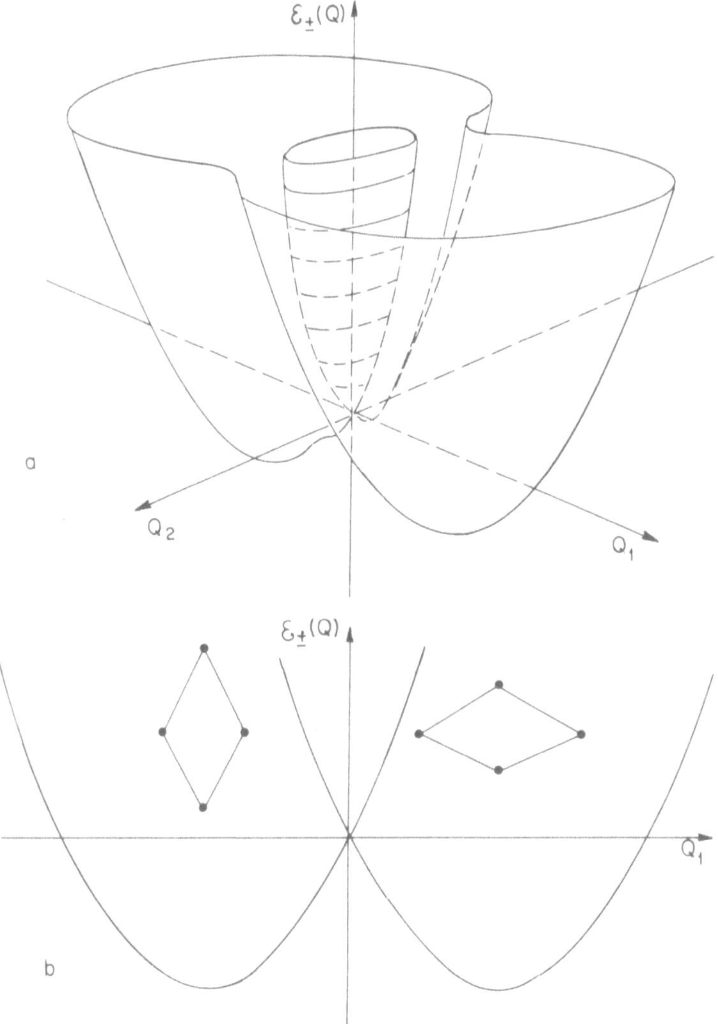

Figure 2.6. AP surface of a quadratic X_4 molecule in an E state linearly coupled to B_1 and B_2 displacements (E–(b_1+b_2) problem) when $E_{JT}^{(1)} > E_{JT}^{(2)}$: (a) general view; (b) cross-section along the Q_1 coordinate with illustration of distortions of the square at the minimum points.

Its extreme points and JT stabilization energies are

$$Q_1^{(0)} = \pm F_1/K_1, \quad Q_2^{(0)} = 0; \qquad E_{JT}^{(1)} = F_1^2/2K_1 \tag{2.16a}$$

$$Q_1^{(0)} = 0, \quad Q_2^{(0)} = \pm F_2/K_2; \qquad E_{JT}^{(2)} = F_2^2/2K_2 \tag{2.16b}$$

The new curvatures K_1' and K_2' (compared with K_1 and K_2 without the VI) are given by the relations

$$K_1' = K_1 \quad \text{and} \quad K_2' = K_2[1 - (E_{JT}^{(2)}/E_{JT}^{(1)})] \tag{2.17}$$

at the points (2.16a), and

$$K_1' = K_1[1 - E_{JT}^{(1)}/E_{JT}^{(2)}] \quad \text{and} \quad K_2' = K_2 \tag{2.18}$$

at the points (2.16b).

It follows that if $E_{JT}^{(1)} > E_{JT}^{(2)}$, the points (2.16a) are minimal and those of (2.16b) are saddle points, and if $E_{JT}^{(1)} < E_{JT}^{(2)}$ the opposite is true; the points (2.16b) are minimal and those of (2.16a) are saddle points. If $E_{JT}^{(1)} = E_{JT}^{(2)}$, a trough of equal energy minima is realized, and if $F_1/K_1 = F_2/K_2$ the trough is circular, similar to the E–e problem. If vibronic coupling with one of the JT active modes is negligible (one of the VC is small compared with the other), we obtain the E–b_1 problem, its AP shape being shown in Figure 2.6b. Upon taking into account the quadratic terms of the VI, coexistence of both types of minima, (2.16a) and (2.16b), becomes possible.[40]

2.2. Triplet and Quadruplet Terms

Threefold orbitally degenerate terms are possible for molecular systems containing cubic or icosahedral symmetry point groups (T, T_d, T_h, O, O_h, I, I_h). There are two types of orbital triplets, T_1 and T_2. Since the vibronic effects in these two cases are similar, consideration of only one of them, say T_2, is sufficient. Denote the three wave functions, which transform as the coordinate products yz, xz, and xy, by $|\xi\rangle$, $|\eta\rangle$, and $|\zeta\rangle$, respectively. In the T term case as distinguished from E, there are five JT active nontotally symmetric coordinates (Table 1.3): two which are tetragonal, Q_θ and Q_ε (E type), and three trigonal, Q_ξ, Q_η, and Q_ζ (T_2 type) (see Figure 1.1 and Table 1.2). For all the other coordinates the AP remains parabolic (i.e., they provide no vibronic contributions ε_k^{v} in equation (2.1)). The problem is T–$(e + t_2)$.

Secular equation (2.2) determining the vibronic $\varepsilon_k^{\text{v}}(Q)$ parts of the AP is of third order here. The matrix elements $W_{\gamma\gamma'}^{\text{v}}$ contain two linear vibronic constants F_E and F_T and several quadratic $G_{\bar{\Gamma}}(\Gamma_1 \times \Gamma_2)$ depending on the number of nontotally symmetric tensor convolutions, which can be prepared by different quadratic combinations of $Q_{E\gamma}$ and $Q_{T\gamma}$ (see Section 1.3). It can be shown[13] that for a T term there may be four quadratic vibronic constants: $G_E(E \times E)$, $G_E(T \times T)$, $G_T(T \times T)$, and $G_T(E \times T)$.

Consider first the linear approximation for which $G_\Gamma = 0$. Denote

$$F_E = \left\langle \zeta \left| \left(\frac{\partial V}{\partial Q_\theta} \right)_0 \right| \zeta \right\rangle, \qquad F_T = \left\langle \eta \left| \left(\frac{\partial V}{\partial Q_\xi} \right)_0 \right| \zeta \right\rangle \tag{2.19}$$

According to equation (1.16) all the matrix elements of the linear terms of the VI can be expressed by means of these two constants, and secular equation (2.2) for $\varepsilon_k^{\mathrm{v}}(Q)$ takes the form

$$\begin{vmatrix} F_E\left(-\frac{1}{2}Q_\theta + \frac{\sqrt{3}}{2}Q_\varepsilon\right) - \varepsilon^{\mathrm{v}} & F_T Q_\zeta & F_T Q_\eta \\ F_T Q_\zeta & F_E\left(-\frac{1}{2}Q_\theta - \frac{\sqrt{3}}{2}Q_\varepsilon\right) - \varepsilon^{\mathrm{v}} & F_T Q_\xi \\ F_T Q_\eta & F_T Q_\xi & F_E Q_\theta - \varepsilon^{\mathrm{v}} \end{vmatrix} = 0 \tag{2.20}$$

The three roots of this equation $\varepsilon_k^{\mathrm{v}}(Q)$, $k = 1, 2, 3$, are the surfaces in the five-dimensional space of the coordinates $Q_{\Gamma\gamma}$, $\Gamma\gamma = E\theta$, $E\varepsilon$, $T\xi$, $T\eta$, $T\zeta$. Together with the parabolic (nonvibronic) parts in equation (2.1) they determine the three sheets of the AP (in the space of these coordinates), crossing at $Q_{\Gamma\gamma} = 0$:

$$\varepsilon_k(Q) = \tfrac{1}{2}K_E(Q_\theta^2 + Q_\varepsilon^2) + \tfrac{1}{2}K_T(Q_\zeta^2 + Q_\xi^2 + Q_\eta^2) + \varepsilon_k^{\mathrm{v}}(Q), \quad k = 1, 2, 3 \tag{2.21}$$

However, the analytical solution of equation (2.20) is difficult. Öpik and Pryce[15] have worked out a procedure to determine the extremal points of the surface (2.21) without requiring a solution of equation (2.20).

Consider first the particular case when $F_T = 0$, $F_E \neq 0$ (the T–e problem). In this case equation (2.20) can be solved immediately:

$$\begin{aligned} \varepsilon_1^{\mathrm{v}}(Q_\theta, Q_\varepsilon) &= -F_E Q_\theta \\ \varepsilon_2^{\mathrm{v}}(Q_\theta, Q_\varepsilon) &= \tfrac{1}{2}F_E Q_\theta + (\sqrt{3}/2)F_E Q_\varepsilon \\ \varepsilon_3^{\mathrm{v}}(Q_\theta, Q_\varepsilon) &= \tfrac{1}{2}F_E Q_\theta - (\sqrt{3}/2)F_E Q_\varepsilon \end{aligned} \tag{2.22}$$

Substitution of these solutions into equation (2.21) yields the AP surface consisting of a set of paraboloids, of which only those containing the tetragonal Q_θ and Q_ε coordinates have minima displaced from the origin. In these coordinates the surface has the shape of three equivalent paraboloids intersecting at the point $Q_\theta = Q_\varepsilon = 0$ (Figure 2.7). The positions of the three minima are given by the coordinates

$$(Q_0^E, 0), \quad (\tfrac{1}{2}Q_0^E, (\sqrt{3}/2)Q_0^E), \quad (\tfrac{1}{2}Q_0^E, -(\sqrt{3}/2)Q_0^E) \tag{2.23}$$

where $Q_0^E = F_E/K_E$. For the depth of the minima and the energy of JT stabilization, we have

$$E_{JT}^E = F_E^2/2K \tag{2.24}$$

Note that the relief of the surface sheets near the point of degeneracy (Figure 2.7) differs from that of the E term (Figures 2.1 and 2.2): in the case of the T term there is a real intersection of the surface sheets at the point $Q_\theta = Q_\varepsilon = 0$, whereas for the E term AP branching occurs at this point. The wave function for the three paraboloids $|\xi\rangle \sim yz$, $|\eta\rangle \sim xz$, and $|\zeta\rangle \sim xy$, as distinct from the E term case, are mutually orthogonal and are not mixed by the tetragonal displacements.

In the other particular case when $F_E = 0$, $F_T \neq 0$ (the T–t_2 problem) the third-order equation (2.20) cannot be solved directly. Using the method of Öpik and Pryce[15] mentioned above, one can determine the extremal points of the AP without solving equation (2.20). For the case in question the surface $\varepsilon(Q_\xi, Q_\eta, Q_\zeta)$ in the space of the trigonal coordinates has four minima lying on the C_3 axes of the cubic system at the points $(m_1Q_0^T, m_2Q_0^T, m_3Q_0^T)$, where the four sets (m_1, m_2, m_3) are $(1,1,1)$, $(-1,1,-1)$, $(1,-1,-1)$, and $(-1,-1,1)$, and

$$Q_0^T = -2F_T/3K_T \tag{2.25}$$

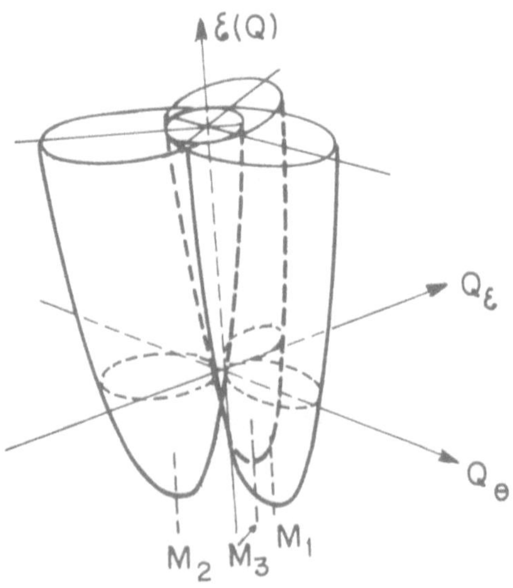

Figure 2.7. AP surface for the T–e problem.

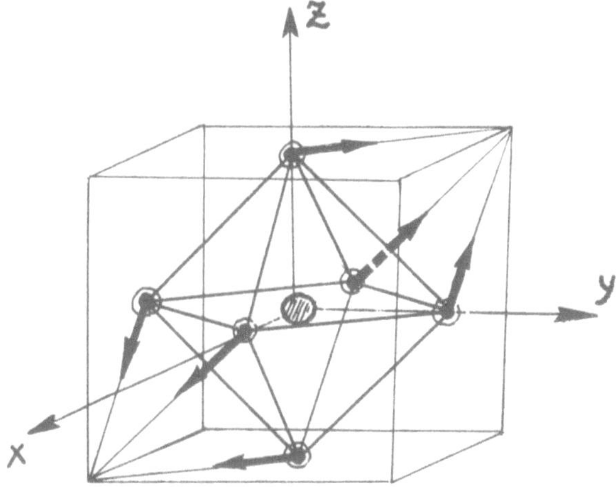

Figure 2.8. Illustration of the trigonal distortion of an octahedron in an electronic T state (cf. the $Q_\xi + Q_\eta + Q_\zeta$ displacement shown in Figure 1.1a).

At these minima the system is distorted along the trigonal axes. The displacements of the atoms corresponding to this distortion for an octahedral system are illustrated in Figure 2.8. The six ligands, in two sets of three ligands each, move on the circumscribed cube toward two apexes lying on the corresponding C_3 axes.[15,36,39] The depth of the minima E_{JT} is

$$E_{JT}^T = 2F_T^2/3K_T \tag{2.26}$$

In this case the frequency ω_T of the trigonal T_2 vibration splits into two[13]:

$$\omega_A = \omega_T \quad (K_A = K_T) \quad \text{and} \quad \omega_E = (2/3)^{1/2}\omega_T \quad (K_E = \tfrac{2}{3}K_T)$$

The electronic wave functions at the minima are given by the relationship

$$\Psi = \frac{1}{\sqrt{3}}(m_1|\zeta\rangle + m_2|\eta\rangle + m_3|\xi\rangle) \tag{2.27}$$

with the values of m_i, $i = 1, 2, 3$ given above.

In the general case of the T–$(e + t_2)$ problem, when simultaneous interaction with both the tetragonal ($F_E \neq 0$) and trigonal ($F_T \neq 0$) displacements is taken into account, the AP surface in the five-dimensional space of the five coordinates $Q_{\Gamma\gamma}$ is rather complicated, but the extremal points may be obtained using the procedure of Öpik and Pryce.[15] In general, the AP surface for the linear T–$(e + t_2)$ problem has three types of extrema summarized in Table 2.1: (1) Three equivalent tetragonal points, at which only tetragonal coordinates Q_θ and Q_ε are displaced (from the origin $Q_{\Gamma\gamma}=0$). Here the coordinates of the minima and their depths are the same as in the linear T–e problem and are given by equations (2.23) and (2.24). (2) Four equivalent trigonal points at which only the trigonal coordinates Q_ξ, Q_η, and Q_ζ are displaced (Figure 2.8). The minimum coordinates coincide with those obtained in the T–t_2 problem (see equation (2.26) and Table 2.1). (3) Six equivalent orthorhombic points at each of which one trigonal and one tetragonal coordinate is displaced (Table 2.1), their depth being given by

$$E_{\mathrm{JT}}^{0}=\tfrac{1}{4}E_{\mathrm{JT}}^{E}+\tfrac{3}{4}E_{\mathrm{JT}}^{T} \tag{2.28}$$

In this case, if $E_{\mathrm{JT}}^{E}>E_{\mathrm{JT}}^{T}$, the tetragonal extremal points are deepest and hence absolute minima, while the trigonal ones are saddle points. If, however, $E_{\mathrm{JT}}^{T}>E_{\mathrm{JT}}^{E}$, the trigonal points are minima and the tetragonal ones are saddle points. The orthorhombic extremal points are always saddle points in the linear approximation.

In particular, when $E_{\mathrm{JT}}^{E}=E_{\mathrm{JT}}^{T}$ all the extremal points (including the orthorhombic ones) have the same depth. It has been shown by O'Brien[41] that in this case a continuum of minima is realized forming a two-dimensional trough on the five-dimensional surface of the AP (in a sense similar to the "mexican hat" in the E term case).

Table 2.1. Extremal Points of the Adiabatic Potential for an Electronic T Term in the Linear T–$(e+t_2)$ Problem

The number of equivalent points	The nature of the extremum	Coordinates of the extremum in five-dimensional space $(Q_\theta, Q_\varepsilon, Q_\xi, Q_\eta, Q_\zeta)$
3	Tetragonal minima or saddle points	$(F_E/K_E, 0, 0, 0, 0)$
		$(-3F_E/2K_E, F_E/2K_E, 0, 0, 0)$
		$(-3F_E/2K_E, -F_E/2K_E, 0, 0, 0)$
4	Trigonal minima or saddle points	$(0, 0, 2F_T/3K_T, 2F_T/3K_T, 2F_T/3K_T)$
		$(0, 0, -2F_T/3K_T, -2F_T/3K_T, 2F_T/3K_T)$
		$(0, 0, -2F_T/3K_T, 2F_T/3K_T, -2F_T/3K_T)$
		$(0, 0, 2F_T/3K_T, -2F_T/3K_T, -2F_T/3K_T)$
6	Orthorhombic saddle points	$(-F_E/2K_E, 0, 0, 0, F_T/K_T)$
		$(-F_E/2K_E, 0, 0, 0, -F_T/K_T)$
		$(3F_E/4K_E, -F_E/4K_E, 0, F_T/K_T, 0)$
		$(3F_E/4K_E, -F_E/4K_E, 0, -F_T/K_T, 0)$
		$(3F_E/4K_E, F_E/4K_E, F_T/K_T, 0, 0)$
		$(3F_E/4K_E, F_E/4K_E, -F_T/K_T, 0, 0)$

When $K_E = K_T$, the classical motion of an octahedral system along the trough corresponds to motions of the ligands around identical spheres centered at the apexes of the octahedron.[13] The displacements of different ligands are correlated: at every instant their radius vectors drawn from the center of the sphere, if shifted to a common origin, form a star, the apexes of which produce a regular octahedron rotating around its geometric center.

In the quadratic approximation the vibronic T–$(e + t_2)$ problem is very complicated and has been solved only recently[42–46] (cubic terms have also been taken into account[46]). Similar to the E term case, the quadratic terms of the vibronic interaction W according to equation (1.7) produce some changes in the shape of the AP for a T term. Taking these quadratic terms into account, the secular equation (2.20), containing the two linear and four quadratic $E \times E$, $T_2 \times T_2(E + T_2)$, and $E \times T_2$ type vibronic constants mentioned above, becomes rather complicated. From the quadratic VC, the $G_E(E \times T_2)$ term which mixes the E and T_2 type displacements seems to be most important in changing the shape of the AP. Therefore, certain qualitative and semiquantitative results can be obtained by retaining only one quadratic VC, $G_E(E \times T_2)$, where

$$G = G_E(E \times T) = \frac{1}{2}\left\langle \xi \left| \left(\frac{\partial^2 V}{\partial Q_\xi \partial Q_\eta} \right)_0 \right| \eta \right\rangle \tag{2.29}$$

and disregarding all the other vibronic constants.

In this case the AP surface contains five parameters: F_E, F_T, K_E, K_T, and G. It has been shown[42,43] that only two dimensionless combinations A and B of these five parameters are necessary to describe the main features of the system:

$$A = G[K_E K_T]^{-1/2}, \qquad B = F_T \cdot G/F_E \cdot K_E \tag{2.30}$$

The case of linear vibronic coupling $G = 0$ correlates with the origin $(0, 0)$ on the plane (A, B) shown in Figure 2.9. Along the two lines $A = \pm(\sqrt{3}/2)B$, the energies of the tetragonal and trigonal extremal points are the same (the positions and depths of these points are not affected by the $E \times T_2$ type quadratic VI). In the cross-hatched area the trigonal extrema are deeper than the tetragonal extrema, whereas in the hatched area the opposite is true. In all cases the stability of the system requires that $|A| < 1$.

An account of the $E \times T_2$ type quadratic terms is most essential in determining the positions and pattern of the six equivalent orthorhombic extremal points which, in the linear approximation, are only saddle points. Their depth and the coordinates of one of them are as follows (the others can be easily derived from symmetry considerations):

$$
\begin{aligned}
E_{\mathrm{JT}}^{0} &= F_T^2(4A^2 - 4A^2B + B^2)/8K_E B^2(1 - A^2) \\
Q_\xi^{(0)} &= Q_\eta^{(0)} = Q_\varepsilon^{(0)} = 0 \\
Q_\zeta^{(0)} &= F_E(2 - B)/2K_T(1 - A^2) \\
Q_\theta^{(0)} &= -G(B - 2A^2)/2K_E B(1 - A^2)
\end{aligned}
\tag{2.31}
$$

When approaching the point $G = 0$ ($A = B = 0$) along the lines $A = \pm(\sqrt{3}/2)B$ on the plane (A, B) (Figure 2.9), the depth of the extremal points in question, E_{JT}^{0}, becomes equal to that of the trigonal and tetragonal cases ($E_{\mathrm{JT}}^{0} = E_{\mathrm{JT}}^{T} = E_{\mathrm{JT}}^{E}$), and the two-dimensional trough of the minima on the five-dimensional AP surface results. Away from this point $E_{\mathrm{JT}}^{0} \neq E_{\mathrm{JT}}^{E,T}$, and the trough is warped; as in the E term case, wells and humps emerge regularly alternating along the line of the minima. It is important that for a large area of the A and B parameters shown in Figure 2.10, the orthorhombic extremal points become absolute minima.

Alongside these three types of extremal points, an accounting of quadratic terms of the vibronic interaction reveals three new types with 12,

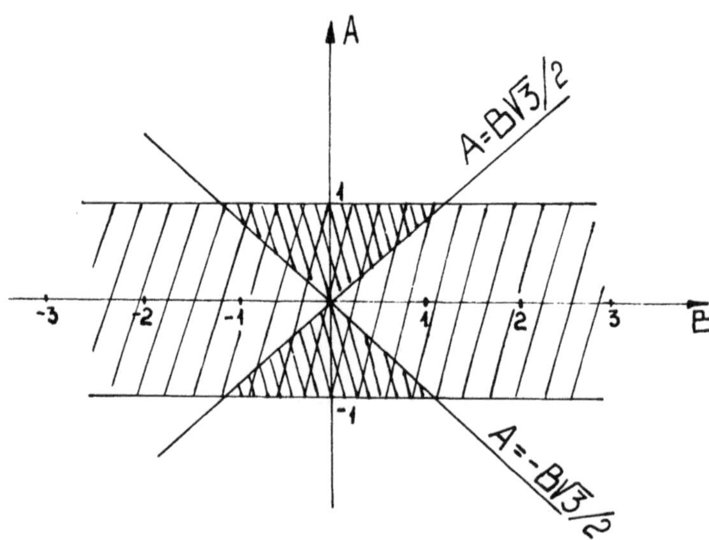

Figure 2.9. Stability areas in the plane (A, B), where $A = G(K_E K_T)^{-1/2}$, $B = GF_E/K_E F_T$, for a molecule in an electronic degenerate T term (quadratic T–$(e + t_2)$ problem). The molecule is stable when $|A| < 1$. Along the lines $A = \pm B\sqrt{3}/2$ the tetragonal and trigonal extremal points have the same depths. In the cross-hatched area the tetragonal extremal points possess lower energy, whereas in the remaining hatched area the trigonal extremal points possess lower energy.

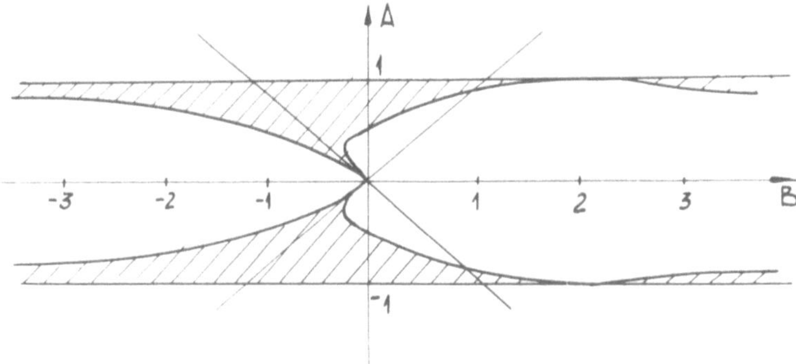

Figure 2.10. Area of existence (hatched) of orthorhombic absolute minima of the AP for a quadratic T–$(e + t_2)$ problem.

12, and 24 equivalent extremal points, respectively. Calculations carried out numerically[43] show that the latter type (24 points) might occur for parameters for which the system is unstable, while the first two types (with 12 points each) never become absolute minima. (See Figure 2.11, in which the areas of coexistence of various types of minima are also shown.) A group-theoretical account of possible extremal points in the AP is given elsewhere.[47] Some features of the classical motion of atoms along the AP have been discussed by Judd.[48]

An important feature of orbital T terms, as distinct from E terms, is the large splitting caused by spin–orbit interaction (in first-order perturbation theory). Therefore, if the system in question has unpaired electrons or, in general, if the total spin S of the state under consideration is nonzero, we do not deal with T terms, but with their components which result from spin–orbit splitting. For instance, the spin doublet ^2T (one unpaired electron) under spin–orbit interaction splits into two components, $^2T = \Gamma_8 + \Gamma_6$, from which the first ($\Gamma_8$) is a spin quadruplet and the second (Γ_6) a spin doublet. The latter can be treated like the doublet considered above (Section 2.2), whereas the quadruplet term Γ_8 requires additional treatment.

Since $[\Gamma_8 \times \Gamma_8] = A_1 + E + T_2$ (Table 1.3), the JT active displacements for the Γ_8 state are E and T_2, and the problem is Γ_8–$(e + t_2)$, like the T–$(e + t_2)$ case. The secular equation (2.2) in this case is of fourth order. The matrix elements $W^{v}_{\bar{\gamma}\bar{\gamma}'}$ as functions of the five coordinates $Q_{\Gamma\gamma}$, $\Gamma_\gamma = E\theta$, $E\varepsilon$, $T\xi$, $T\eta$, $T\zeta$, in the linear approximation contain two vibronic constants, F_E and F_T. The solution of equation (2.2) results in two sheets of the AP surface (two spin doublets) given by the following expression[39,13]:

$$\varepsilon_\pm(\rho, Q) = \tfrac{1}{2}(K_E\rho^2 + K_T Q^2) \pm [F_E^2\rho^2 + F_T^2 Q^2]^{1/2} \qquad (2.32)$$

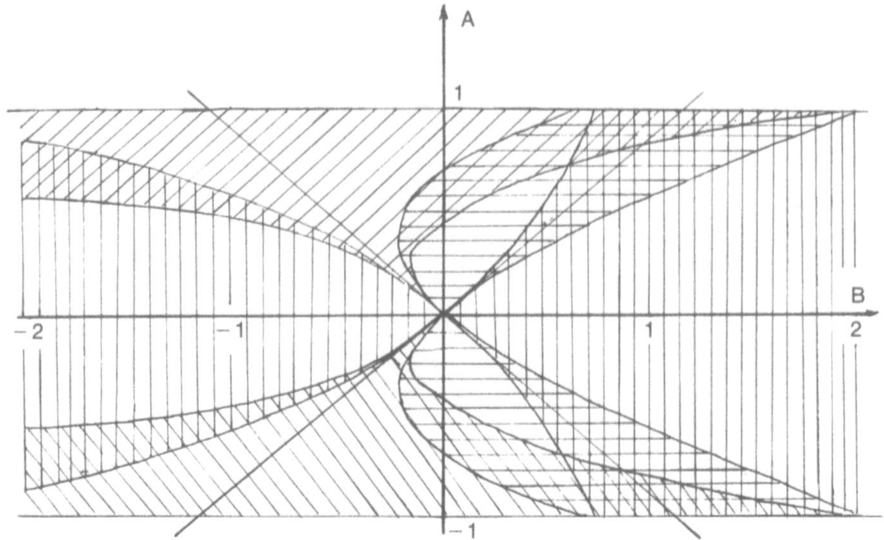

Figure 2.11. Areas of existence and coexistence of different types of minima of the AP for the quadratic T–$(e + t_2)$ problem (in A and B coordinates): vertical, horizontal and oblique shadings correspond to areas of occurrence of tetragonal, triagonal, and orthorhombic minima, respectively. Overlap shading shows areas of coexistence of corresponding minima.

where $\rho^2 = Q_\theta^2 + Q_\varepsilon^2$ and $Q^2 = Q_\xi^2 + Q_\eta^2 + Q_\zeta^2$. From this it is clear in the first place that the surface (2.32) depends on only two (ρ and Q) of five coordinates, being a surface of rotation for the remaining three (the angle ϕ in polar coordinates $Q_\theta = \rho \cos \phi$ and $Q_\varepsilon = \rho \sin \phi$ and similar angles α and β in the corresponding spherical coordinates for Q_ξ, Q_η, and Q_ζ). On the other hand, expression (2.32) formally coincides with equation (2.15) for the AP of the E–$(b_1 + b_2)$ problem, which means that in (ρ, Q) space the AP (2.32) has the shape illustrated in Figure 2.6, provided only positive values of $Q_1 = \rho$ and $Q_2 = Q$ are taken into account. It follows that, depending on the relation between the JT stabilization energies E_{JT}^E and E_{JT}^T,

$$E_{JT}^F = F_E^2/2K_E, \qquad E_{JT}^T = F_T^2/2K_T \tag{2.33}$$

the minimum of the surface will be either tetragonal:

$$\rho^{(0)} = F_E/K_E, \qquad Q^{(0)} = 0,$$

or trigonal:

$$\rho^{(0)} = 0, \qquad Q^{(0)} = F_T/K_T$$

The first case (tetragonal distortion) occurs if $E_{JT}^{E} > E_{JT}^{T}$; the second case (trigonal distortion) is specific for the opposite inequality. Since the other three coordinates remain arbitrary at the minimum points, these points in fact form a three-dimensional continuum — a three-dimensional trough (formally analogous to the one-dimensional trough in the E–e problem and the two-dimensional trough in the T–$(e+t_2)$ problem[49]). When $E_{JT}^{E} = E_{JT}^{T}$, an additional one-dimensional trough in the (ρ, Q) space occurs, resulting in a four-dimensional trough in the five-dimensional space of E and T_2 displacements.

Alongside the above doublet, triplet and quadruplet terms, there are quintet and sextet terms in icosahedral systems which are also subject to the Jahn–Teller effect. Some of these problems have been considered elsewhere.[50,51]

2.3. Pseudodegenerate States

Consider first an easy case of two nondegenerate states Γ and Γ' separated by an energy interval of 2Δ.[15] In order to obtain the adiabatic potential of these states the vibronic contributions ε_k^{v} must be calculated as solutions of secular equation (2.2). Assuming that only one coordinate $Q = Q_{\bar{\Gamma}}$, $\bar{\Gamma} = \Gamma \times \Gamma'$, mixes the two states (in principle, there may be more than one coordinate of type $\bar{\Gamma}$) and taking into account only linear terms in the vibronic interaction W after equation (1.14), we obtain (the energy is read off the middle of the 2Δ interval between the initial levels)

$$\begin{vmatrix} -\Delta - \varepsilon^{\text{v}} & FQ \\ FQ & \Delta - \varepsilon^{\text{v}} \end{vmatrix} = 0 \tag{2.34}$$

where $F = \langle \Gamma | (\partial V/\partial Q_{\bar{\Gamma}})_0 | \Gamma' \rangle$ is the nondiagonal linear vibronic constant. Inserting the solutions of equation (2.34)

$$\varepsilon_{\pm}^{\text{v}} = \pm[\Delta^2 + F^2Q^2]^{1/2} \tag{2.35}$$

into equation (2.1) and assuming that the force constant is the same in both states, $K = K_{\Gamma} = K_{\Gamma'}$, we obtain

$$\varepsilon_{\pm} = \tfrac{1}{2}KQ^2 \pm [\Delta^2 + F^2Q^2]^{1/2} \tag{2.36}$$

or, after expanding the second term in Q,

$$\varepsilon_{\pm}(Q) = \tfrac{1}{2}[K \pm F^2/\Delta]Q^2 \pm (F^4/\Delta^3)Q^4 \mp \cdots \tag{2.37}$$

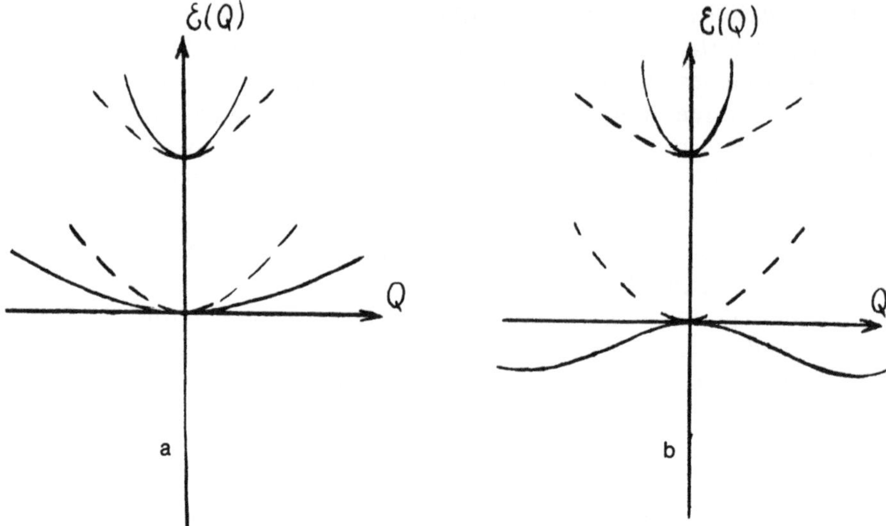

Figure 2.12. Vibronic mixing (repulsion) of two near-lying terms — the pseudo-Jahn–Teller effect: (a) weak PJTE: the ground state is softened but remains stable; (b) strong PJTE: the ground state becomes unstable in the initial configuration. Terms without mixing are shown by dashed lines.

It is seen from these expressions that, on taking into account the VI, the two AP curves change in different ways: in the upper sheet the curvature (force constant) increases, whereas in the lower one it decreases. However, until $\Delta > F^2/K$, the minima of both states correspond to the point $Q = 0$, as in the absence of vibronic mixing. This is the case of the weak pseudo-Jahn–Teller effect (PJTE) (Figure 2.12a).

However, if

$$\Delta < F^2/K \tag{2.38}$$

the curvature of the lower sheet of the AP becomes negative and the system unstable with respect to the Q displacements. This is the strong pseudo-Jahn–Teller effect (Figure 2.12b). The minima of the AP in this case are given by

$$\pm Q_0 = [(F^2/K^2) - (\Delta^2/F^2)]^{1/2} \tag{2.39}$$

For $\Delta = F^2/K$ the curvature is zero everywhere.

Similar to distortions in the JTE (see Section 1.6), the origin of instability of the high symmetry configuration and its distortion due to the strong PJTE can be illustrated in simple terms. Consider, for example, a planar square complex of type MA_4 (D_{4h} symmetry) and suppose that the highest occupied MO (HOMO) is the d_{z^2} orbital of the metal M (the A_{1g} orbital), while the lowest unoccupied orbital (LUMO) of A_{2u} symmetry is formed by the four p_z orbitals of the ligands A (Figure 2.13), or vice versa (i.e., the d_{z^2} orbital is the LUMO and the A_{2u} orbital is the HOMO). In the planar configuration these two MO do not mix, since the overlap integral S is zero. However, if the atom M is displaced out of and transversely to the plane (in the A_{2u} direction), a nonzero overlap occurs and the two orbitals mix to form a π bond between the metal and the ligands. In this case the energy of the ground state is lowered (as compared with the one expected without the above mixing) and the curvature of the AP in the direction of the A_{2u} displacement decreases (weak PJTE, Figure 2.12a), or becomes negative (strong PJTE, Figure 2.12b), provided the covalency of the π bond formed is superior to the changes of the elasticity forces.

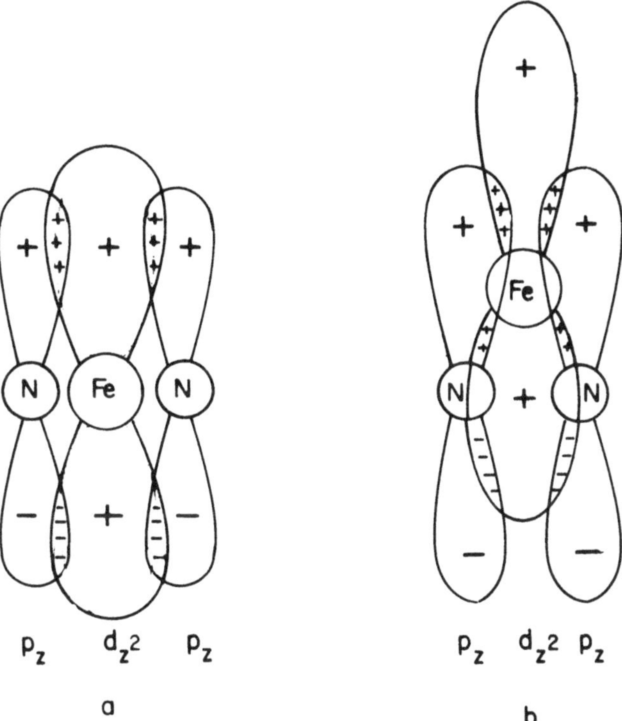

Figure 2.13. Illustration of the visual treatment of the origin of the PJTE using the N–Fe–N fragment of the square-planar FeN_4 molecular group as an example. Orbitals of the nitrogen atoms form the HOMO (ground state), while the d_{z^2} orbital of the Fe atom is the LUMO, or vice versa. (a) When the Fe atom is in the N_4 plane (on the N–N line) the d_π–p_π overlap is zero and there are no π bonds Fe–N. (b) The out-of-plane displacement of the Fe atom results in nonzero d_π–p_π overlap and π bond formation, which lowers the curvature of the AP in the direction of such a displacement.

It is seen from the above example that the PTJE is due to the occurrence of additional bonding in the distorted configuration which is absent in the high symmetry one. An analogous explanation can be given for the Renner effect in which, similar to the PJTE, the distortion of the high symmetry (linear) configuration is quadratic in the nuclear displacements.

Consider now a more complicated case of a transition metal complex with electron configuration d^0, such as four-valent titanium in an octahedral coordination of oxygen atoms as in the cluster $[TiO_6]^{8-}$.[52,53] For an approximate treatment of the problem, consider vibronic mixing only within the group of electronic terms close in energy and well separated from the other terms. The typical qualitative scheme of the MO energy levels and electron population numbers for such a complex is shown in Figure 2.14. It is seen that, for a full consideration, at least nine MO, t_{2g}, t_{1u}, and t^*_{2g} (occupied by six electrons in the TiO_6^{8-} cluster), have to be taken into account. With allowance for interelectron repulsion, these states form a ground state $^1A_{1g}$, and excited states (of the same multiplicity) $^1A_{2u}$, ^1E_u, $^1T_{1u}$, $^1T_{2u}$, $^1A_{2g}$, ^1E_g, $^1T_{1g}$, and $^1T_{2g}$.

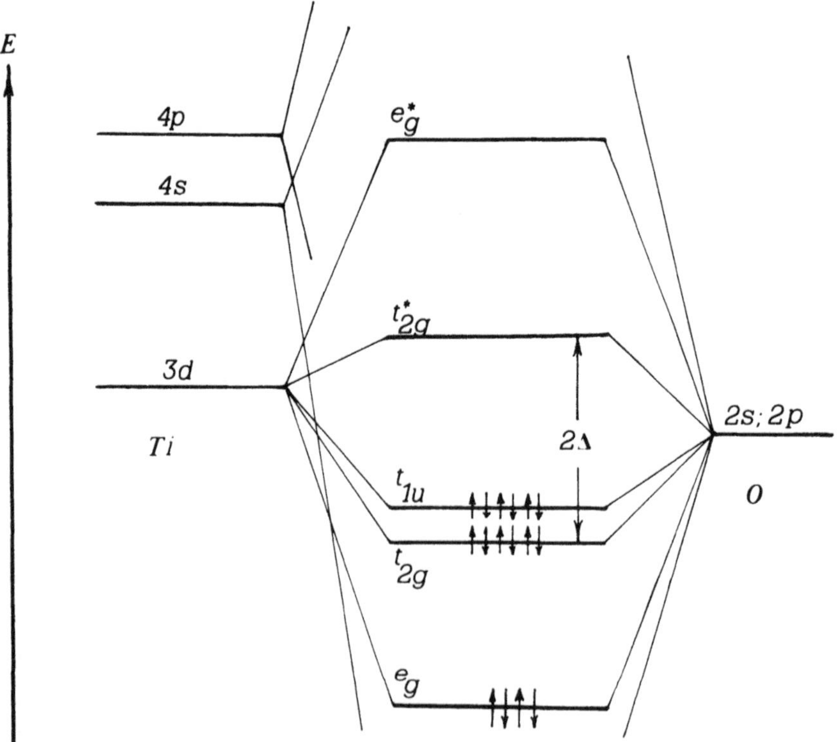

Figure 2.14. HOMO and LUMO for the TiO_6^{8-} cluster in the $BaTiO_3$ crystal.

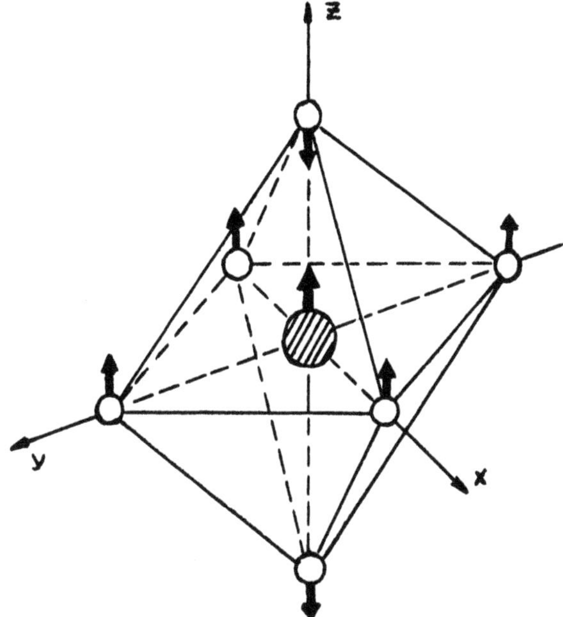

Figure 2.15. One component of the T_{1u} displacement for the octahedral TiO_6 cluster.

In most cases of practical use the energies and wave functions, as well as covalency parameters for the MO, are unknown. Therefore, and for a qualitative understanding of the vibronic effects, the VI may be considered at an earlier stage before covalency and multielectron term formation. Then the nine atomic functions, three $3d_\pi$ functions of the Ti^{4+} ion (d_{xy}, d_{xz}, d_{yz}) and six combinations of the $2p_\pi$ functions (p_x and p_y) of the O^{2-} ions (forming the above nine MO, t_{2g}, t_{1u}, and t_{2g}^*) may be taken as a basis of the vibronic treatment. These states mix by the T_{1u} type displacements, which have three components, Q_x, Q_y, and Q_z, one of them shown in Figure 2.15.

According to the Wigner–Eckart theorem (Section 1.2), the matrix elements of the ninth-order secular equation of the VI perturbation theory in the linear approximation contains only one VC:

$$F = \langle 2p_y|(\partial V/\partial Q_x)_0|3d_{xy}\rangle \tag{2.40}$$

Omitting the equation itself we present its solutions (2Δ is the energy interval between the $3d_{xy}$ and $2p_y$ states):

$$\begin{aligned} \varepsilon_{1,2}^{\text{v}} &= \pm[\Delta^2 + F^2(Q_x^2 + Q_y^2)]^{1/2}; \qquad \varepsilon_{3,4}^{\text{v}} = \pm[\Delta^2 + F^2(Q_y^2 + Q_z^2)]^{1/2} \\ \varepsilon_{5,6}^{\text{v}} &= \pm[\Delta^2 + F^2(Q_x^2 + Q_z^2)]^{1/2}; \qquad \varepsilon_{7,8,9}^{\text{v}} = -\Delta \end{aligned} \tag{2.41}$$

Of these nine levels, only the lower three are populated by the above six electrons in the ground state. Substitution of the expressions $\varepsilon_1^{\mathrm{v}}$, $\varepsilon_3^{\mathrm{v}}$, and $\varepsilon_5^{\mathrm{v}}$ from (2.41) into equation (2.1) yields the following expression for the ground state AP:

$$\begin{aligned}\varepsilon(Q_x, Q_y, Q_z) = \tfrac{1}{2}K(Q_x^2 + Q_y^2 + Q_z^2) \\ - 2\{[\Delta^2 + F^2(Q_x^2 + Q_y^2)]^{1/2} + [\Delta^2 + F^2(Q_y^2 + Q_z^2)]^{1/2} \\ + [\Delta^2 + F^2(Q_x^2 + Q_z^2)]^{1/2}\}\end{aligned} \tag{2.42}$$

The shape of this surface depends on the relation between the constant Δ, F, and K. If $\Delta > 4F^2/K$ the surface has one minimum at the point $Q_x = Q_y = Q_z = 0$, for which the system remains undistorted. This case corresponds to the weak PJTE (Figure 2.12a). However, if

$$\Delta < 4F^2/K \tag{2.43}$$

the surface (2.42) acquires a rather complicated shape with four types of extremal points:

(1) One maximum at $Q_x = Q_y = Q_z$ (dynamic instability).

(2) Eight minima at the points $|Q_x| = |Q_y| = |Q_z| = Q_0^{(1)}$, where

$$Q_0^{(1)} = [(8F^2/K^2) - (\Delta^2/2F^2)]^{1/2} \tag{2.44}$$

with JT stabilization energy

$$E_{\mathrm{JT}}^{(1)} = 3[(4F^2/K) + (\Delta^2 K/4F^2) - 2\Delta] \tag{2.45}$$

At these minima the Ti atom is displaced along the trigonal axis, equally approaching three oxygen atoms and removed from the other three.

(3) Twelve saddle points at $|Q_p| = |Q_q| \neq 0$, $Q_r = 0$, $p, q, r = x, y, z$ (with a maximum in section r and minima along p and q). At these points the Ti atom is displaced toward two oxygen atoms lying on the p and q axes, respectively.

(4) Six saddle points at $Q_p = Q_q = 0$, $Q_r = Q_0^{(2)}$, where

$$Q_0^{(2)} = [(16F^2/K^2) - (\Delta^2/F^2)]^{1/2} \tag{2.46}$$

with depth given by

$$E_{\mathrm{JT}}^{(2)} = 2[(4F^2/K) - (\Delta^2 K/4F^2) - 2\Delta] \tag{2.47}$$

When taking account of covalency and multielectron term formation these results, especially their quantitative expression, will certainly be modified.

A similar treatment is possible for tetrahedral complexes of type MA_4.[54] In this case, in the strong vibronic coupling limit and under certain vibronic mixing conditions, four equivalent minima are expected, at each of which one M–A bond is longer or shorter than the other three which remain identical.

In stereochemistry and reactivity problems (as well as in spectroscopy and crystal chemistry) other cases of the PJTE may be significant. In particular, the vibronic mixing of an E term with an A_1 term under E type displacements is commonly used. We consider here such mixing for a system with C_{4v} symmetry.[55] The two wave functions of the E term transform as x and y and the two components of the E mode may be denoted by Q_x and Q_y, the energy gap between the E and A_1 levels being 2Δ. Then secular equation (2.2) in the linear approximation takes the form

$$\begin{vmatrix} -\Delta - \varepsilon^{\mathrm{v}} & 0 & FQ_x \\ 0 & -\Delta - \varepsilon^{\mathrm{v}} & FQ_y \\ FQ_x & FQ_y & \Delta - \varepsilon^{\mathrm{v}} \end{vmatrix} = 0 \tag{2.48}$$

In polar coordinates

$$Q_x = \rho \cos \phi, \qquad Q_y = \rho \sin \phi \tag{2.49}$$

the roots of this equation are

$$\varepsilon^{\mathrm{v}}_{1,3} = \pm [\Delta^2 + F^2\rho^2]^{1/2} \tag{2.50}$$

and the AP of the ground state is given by

$$\varepsilon(\rho, \phi) = \tfrac{1}{2}K_E\rho^2 - [\Delta^2 + F^2\rho^2]^{1/2} \tag{2.51}$$

In the above linear approximation the AP is independent of angle ϕ, i.e., it is a figure of rotation. Similar to the two-level case, considered at the beginning of this section, if $\Delta < F^2/K_E$ the surface has a maximum at the point $\rho = 0$ (dynamic instability), and a circular trough at $\rho = \rho_0 = [(F^2/K_E^2) - (\Delta^2/F^2)]^{1/2}$ with depth (read off at the point $\varepsilon(0) = -\Delta$)

$$E_{\mathrm{PJT}} = (F^2/2K_E) + (\Delta^2 K_E/F_E^2) - \Delta$$

If quadratic terms of the VI are taken into account, the secular equation becomes complicated, since the $|x\rangle$ and $|y\rangle$ states of the E term mix by quadratic coordinates (whereas the E and A_1 ones do not). In accordance with the Wigner–Eckart theorem (1.16) this mixing can be described by means of two VC (for the C_{4v} group the symmetric product $[E \times E]$ contains two nontotally symmetric representations B_1 and B_2).

Denote

$$G_1 = \frac{1}{2}\left\langle x \left| \left(\frac{d^2V}{dQ_x^2}\right)_0 \right| x \right\rangle, \qquad G_2 = \frac{1}{2}\left\langle x \left| \left(\frac{d^2V}{dQ_x dQ_y}\right)_0 \right| y \right\rangle \tag{2.52}$$

Then secular equation (2.2) for the case in question takes the form

$$\begin{vmatrix} -\Delta + G_1(Q_x^2 - Q_y^2) - \varepsilon^v & 2G_2Q_xQ_y & FQ_x \\ 2G_2Q_xQ_y & -\Delta - G_1(Q_x^2 - Q_y^2) - \varepsilon^v & FQ_y \\ FQ_x & FQ_y & \Delta - \varepsilon^v \end{vmatrix} = 0 \tag{2.53}$$

Turning to polar coordinates (2.49), it can be shown that the AP corresponding to equation (2.53), as distinct from the linear case, is dependent on angle ϕ (acquiring initial symmetry C_{4v}); four minima regularly alternating with four saddle points occur on the AP as a function of ϕ (along the trough (2.51)). At the extremal points, $\phi_0 = n\pi/4$, $n = 0, 1, 2, \ldots, 7$. If $G_1 > G_2$, the minima are given by

$$\phi_0(\min) = n\pi/4, \qquad n = 1, 3, 5, 7 \tag{2.54}$$

and the saddle points are at $n = 0, 2, 4, 6$. In the opposite case $G_1 < G_2$, the minima and saddle points interchange, and if $G_1 = G_2$ the surface preserves the shape of a rotating body (trough) of linear approximation.

Analytic consideration of s–p mixing by T_{1u} displacement ($s \times p - t_{1u}$ problem) is given by Pooler.[56] Simultaneous mixing of many electronic states may lead to an AP with six equivalent minima of elongated and compressed octahedra.[56a]

The above examples show that in the case of relatively close electronic terms (pseudodegeneracy), *the AP as a result of the PJTE acquires features similar to that obtained in the JTE: instability of the high symmetry configuration and several equivalent (or a continuum of) minima.* The criterion of the strong PJTE (or of PJT instability), (2.38) or (2.43) (in every other case this condition looks similar), contains three parameters, Δ, K, and F, and therefore it may be "soft" for any one of them taken on its own. In particular, a strong PJTE may occur for large values of the energy gap Δ, if the force constant K is small enough and the vibronic constant F large enough.

The cases when the criterion of strong PJTE is not fulfilled, i.e., $\Delta \gtrsim (F^2/K)$, are also very important. Indeed, although in this case there is no instability (in the sense of absence of a minimum of the AP for the nuclear configuration in question), the curvature of the ground state AP as indicated above is lowered, the nuclear configuration is softened in the Q direction, and it is more softened, the larger F and the smaller Δ (for a given value of K). This situation is significant, e.g., in the investigation of chemical reaction mechanisms (Chapter 6).

Along with the strong similarity, there are essential differences between the JTE and PJTE. An important feature of the PJTE is that the two (or more) mixing electronic states Γ and Γ' may pertain to different irreducible representations of the point group of the system, whereas in the JT case they belong to the same representation $\Gamma = \Gamma'$. Consequently, in the PJTE the direction of distortion may possess any symmetry possible in the symmetry group of the system under consideration, whereas in the JTE the directions are limited by JT active modes. In particular, for systems with an inversion center the two states mixing Γ and Γ' in the PJTE may possess opposite parity. In this case the VC $F_{\bar{\Gamma}}^{(\Gamma\Gamma')}$ is nonzero for odd nuclear displacements Q_u which remove the inversion center and form a dipole moment (dipolar instability[57,53]; see Chapters 4 and 5). In this case the system in its minima configurations has a dipole moment. It is clear that this effect of dipolar instability is impossible in the JTE where mixing electronic terms have the same symmetry $\Gamma = \Gamma'$ and hence active modes can only be even.

2.4. The Multimode Problem

It is known that in a nonlinear molecular system with N atoms ($N > 2$) there are $2N - 6$ degrees of freedom, which in the harmonic approximation are conveniently described by normal coordinates transforming according to corresponding irreducible representations of the symmetry point group of the system (Section 1.2). If N is not very small there may be two or more (or an infinite number for a crystal lattice) normal modes of the same symmetry, which are active in the JTE. For instance, already in a simple tetrahedral system of type MA_4 there are two types of normal coordinates of T_2 symmetry, T_2' and T_2'' (Table 1.1), both JT active. In this case the AP instability must be considered in the space of not three but six coordinates transforming as T_2' and T_2'', and the problem will not be T–$(e + t_2)$, but T_2–$(e + t_2' + t_2'')$. Such problems, in which there are more than one set of JT active coordinates of given symmetry, are called multimode problems. Unlike the latter problems, those with only one JT active mode of a particular symmetry are called ideal problems.

The multimode problems are very important in the investigation of polyatomic molecules, especially crystals, for which the JT center in question and its nearest neighbors cannot be isolated from the remaining lattice. The number of JT active modes with a particular symmetry increases with the number of coordination spheres around the JT center. On the other hand, the multimode problem is much more complicated owing to the multidimensional space of JT active displacements. For the above example of the MA_4 molecule, the number of JT active coordinates is eight instead of five in the ideal problem.

In addition to the increasing dimensionality of the JT active space, the number of vibronic constants is also increasing since, strictly speaking, the vibronic coupling to each of the modes has to be described by its own VC (however, see below).

Fortunately, there is an important feature of the multimode problem which allows some simplification, at least at the stage of qualitative consideration. This simplification is based on the statement that the extremal properties of the multimode problem AP are similar to those of the corresponding ideal problem. As an example, we shall illustrate it for the two-mode E–$(b_1' + b_1'')$ problem. The appropriate ideal problem is E–b_1, which leads to a simple AP illustrated in Figure 2.6b.

Set $Q_{b1}' = Q_1$ and $Q_{b1}'' = Q_2$, and the corresponding constants of linear coupling with b_1' and b_1'' displacements by F_1 and F_2, respectively. Then secular equation (2.2) takes the form

$$\begin{vmatrix} F_1Q_1 + F_2Q_2 - \varepsilon^{\text{v}} & 0 \\ 0 & -(F_1Q_1 + F_2Q_2) - \varepsilon^{\text{v}} \end{vmatrix} = 0 \tag{2.55}$$

with roots

$$\varepsilon^{\text{v}} = \pm(F_1Q_1 + F_2Q_2) \tag{2.56}$$

which, upon substitution in equation (2.1), yield the following AP (the force constants are denoted by $K_{b1} = K_1$ and $K_{b1} = K_2$):

$$\varepsilon_{\pm}(Q_1, Q_2) = \tfrac{1}{2}(K_1Q_1^2 + K_2Q_2^2) \pm (F_1Q_1 + F_2Q_2) \tag{2.57}$$

with minima at the points

$$Q_1^{\pm} = \pm(F_1/K_1), \qquad Q_2^{\pm} = \pm(F_2/K_2) \tag{2.58}$$

at depth

$$E_{\text{JT}} = \tfrac{1}{2}[(F_1^2/K_1) + (F_2^2/K_2)] \tag{2.59}$$

It is seen that in the E–$(b_1' + b_1'')$ problem, as in the appropriate ideal E–b_1 problem, there are two equivalent minima of the AP. Now, in the multimode case both coordinates Q_1 and Q_2 are displaced at the minima (whereas in the ideal problem only one coordinate is shifted). However, the expression (2.57) allows a linear transformation to other coordinates, in which only one mode will be displaced at the minima (as in the ideal case). These new coordinates are given by

$$\begin{aligned} q_1 &= F^{-1}(K_1^{-1}F_1Q_1 + K_2^{-1}F_2Q_2) \\ q_2 &= F^{-1}(K_1K_2)^{-1/2} \cdot (-F_2Q_1 + F_1Q_2) \\ F &= [(F_1^2/K_1) + (F_2^2/K_2)]^{1/2} \end{aligned} \tag{2.60}$$

In terms of coordinates q_1 and q_2, the AP (2.57) takes the form

$$\varepsilon_\pm(q_1, q_2) = \tfrac{1}{2}(q_1^2 + q_2^2) \pm Fq_1 \tag{2.61}$$

with minima displaced only in q_1:

$$q_1^\pm = \pm F, \qquad q_2^\pm = 0 \tag{2.62}$$

The depth of the minima remains as in (2.59).

In other words, by means of a coordinate transformation the AP of the two-mode problem has been reduced to qualitatively the same shape as that of the ideal problem, i.e., with the same vibronic features (the same number and types of minima and only one coordinate displaced in them). This result, illustrated here by the E–$(b_1' + b_1'')$ problem, is valid in every case.[13]

Note that the reduction of the multimode problem to the form in which vibronic effects are expressed by only one mode, does not mean that some of the atoms of the polyatomic system are excluded from participation in the JTE. Actually, the coordinates of all the atoms are present via the new coordinates $q_{1,2}$ given by equation (2.60). However, the displacements of different atoms in the JT distortion are not independent but concerted (coherent), and therefore they can be taken into account by means of an essentially smaller number of coordinates.

For instance, in case of the linear multimode E–$(e_1 + e_2 + \cdots)$ problem the AP can be reduced to that of the "mexican hat" for the ideal E–e problem (Figure 2.1). When the system is moving along the trough, *the circular motions of the atoms of the first coordination sphere* (ideal system), described in Section 2.1, *involve in these motions the atoms of all the coordination spheres through the vibronic and vibrational coupling. This produces a wave of deformations which rotate around the JT center, drawing into this rotation all the nearby atoms.*

As indicated above, the number of JT active modes and vibronic constants increases with the number of next-neighbor coordination spheres involved in the vibronic interaction. There are cases, however, when the electronically degenerate term is strongly localized and therefore the proper vibronic coupling (of the electrons with the nuclear displacements) is negligible beyond the first, or first few, coordination spheres. In this case there is no need to introduce as many VC as in JT active modes; the number of VC may be greatly reduced, in many cases to one VC, which characterizes the vibronic interaction with the atomic displacements of the first coordination sphere. This does not mean, however, that the problem ceases to be multimodal and becomes ideal, since the displacements of the first coordination sphere enter into all the symmetrized displacements $e_1, e_2, \ldots$, etc. *The coupling to even one (the first) coordination sphere involves all the modes in the JTE*, but the fewer the number of nonzero VC, the easier the solution of the problem.

Consider, for instance, the n-mode E–e problem (the E–$(e_1 + e_2 + \cdots + e_n)$ problem) with only one VC, F_E. The AP, as indicated above, can be reduced by means of a coordinate transformation to the shape of a trough, similar to that illustrated in Figure 2.1. In this trough there is a free "rotation" of the system (a rotation of the wave of JT deformation) around the JT center, and harmonic vibrations along the remaining $2n-1$ coordinates. The dimensions of the trough, its radius and depth, are determined by the contributions of all the modes E_x, proportional to the common VC, F_E, multiplied by the coefficients a_{E_x}, with which the displacements of the first coordination sphere enter the E_x mode in question. The numerical values of the coefficients a_{E_x} are known, provided the dynamic problem of molecular vibrations is solved; for crystals, the coefficients a_{E_x} are called Van Vleck coefficients[58] (see also Perlin and Tsukerblat[59]). As a result, the JT stabilization energy E_{JT} equals the sum of the contributions of each JT active mode E_{JT}^x:

$$E_{JT}^x = F_{E_x}^2/2K_{E_x} = F_E^2 a_{E_x}^2/2K_{E_x}$$
$$F_{JT} = \tfrac{1}{2}F_E^2 \sum_x (a_{E_x}^2/K_{E_x}) \tag{2.63}$$

Thus with known coefficients a_{E_x} and appropriate force constants $K_{E_x} = M_{E_x}\omega_{E_x}^2$ (for a crystal lattice, the phonon density of states $\rho_E(\omega)$ is used instead of K_{E_x}) the shape of the AP for the multimode problem can be determined. It must be emphasized, however, that the transformation (2.60), essentially simplifying the expression for the AP and allowing separation of a single active JT mode, does not simplify the multimode problem as a whole, especially calculations of the energy spectra and wave functions. In general, this transformation does not separate the variables. As a result of the potential energy simplification, the expression for the kinetic energy will become more complicated.[13]

3

Energy Spectra. Dynamic Distortions of Nuclear Configurations

The main effect of vibronic interactions resolves itself into peculiar dynamics of nuclear configurations. The resulting energy levels and wave functions have been determined for the most important cases.

3.1. Some Limiting Cases. Free and Hindered Rotations of the Distortion

Calculation of the energy spectrum and wave functions of a Jahn–Teller or pseudo-Jahn–Teller molecule as solutions of the coupled equations (1.6) is a very complicated problem, and one cannot expect to solve it in general form. However, as in similar quantum-mechanical situations, *analytic solutions for some limiting cases in combination with exact numerical solutions of some particular cases yields general trends, the origin and mechanism of the phenomenon as a whole, and prediction of new effects.*

For vibronic problems, such limiting cases as weak and strong vibronic coupling (with relatively small and large vibronic constants, respectively) can be solved analytically. A quantitative criterion of weak and strong coupling can be defined by comparing the JT stabilization energy E_{JT}^{Γ} with the energy of zero-point vibrations $n_\Gamma \hbar\omega_\Gamma/2$ of n-fold degenerate vibration. Denote $\lambda_\Gamma = 2E_{JT}^{\Gamma}/n_\Gamma\hbar\omega_\Gamma$. Then, if $\lambda_\Gamma \ll 1$ $(2E_{JT}^{\Gamma} \ll n_\Gamma\hbar\omega_\Gamma)$ the vibronic coupling will be regarded as weak, and if $\lambda_\Gamma \gg 1$ $(2E_{JT}^{\Gamma} \gg n_\Gamma\hbar\omega_\Gamma)$ the coupling is strong; λ_Γ is the dimensionless vibronic constant.

In the limiting case of weak vibronic coupling the depths of the adiabatic potential minima, formed by the vibronic interactions, are much smaller than the zero vibration energies, and therefore there are no local states in the minima. In the strong coupling case, however, such local states exist. Nevertheless, in both cases the system is delocalized into all the equivalent minima, provided stationary states of the free system (and not instantaneous or specially prepared ones) are considered. Therefore the terms "dynamic JTE" for weak coupling and "static JTE" for strong coupling are unsuitable since, strictly speaking, both cases are dynamic. However, the term "static JTE" may still be meaningful if used to indicate the situation when, in the limiting case of very deep minima, the one-minimum state for the given process of measurement may be regarded as a quasi-stationary one (see the relativity rule in Section 5.1).

Consider first the approximate solutions of the vibronic equations for weak coupling, when $\lambda \ll 1$. General considerations[60-62] show that in this case the angular momenta of the electrons and nuclei are not preserved separately but as their vector sum. An additional special quantum number, comprising the quantum numbers of the projections of the angular momenta of electrons and nuclei, characterizes the energy levels of the system as a whole. This situation is similar to that in atomic theory, where orbital and spin angular momenta in the weak coupling limit are summarized by the LS coupling scheme, so we can use some results of this theory.

When $\lambda \ll 1$, the energy levels can be obtained, including VI, by means of perturbation theory. For the linear E–e problem, the unperturbed system is a dimeric oscillator with energies $E_n = \hbar\omega_E(n+1)$, $n = 0, 1, 2, \ldots$. Here, the angular momentum of the nuclei defined by the quantum number $l = n, n-2, \ldots, -n+2, -n$ is preserved, and therefore each energy level is $(n+1)$-fold degenerate with respect to the vibrational quantum numbers; taking into account the twofold electronic degeneracy, it results in a total $2(n+1)$ degeneracy.

If the VI is weak enough, the quantum number l remains "good". In addition, the total angular momentum of the electrons and nuclei defined by the quantum number $m = l \pm \frac{1}{2}$ (the effective angular momentum of the electrons in the doubly degenerate state is defined by quantum numbers $\pm\frac{1}{2}$) must be preserved. As a result the following expression for the energy levels has been obtained using second-order perturbation theory (with respect to linear VI terms)[60,13] (Figures 3.1 and 3.2):

$$E_{nlm} = \hbar\omega_E[n + 1 + 2\lambda_E(l^2 - m^2 - \tfrac{3}{4})]$$

$$l = n, n-2, \ldots, -n+2, -n, \qquad m = \pm\tfrac{1}{2}, \pm\tfrac{3}{2}, \ldots, \pm(n+\tfrac{1}{2}) \qquad (3.1)$$

where $\lambda_E = (E^E_{\text{JT}}/\hbar\omega_E) = F_E^2/2\hbar\omega_E K_E$ (see equation (2.7)).

Since l and m are not independent, the energy E_{nlm} depends on only two quantum numbers, say n and l, so $E_{nl} = \hbar\omega_E[n + 1 + 2\lambda_E(\pm l - 1)]$. It is seen that the $(2n+1)$-degenerate level with given n splits into $n+1$ components (since l may have $n+1$ values). Each level remains twofold degenerate because the energy (3.1) does not depend on the sign of m. The ground state is also twofold degenerate: $E_{00,\pm 1/2} = \hbar\omega_E(1 - \lambda_E) =$

$\hbar\omega_E - 2E_{JT}$. However, some of these levels, namely, those with quantum numbers $m = \pm\frac{3}{2}, \pm\frac{9}{2}, \pm\frac{15}{2}, \pm\ldots$, pertain to the symmetry $A_1 + A_2$, which indicates that they are incidentally degenerate (the others being of E symmetry and therefore regularly degenerate).

The energy distance between the extreme component doublets for a given value of n equals $4n\hbar\omega_E\lambda_E = 4nE_{JT}$, and according to the criterion of perturbation theory, equation (3.1) is valid if and only if this value $4nE_{JT}$ is much smaller than the distance to the next vibrational level, i.e., if $E_{JT} \ll \hbar\omega_E/4n$.

The wave functions of the unperturbed dimeric oscillator may be written in the form

$$\Psi_{\pm n_1 n_2} = \psi_{\pm} |n_1 n_2\rangle \tag{3.2}$$

where $|n_1 n_2\rangle$ is the vibrational harmonic function, in which one of the components of the twofold degenerate E vibration (say, the E_θ component) is in the n_1 state, the other in the n_2 state, and the electronic functions $\psi_{\pm}$ are given by

$$\begin{aligned} \psi_+ &= |\theta\rangle + i|\varepsilon\rangle \\ \psi_- &= |\theta\rangle - i|\varepsilon\rangle \end{aligned} \tag{3.3}$$

Then, the perturbed functions of the ground state assume the form[13]

$$\begin{aligned} \Psi_{+00}(r, Q) &= \psi_+ |00\rangle + [F_E/(\hbar\omega_E)^{3/2}]\psi_- |01\rangle \\ \Psi_{-00}(r, Q) &= \psi_- |00\rangle - i[F_E/(\hbar\omega_E)^{3/2}]\psi_+ |10\rangle \end{aligned} \tag{3.4}$$

It is seen that a weak linear VI mixes the ground state with the first vibrational excited state.

When the T term is coupled to only E vibrations (the T–e problem), the energy levels of the initial three-dimensional oscillator $E_n = \hbar\omega_T(n + \frac{3}{2})$ under the influence of VI are not split, but shifted down by an amount E^E_{JT}. This result has a general meaning and remains valid in the strong coupling case, too.[41]

When coupling with only trigonal T_2 vibrations is taken into account (the T–t_2 problem), the initial zero-approximation system is also a three-dimensional oscillator with energy levels $E_n = \hbar\omega_T(n + \frac{3}{2})$ but, unlike the T–e problem, the VI mixes the electronic and nuclear motions. The quantum number of the vibrational angular momentum $l = n, n-2, n-4, \ldots, 1$, or 0, in case of weak coupling, remains a "good" one. Besides, the energy is classified by the quantum number m of the square of the total angular momentum, equal to the sum of the orbital and vibrational momenta: $m = 1, l \pm 1$ for $l \geqslant 1$, and $m = 1$ for $l = 0$.

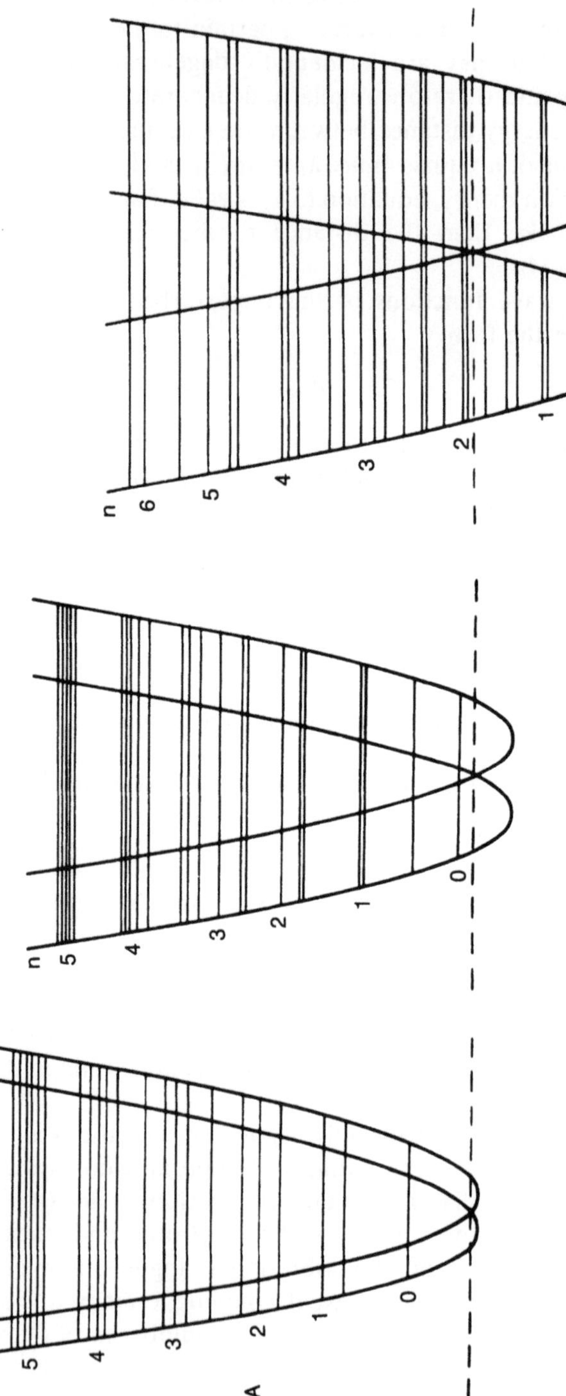
A
n
5
4
3
2
1
0
n
5
4
3
2
1
0
n
6
5
4
3
2
1
0

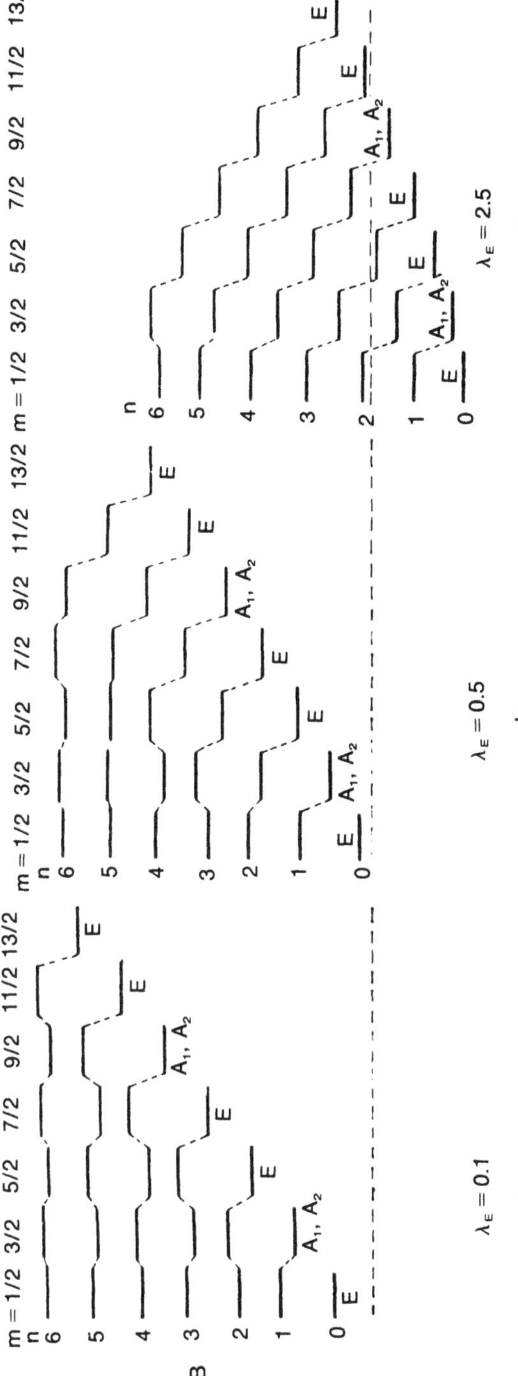

Figure 3.1. Cross-sections of the AP for the linear E–e problem (Figure 2.1) with corresponding vibronic energy levels (A), and the same levels with indication of the quantum numbers and symmetries (B), for three cases of the dimensionless coupling parameters: (a) $\lambda_E = 0.1$; (b) $\lambda_E = 0.5$; (c) $\lambda_E = 2.5$. Energy levels with the same vibrational quantum number n are linked by dashed lines. All levels with a given value m have the same symmetry indicated at the lowest level.

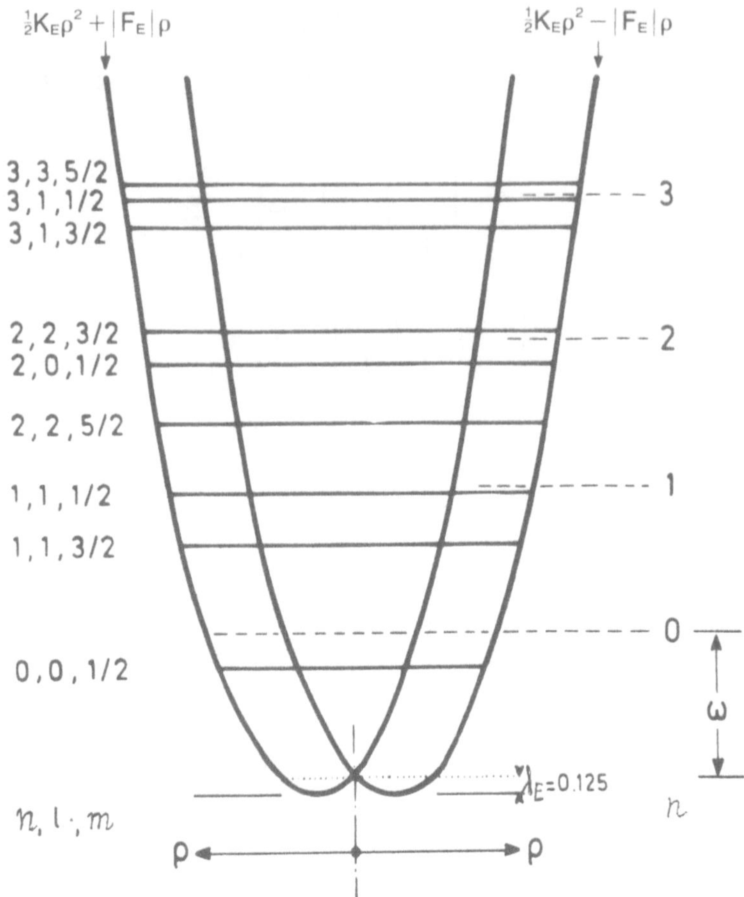

Figure 3.2. More detailed illustration of the cross-section of the AP for the linear E–e problem in the case of weak vibronic coupling ($\lambda_E = 0.125$) with indication of the quantum numbers n, l, m. Positions of vibrational levels without vibronic coupling are shown by dashed lines.

The resulting expression for the energy in the approximation of second-order perturbation theory is[62]

$$F_{nlm} = \hbar\omega_T\{n + \tfrac{3}{2} + \tfrac{9}{16}\lambda_T[m(m+1) - l(l+1) - 6]\} \quad (3.5)$$

where $\lambda_T = 2E^T_{JT}/3\hbar\omega_T$. Each of these levels is $(2m+1)$-fold degenerate with respect to the quantum number of the projection of the angular momentum, which is also preserved. In particular, for the ground state

$n = 0, l = 0, m = 1, 2m + 1 = 3$, i.e., the ground state is three-fold degenerate, as is the initial electronic T term.

The energy gap between the extreme components of the split level with a given value of n equals $\frac{3}{4}(2n + 1)E_{JT}^{T}$. Therefore, the perturbation theory formula (3.5) is valid until $E_{JT}^{T} \ll 4\hbar\omega_T/3(2n + 1)$.

If the coupling with both E and T_2 nuclear displacements is taken into account simultaneously (the T–$(e + t_2)$ problem), the solution becomes considerably more complicated even in the limiting case of weak vibronic coupling. Analytical expressions for splitting of several first vibronic levels in the second-order perturbation theory approximation have been obtained by means of special functions.[63] Important methods of vibronic effect calculations were suggested by Wagner.[63a]

In the other limiting case of strong vibronic coupling, when $2E_{JT}^{\Gamma} > n_{\Gamma}\hbar\omega_{\Gamma}$, some analytical solutions have been obtained, too. Consider first the linear E–e problem. The corresponding AP has the shape given in equation (2.1). If the JT stabilization energy E_{JT} is large enough, as assumed in the strong coupling case, the energy separation from the ground state sheet of the AP to the upper one in regions far from the branching point is also large (near the bottom of the trough this distance equals $4E_{JT}$). It follows that, for the nuclear motion in these areas, the vibronic equations (1.6) may be approximately decoupled, and for a small number of states near the bottom of the trough the Hamiltonian of the system takes the form (in polar coordinates)

$$H = -\frac{\hbar^2}{2M}\left[\frac{1}{\rho}\frac{\partial}{\partial\rho}\left(\rho\frac{\partial}{\partial\rho}\right) + \frac{1}{\rho^2}\frac{\partial^2}{\partial\phi^2}\right] + \frac{1}{2}K_E\rho^2 - |F_E|\rho \tag{3.6}$$

It can be shown that in the Schrödinger equation $(H - E)\Psi = 0$ with this Hamiltonian, the variables ρ and ϕ separate. The motions along the trough described by the ϕ coordinate are free rotations with energy $m^2(\hbar\omega_E)^2/4E_{JT}$ and wave function $(2\pi)^{1/2}\exp(im\phi)$, where m is the rotational quantum number, while radial motions (along ρ) are described well by harmonic vibrations with energies $\hbar\omega_E(n + \frac{1}{2})$ and oscillator wave functions $\chi_n(\rho - \rho_0)$. The resulting energy and wave function are[64] (the factor $1/\sqrt{\rho}$ is introduced for normalization)

$$\begin{aligned} \Psi_{nm}(r, \rho, \phi) &= \psi_-(r, \phi)\chi_n(\rho - \rho_0)\exp(im\phi)/(2\pi\rho)^{1/2} \\ E_{nm} &= \hbar\omega_E(n + \tfrac{1}{2}) + m^2(\hbar\omega_E)^2/4E_{JT} - E_{JT} \end{aligned} \tag{3.7}$$

$n = 0, 1, 2, \ldots;\ m = \pm\frac{1}{2}, \pm\frac{3}{2}, \pm\frac{5}{2}, \ldots$, where $\psi_-(r, \phi)$ is the electronic function of the lower sheet of the AP determined by equation (2.12). The half-integer values of the quantum number m of the rotational motions follow from the special conditions of the problem, namely, from the symmetry properties of the electronic function $\psi_-(r, \phi)$, which changes sign every 2π rotation (for which the total wave function must remain unchanged). The ρ_0 and E_{JT} values are given by equation (2.7).

It is seen from equation (3.7) that in this limiting case of strong vibronic linear coupling, every level of radial vibration (with frequency $\hbar\omega_E$) is accompanied by a series of doublet levels of rotational motion with energy intervals $\Delta E_{rot} = (3|m|+1)(\hbar\omega_E)^2/4E_{JT}$ equal to the rotational frequencies. Since $\hbar\omega_E \ll E_{JT}$, $\Delta E_{rot} \ll \hbar\omega_E = \Delta E_{vib}$, and the usual relation between the frequencies of molecular motion is maintained: $\Delta E_{el} \gg \Delta E_{vib} \gg \Delta E_{rot}$.

The general picture of nuclear motions described by equation (3.7) is as follows. At any point of the bottom of the trough, the distorted (as shown in Figure 2.5) system performs radial (E type) vibrations with frequency ω_E simultaneously moving over ϕ along the bottom of the trough with a rotational frequency much less than ω_E. Unlike the vibrations, the rotation over ϕ is not a real rotation of the system, but rather a kind of internal rotation. In real rotations, all the atoms bend synchronously around a common axis of rotation, whereas in motion along the trough each atom bends around its individual axis, the angular displacements of the latter being correlated (see Section 2.1). As a result of this "rotation," a slow (compared with the vibrations) change of the nuclear configuration in the space of E coordinates takes place (Figure 2.5); actually *the molecule does not itself rotate, but the wave of its tetragonal deformations does rotate.*

When the quadratic terms of VI are taken into account the previously indicated (Section 2.1) regularly alternating wells and humps occur along the bottom of the trough every 120° of the ϕ angle (Figures 2.2 and 2.3). The bigger the quadratic barrier height Δ in equation (2.10), the more the motion along such a warped trough differs from free rotation. Suppose the values of Δ are small compared with the quantum of radial vibrations ($\Delta \ll \hbar\omega_E$), the criterion of strong linear vibronic coupling $E_{JT} \gg \hbar\omega_E$ being preserved (the case $\Delta > \hbar\omega_E$ is treated in the next section). Here, the expansion of the AP (2.5) with respect to small quadratic VC is valid, and the approximate Hamiltonian for motion near the bottom of the trough takes the form

$$H = -\frac{\hbar^2}{2M}\left[\frac{1}{\rho}\frac{\partial}{\partial\rho}\left(\rho\frac{\partial}{\partial\rho}\right) + \frac{1}{\rho^2}\frac{\partial^2}{\partial\phi^2}\right] + \frac{1}{2}K_E\rho^2 - |F_E|\rho + G_E\rho^2\cos 3\phi \qquad (3.8)$$

Separation of the variables ρ and ϕ in equation (3.8) is impossible due to the additional term $G_E\rho^2 \cos\phi$, not present in equation (3.6). However, if this term is sufficiently small (as was actually assumed) and the difference in the aforementioned frequencies of the radial and rotational motion is large enough, an approximate separation of these variables is possible in the adiabatic approach. Taking into account the above parameter differences, stationary radial motions along ρ are adjusted for every instantaneous configuration of the system corresponding to a fixed angle ϕ, and hence the kinetic energy of the motions along ϕ may be neglected when considering fast radial motions.

Following the adiabatic approach (Section 1.1) the total wave function may be sought in the form $\chi'_n(\rho, \phi) \cdot \Phi(\phi)$, where $\chi'_n(\rho, \phi)$ is determined by the solution of the equation for the fast subsystem (3.8) in which the term $\partial^2/\partial\phi^2$ is neglected. If the last term of this equation, regarded as small, is also neglected, the remaining equation is independent of ϕ and yields the same solution $\chi_n(\rho - \rho_0)$ (with normalizing factor $1/\sqrt{\rho}$) as in the linear case. Substituting this solution into equation (3.8) and averaging over the variables ρ with ground state function χ_0, the following equation of motion along ϕ is derived[64]:

$$\left(-\alpha \frac{\partial^2}{\partial\phi^2} + \beta \cos 3\phi - E_m\right) \Phi(\phi) = 0 \tag{3.9}$$

where $\alpha = \hbar^2/3M\rho_0^2$, $\beta = G_E\rho_0^2$.

This equation has been solved numerically by means of the expansion $\Phi = \Sigma_m C_m \exp(im\phi)$ in the free rotation $\exp(im\phi)$. The energy levels obtained as functions of β/α are illustrated in Figure 3.3 (cf. Figures 3.1 and 3.2). It is seen that some of the doublet rotational levels (3.7) for $\beta = 0$, namely those transforming after the irreducible representations $A_1 + A_2$ (i.e., accidentally degenerate), become split when the quadratic VI is taken into account, i.e., when $\beta \neq 0$.

The total wave function in this case is

$$\Psi(r, \rho, \phi) = (2\pi\rho)^{-1/2}\psi_-(r, \Omega)\chi_0(\rho - \rho_0) \sum_m C_m \exp(im\phi) \tag{3.10}$$

where $m \pm \frac{1}{2}, \pm \frac{3}{2}, \ldots, \pm k$ correspond to the finite number of functions of free rotations in the trough chosen as a basis of the calculations, and C_m are the expansion coefficients determined by equation (3.9) (the electronic function is given by equation (2.12)).

In the case of small quadratic VI under consideration, the motions in the circular trough, as distinct from the linear case, cease to be free rotations. At every point of the trough (as in the linear case) the distorted system performs fast vibrations with frequency ω_E, and the distortion of the nuclear configuration changes slowly while assuming a continuous set of geometric figures in the space of E displacements, illustrated in Figure 2.5. However, unlike the linear case for which the "motion" of the

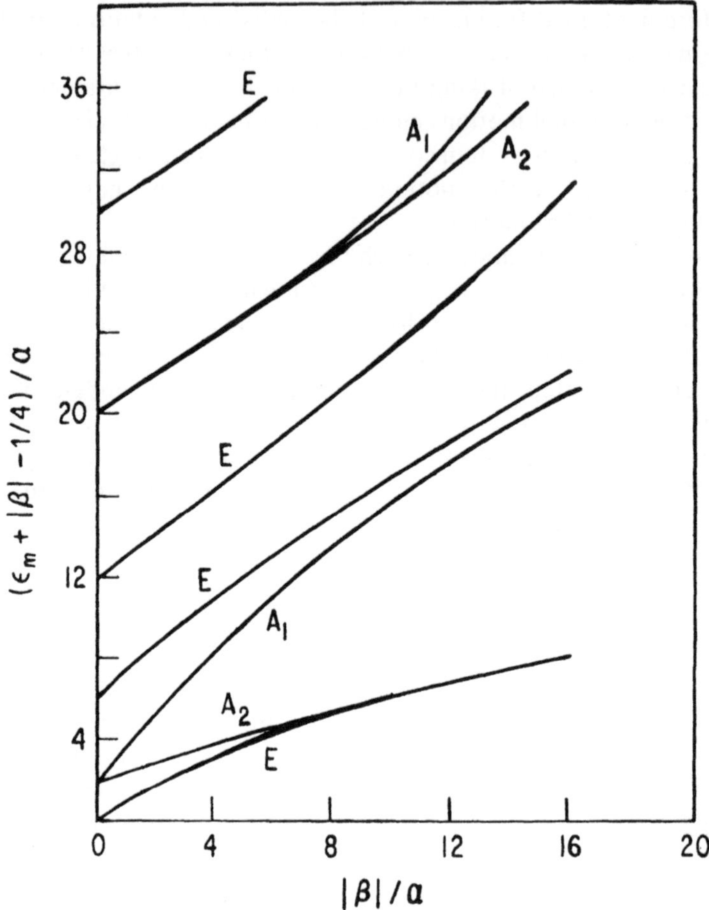

Figure 3.3. Energy levels for the quadratic E–e problem, solutions of equation (3.6), as functions of the ratio between the quadratic and linear VC parameters, β/α (after O'Brien[64]).

distorted configuration along the bottom of the trough is uniform, in the case of the quadratic VI the above changes in nuclear configuration are hindered (or even reflected) by the quadratic barriers. As a result, the system remains longer at the minima (Figure 2.3) than at the barrier maxima. *The picture as a whole is that of hindered internal rotations of the* JT *distortions.*

3.2. Tunneling Splitting. Pulsating Deformations

In the case of sufficiently large quadratic VI terms, when the quadratic barrier Δ in equation (2.10) is of the order of, or greater than the vibrational quanta $\hbar\omega_E$, the E–e problem becomes complicated with three minima of the AP and high barriers between the minima. For a T term, the minima of the AP appear in the linear approximation (Sections 2.1 and 2.2).

In the E–e problem with $\Delta \gtrsim \hbar\omega_E$, the above expansion of the AP and separation of variables ρ and ϕ are not valid, so the problem must be solved by another technique. Since the minima of the AP are deep enough, a local state arises at each of them which, in the limiting case of high barriers between the minima, may be considered as a quasi-stationary state. Therefore, as far as the lowest vibronic states are concerned, the local states with three minima can be taken as a starting approximation in the calculations, and then modified due to the minima state interactions (finite barrier height Δ) by means of perturbation theory.[65,66,9] A similar approach is applicable to all JT systems with equivalent minima sufficiently deep to have high barriers between them.

Suppose that, in the zero approximation, the states at the equivalent minima of the AP are independent, and the electronic function of the ith minimum is denoted by φ_i, its vibrational function by $\chi_{i\kappa}$, and the total function by $\Phi_{i\kappa} = \varphi_i\chi_{i\kappa}$. Since, in general, there are r equivalent minima, the system is r-fold degenerate in this approximation. We can now determine which terms in the Hamiltonian should be neglected in order to solve $\Phi_{i\kappa}$. If these terms are assumed to be small perturbations, one can look for the total wave function in the form of a linear combination of local minima functions:

$$\Psi_{\alpha\kappa} = \sum_{i=1}^{r} C_{i\alpha}\Phi_{i\kappa}, \qquad \alpha = 1, 2, \ldots, r \tag{3.11}$$

The energy levels E_κ and coefficients $C_{i\kappa}$ are determined by the secular equation

$$\| H_{ij}^{\kappa} - E_\kappa S_{ij}^{\kappa} \| = 0, \qquad i, j = 1, 2, \ldots, r \tag{3.12}$$

where H is the total Hamiltonian, and $S_{ij}^{\kappa} = \int \Phi_{i\kappa}\Phi_{j\kappa}d\tau$ is the overlap integral.

It can be easily shown that, due to the equivalence of the minima, $H_{ii} = H_{11}$, $H_{ij}^{\kappa} = H_{12}^{\kappa} = U_\kappa$, and $S_{ij}^{\kappa} = S_{12}^{\kappa} = S_\kappa$. For instance, in the E–e problem with three equivalent minima of the AP, equation (3.12) assumes the form

$$\begin{vmatrix} H_{11}^{\kappa} - E_\kappa & U_\kappa - E_\kappa S_\kappa & U_\kappa - E_\kappa S_\kappa \\ U_\kappa - E_\kappa S_\kappa & H_{11}^{\kappa} - E_\kappa & U_\kappa - E_\kappa S_\kappa \\ U_\kappa - E_\kappa S_\kappa & U_\kappa - E_\kappa S_\kappa & H_{11}^{\kappa} - E_\kappa \end{vmatrix} = 0 \tag{3.13}$$

Equation (3.12) or (3.13) for several, most important cases can be solved immediately. The E_α and $C_{i\alpha}$ values obtained for $r = 3, 4$, and 6 are listed in Table 3.1. It is seen that if $U_\kappa \neq 0$ the κth vibronic level under consideration, being r-fold degenerate at the minimum, splits into two levels (Figure 3.4): one nondegenerate (A_1 or A_2) and the other twofold E (for $r = 3$), or threefold T (for $r = 4$), or two threefold levels, T_1 and T_2, for $r = 6$. Initially, this splitting was called inversion splitting[66] (similar to inversion splitting in ammonia). Now the term *tunneling splitting* is more often used. We denote inversion (tunneling) splitting by δ_κ. Another symbol for this splitting, namely 3Γ, 4Γ, etc. (in general, $r\Gamma$) for three, four, etc. (r) minima problems, is also encountered in the literature, especially in ESR spectroscopy.

The criterion for perturbation theory to be applicable to this problem is

$$\delta_\kappa \ll \hbar\omega \tag{3.14}$$

where $\hbar\omega$ is the vibrational quantum at the minimum.

Table 3.1. Energies and Wave Function Coefficients $C_{i\alpha}$ (after equation (3.11)) of Tunneling States in Systems with Three, Four, and Six Equivalent Minima

Number of equivalent minima	Symmetry of the tunneling states	Energy	Wave function coefficients[a] $N_\alpha(C_{1\alpha}, C_{2\alpha}, \ldots, C_{r\alpha})$
3	E	$-\dfrac{U}{1-S}$	$[6(1-S)]^{-1/2}(2, -1, -1)$ $[2(1-S)]^{-1/2}(0, 1, -1)$
	A	$\dfrac{2U}{1+2S}$	$[3(1+2S)]^{-1/2}(1, 1, 1)$
4	T	$-\dfrac{3U}{1-S}$	$[4(1-S)]^{-1/2}(1, -1, -1, 1)$ $[4(1-S)]^{-1/2}(1, -1, 1, -1)$ $[4(1-S)]^{-1/2}(1, 1, -1, -1)$
	A	$\dfrac{U}{1+3S}$	$[4(1+3S)]^{1/2}(1, 1, 1, 1)$
6	T_1	$\dfrac{2U}{1+2S}$	$[4(1+2S)]^{-1/2}(1, 1, 1, 1, 0, 0)$ $[4(1+2S)]^{-1/2}(1, 0, -1, 0, 1, 1)$ $[4(1+2S)]^{-1/2}(0, -1, 0, 1, 1, -1)$
	T_2	$-\dfrac{2U}{1-2S}$	$[4(1-2S)]^{-1/2}(1, -1, 1, -1, 0, 0)$ $[4(1-2S)]^{-1/2}(-1, 0, 1, 0, 1, 1)$ $[4(1-2S)]^{-1/2}(0, -1, 0, 1, -1, 1)$

[a] N_α is a common factor.

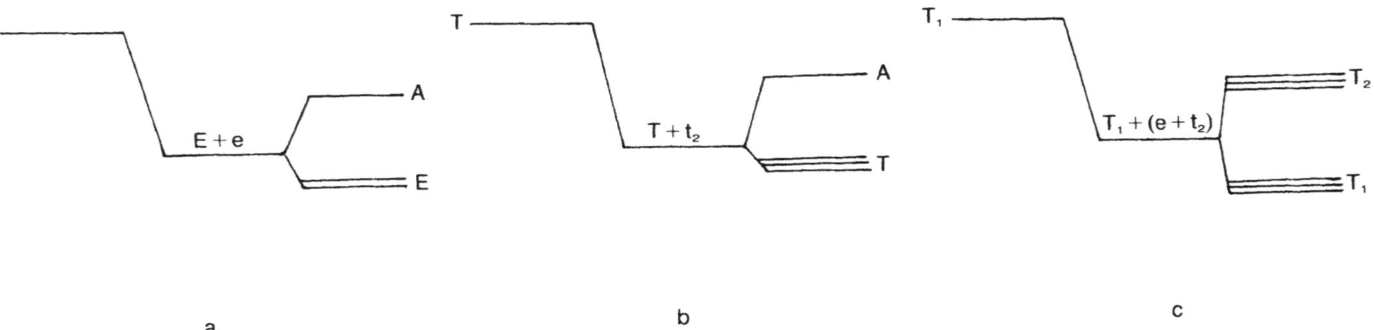

Figure 3.4. Tunneling splitting in case of: (a) three equivalent minima of the quadratic E–e problem; (b) four equivalent minima of the linear T–t_2 problem; (c) six orthorhombic minima of the quadratic T–$(e + t_2)$ problem.

The tunneling splitting δ_κ in the E–e problem (Table 3.1) is given by

$$\delta_\kappa = \frac{2U_\kappa}{1+2S_\kappa} - \frac{U_\kappa}{1-S_\kappa} = \frac{3U_\kappa}{(1+2S_\kappa)(1-S_\kappa)} \tag{3.15}$$

The functions $\chi_{i\kappa}$ are needed in order to calculate the values of U_κ and S_κ. In the approximation of deep minima used above, the vibrations may be treated as harmonic (at least for the lowest vibrational states κ). In this approximation

$$\chi_{i\kappa} = \prod_{\alpha=1}^{N} \chi_{m_\alpha}(Q_\alpha^{(i)}) \tag{3.16}$$

where $\chi_{m_\alpha}(Q_\alpha^{(i)})$ is the harmonic oscillator function for the m_αth vibrational state of the normal coordinate $Q_\alpha^{(i)}$ in the ith minimum configuration. Only the tetragonal coordinates Q_θ and Q_ε are different at the three minima of the E–e problem, the others remaining unchanged. The twofold degenerate E vibration at the tetragonal minimum is split into two components, A_1 and B_1, described by coordinates Q_θ and Q_ε, respectively. The ratio of the corresponding frequencies, $p = \omega_{B_1}/\omega_{A_1}$, characterizes the measure of the tetragonal distortion at the minimum.

The new normal coordinates at the minima, $Q_\theta^{(i)}$ and $Q_\varepsilon^{(i)}$, can be determined using symmetry considerations: they must transform into each other under symmetry transformations, in particular, rotation through $\pm 2\pi/3$ (Figure 3.5).

Hence

$$\begin{aligned} &Q_\theta^{(1)} = Q_\theta - Q^0, && Q_\varepsilon^{(1)} = Q_\varepsilon \\ &Q_\theta^{(2)} = -\tfrac{1}{2}Q_\theta + \frac{\sqrt{3}}{2}Q_\varepsilon - Q_0, && Q_\varepsilon^{(2)} = -\tfrac{1}{2}Q_\varepsilon - \frac{\sqrt{3}}{2}Q_\theta \\ &Q_\theta^{(3)} = -\tfrac{1}{2}Q_\theta - \frac{\sqrt{3}}{2}Q_\varepsilon - Q_0, && Q_\varepsilon^{(3)} = -\tfrac{1}{2}Q_\varepsilon + \frac{\sqrt{3}}{2}Q_\theta \end{aligned} \tag{3.17}$$

where $Q^0 = \rho_0$ if $(F_E/G_E) > 0$, and $Q^0 = -\rho_0$ if $(F_E/G_E) < 0$.

The electronic wave functions at the minima in the simple adiabatic approximation can be obtained immediately by inserting into equation (2.12) the values of Ω corresponding to the minimum positions. For instance, for the case of the elongated octahedron (Section 2.1),

$$\varphi_1 = |\theta\rangle, \quad \varphi_2 = -\tfrac{1}{2}|\theta\rangle - \frac{\sqrt{3}}{2}|\varepsilon\rangle, \quad \varphi_3 = -\tfrac{1}{2}|\theta\rangle + \frac{\sqrt{3}}{2}|\varepsilon\rangle \tag{3.18}$$

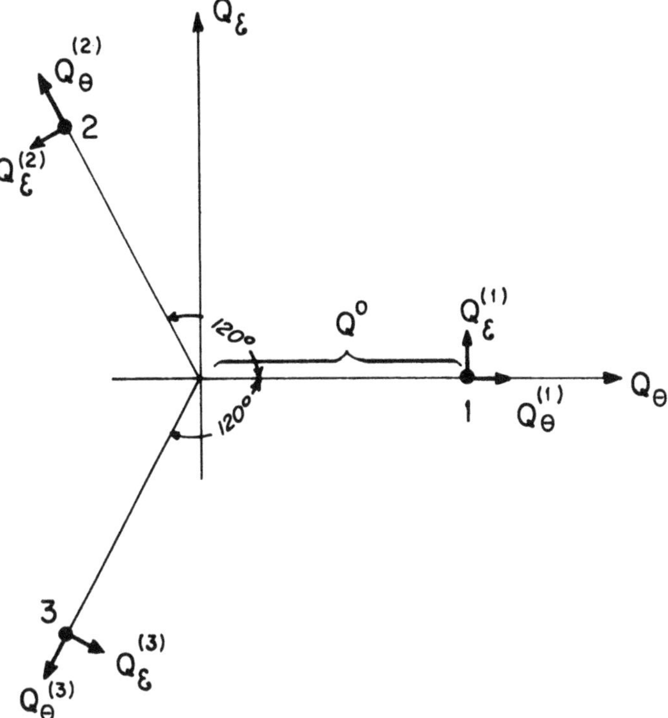

Figure 3.5. General $(Q_\theta, Q_\varepsilon)$ and local $(Q_\theta^{(i)}, Q_\varepsilon^{(i)})$ symmetrized coordinates of E type at the minima in the E–e problem.

The overlap integral is given by

$$S_\kappa = \int \varphi_1 \cdot \varphi_2 d\tau_r \int \chi_{1\kappa}\chi_{2\kappa} d\tau_Q = -\tfrac{1}{2}\gamma_\kappa$$
$$\gamma_\kappa = \int \chi_{m_\theta}(Q_\theta^{(1)})\chi_{m_\theta}(Q_\theta^{(2)})\chi_{m_\varepsilon}(Q_\varepsilon^{(1)})\chi_{m_\varepsilon}(Q_\varepsilon^{(2)})dQ_\theta dQ_\varepsilon \tag{3.19}$$

Without entering into the intermediate mathematical manipulations, we obtain for the ground state $m_\theta = m_\varepsilon = 0$

$$\gamma_0 = \left(\frac{16p}{3p^2 + 10p + 3}\right)^{1/2} \exp\left(-\frac{6p}{1+3p}\frac{E_{JT}}{\hbar\omega_{A_1}}\right) \tag{3.20}$$

$$U_0 = \frac{3\gamma_0}{1+3p} E_{JT}\left(1 - \frac{2p}{1+3p}\frac{|G_E|}{K - 2|G_E|}\right) \tag{3.21}$$

Neglecting terms of order γ^2 compared with unity, we have

$$\delta_0 = \frac{g}{1+3p}\,\gamma_0\left[E_{JT} - \frac{\Delta}{2(1+3p)}\right] \tag{3.22}$$

In the limiting case of $p = 1$, $\delta_0 = \frac{9}{4}\gamma_0[E_{JT} - (\Delta/8)]$, whereas for $p \ll 1$, $p^2 \approx 9|G_E|/K_E$ (see equation (2.11)) and

$$\delta_0 \approx 36(p/3)^{1/2}\exp(-6E_{JT}/\hbar\omega_{A_1}) \tag{3.23}$$

For strong linear coupling, $\Delta < E_{JT}$ and hence $\delta_0 > 0$, thus the ground state remains doubly degenerate (see Table 3.1).

The wave functions of tunneling levels A and E in the approximation under consideration are determined by expression (3.11) with coefficients $C_{i\alpha}$ given in Table 3.1.

The above approximation is not very accurate for numerical estimation of tunneling splitting, since in equation (3.18) the dependence of the electronic functions on the nuclear coordinates within the minimum is neglected (cf. equation (2.12)). The problem has been solved by other methods. In particular, the formulas of the quasi-classic approximation may be used for this purpose.[67] For the E–e problem with weak quadratic VI when $|G_E| \ll K_E$, this approach yields

$$\delta \approx (3\hbar\omega_E/2\pi)p\exp(-0.6pE_{JT}/\hbar\omega_E) \tag{3.24}$$

The variational principle also has been proposed for a solution to this problem. It can be shown [68,8] that variational functions of the form

$$\chi_E(\rho,\phi) = \mathcal{N}\begin{Bmatrix}\cos(\phi/2)\\ \sin(\phi/2)\end{Bmatrix} q^{1/2}\exp[-\tfrac{1}{2}u(q-1)^2 + \nu\cos 3\phi] \tag{3.25}$$

where N is the normalization factor, $q = \rho/\rho_0$, u and ν are the variational parameters, and an analogous function for the A level with factor $\cos(3\phi/2)$, satisfy all the symmetry requirements of the problem, as well as the known asymptotic behavior. Minimization of the total energy calculated by these functions results in the energy and u and ν parameter values as functions of the linear ($\lambda_1 = E_{JT}/\hbar\omega_{A_1}$) and quadratic ($\lambda_2 = \Delta/\hbar\omega_{A_1}$) barrier parameters.

In the T–t_2 problem, the AP has four trigonal minima (Table 2.1) at a depth determined by equation (2.26). The tunneling splitting of the ground level δ_0 is given in Table 2.1 and illustrated in Figure 3.4. With the approximate harmonic functions at the minima, equation (2.27) for the electronic functions, and other relations from Section 2.2, the above approach can be employed to obtain for this case, when splitting of the ω_T frequency is neglected,

$$\delta_\kappa = 4U_\kappa/(1+2S_\kappa - 3S_\kappa^2) \qquad (3.26)$$

$$U_\kappa = \tfrac{2}{3}E_{JT}^T\gamma_\kappa, \qquad S_\kappa = -\tfrac{1}{3}\gamma_\kappa \qquad (3.27)$$

For the ground state

$$\gamma_0 = \exp(-4E_{JT}^T/3\hbar\omega_T) \qquad (3.28)$$

If splitting of the ω_T frequency at the trigonal minimum is taken into account (Section 2.2), the following approximate formula can be obtained:

$$\delta \approx 2E_{JT}^T \exp(-1.24E_{JT}^T/\hbar\omega_T) \qquad (3.29)$$

A somewhat different expression has been deduced from a quasi-classical approximation[67]:

$$\delta \approx (32/3)^{1/2}(\hbar\omega_T/2\pi)\exp(-0.73E_{JT}^T/\hbar\omega_T) \qquad (3.30)$$

In the case of strong vibronic coupling (large values of E_{JT}^T), the difference in the exponential powers in these two expressions is compensated for by the pre-exponential factors.

In the linear T–e problem the electronic functions, as indicated in Section 2.2, are mutually orthogonal and hence $U_\kappa = 0$, i.e., there is no tunneling splitting, provided the influence of the T_2 vibrations (as well as quadratic VI terms) is disregarded. The absence of tunneling splitting in this case also results from general symmetry considerations.

The solution to the problem in the general case of the T–$(e+t_2)$ problem is very difficult, due to the complicated five-dimensional AP discussed in Section 2.2. If the trigonal minima are the lowest, tunneling splitting can be obtained in the quasi-classical approximation[67]:

$$\delta = (32/3)^{1/2}(\hbar\omega_T/2\pi)\exp\left[-0.73\left(\frac{E_{JT}}{\hbar\omega_T}\frac{E_{JT}^T - E_{JT}^E}{\hbar\omega_T}\right)^{1/2}\right] \qquad (3.31)$$

where E_{JT}^T and E_{JT}^E are given by equations (2.24) and (2.26).

Tunneling splitting in the case of six equivalent orthorhombic minima of the AP in the T–$(e+t_2)$ problem, with allowance for the quadratic terms of VI, results in two triplet vibronic levels T_1 and T_2 (Table 3.1 and Figure

3.4) with the splitting given by

$$\begin{aligned}\delta &= \tfrac{3}{2}E_{JT}^{0}\,\frac{24-28B-2B^2+9B^3}{(4-3B)(4-3B^2)}\,\frac{S}{1-S}\\ S &= \exp\left[-\frac{3}{2}\frac{E_{JT}}{\hbar\omega}\,\frac{12-20B-11B^2}{(4-3B)(4-3B^2)}\right]\end{aligned} \tag{3.32}$$

where B and E_{JT}^{0} are determined by equations (2.30) and (2.31), respectively.

For the multimode problem discussed in Section 2.4, the AP in corresponding coordinates can be reduced to a one-mode (ideal) case, but the equations determining the energy spectra and wave function remain multimode. Some solutions to this problem obtained recently[69–73a,13] show that, in the energy spectra, new local resonances occur due to the VI, some of which are considered as resulting from tunneling splitting (although the spectra as a whole is rather complicated).

Tunneling splitting can be estimated by means of the quasi-classic approximation. In the case of an infinite number of JT active modes (a JT center in a crystal), the vibronic levels are subject to strong relaxational broadening $\Delta\Gamma$. At low temperature $\Delta\Gamma \sim \delta^4$. Therefore tunneling splitting can be observed for very small values of δ, when $\Delta\Gamma < \delta$. As the temperature increases the splitting smears and disappears. These results become considerably more complicated when taking into account the dependence of the tunneling splitting itself on temperature, which occurs due to nonharmonic multimode effects. The problem as a whole remains far from being solved.

Tunneling effects in JT systems are not occasional, but occur more often than might be imagined at first sight. The above treatment is based on the assumption that the states at the AP minima are good zero approximations to the problem, in which case tunneling transitions between the minima result in small (tunneling) splitting of the energy levels. This approach is valid if and only if the tunneling splitting δ is small compared with the magnitude of the vibration quantum. This condition is satisfied if (see the expression for δ_0 in equation (3.20) or (3.28))

$$E_{JT} \gtrsim \hbar\omega \tag{3.33}$$

i.e., if the minimum depth is bigger than the value of the quantum of active vibrations. In other words, if there is a local vibrational state at the minimum, then the system performs tunneling transitions between the configurations of the equivalent minima of the AP.

Visually, these transitions between differently oriented distorted configurations may be imagined as pulsating deformations (distortions). Let us assume that as a result of synthesis or some external perturbation the system happens to fall into the JT distorted configuration of one of the equivalent minima of the AP. Then, after a time τ (inversely proportional to the tunneling splitting), during which the system performs ordinary vibrations, it turns out to be in the equivalent state of another minimum, at which the distortion differs from that of the first minimum by the direction of the appropriate nuclear displacements (Figure 2.3). Then again, after time τ, the system moves into the third equivalent minimum configuration (or back to the first), and so on.

For example, if the system is octahedral and it is tetragonally distorted at the minimum (as in the E–e problem), elongated along the fourfold axis, the pulsations of these deformations result in periodic (with period τ) elongations along each of the three fourfold axes, alternatively. Since $\delta \ll \hbar\omega$, the frequency of pulsating distortions of the molecule is much less than the frequency of the vibrations in the distorted configurations, and the lifetime τ of the latter is much greater than the period of vibrations.

In the case of trigonal minima of a T term, the corresponding distortions at the minima are illustrated in Figure 2.8. The pulsating motions here result in periodic alternative changes of the direction of the distortion along each of the four trigonal axes of the system.

With the introduction of pulsating motions, *there are now three types of JT dynamics*: *free rotation, hindered rotation, and pulsating motion of JT deformations.* The phenomenon of pulsating motion and tunneling splitting in polyatomic systems together with relevant ensuing consequences (effects, laws, applications) was registered (in the USSR) as the scientific discovery N 202.[74]

3.3. Vibronic Reduction Factors

The reduction of ground-state physical quantities of electronic origin is one of the important effects of the VI. It originates from the back influence of the JT nuclear dynamics on the electronic structure and properties.

In every solution of the vibronic problems obtained above, the ground vibronic state possesses the same type of symmetry, degeneracy, and multiplicity as the initial electronic term in the highest symmetry configuration. This result may be due to the fact that the VI terms W (equation (1.14)), having the same symmetry as the total Hamiltonian, do not remove the degeneracy of the ground state (in contrast to the rough formulation of the JT theorem which contains an opposite statement; see Section 1.6).

The coincidence of the symmetry of the ground-state terms with and without regard to VI allows us to simplify the calculation of many properties. Already in the first work on calculations of spin–orbit splitting of the ground vibronic state,[75] it was shown that, ignoring mixings with other vibronic levels, this splitting is proportional not only to the spin–orbit coupling constant, as in the usual nonvibronic cases, but to this constant multiplied by the overlap integral between the vibrational functions of different minima, equal to γ_κ after equation (3.13). Since $\gamma_\kappa < 1$, the VI essentially reduces the spin–orbit splitting, sometimes by several orders of magnitude. Ham[76,6] generalized this idea and has shown that such a reduction occurs for any physical quantity, provided its operator depends on electronic coordinates only. This reduction is often called the *Ham effect.*

The following theorem of vibronic reduction can be formulated. Suppose we need to calculate the matrix element of the physical quantity $X_{\bar{\Gamma}\bar{\gamma}}(r)$, transforming according to the $\bar{\gamma}$ line of the $\bar{\Gamma}$ irreducible representation of the symmetry group of the system, with the function of the ground vibronic state $\Psi_{\Gamma\gamma}(r, Q)$. We denote the corresponding wave functions of the initial electronic term by $\psi_{\Gamma\gamma}(r)$. The theorem of vibronic reduction states that

$$\int \Psi^*_{\Gamma\gamma_1} X_{\bar{\Gamma}\bar{\gamma}} \Psi_{\Gamma\gamma_2} d\tau = K_\Gamma(\bar{\Gamma}) \int \psi_{\Gamma\gamma_1} X_{\bar{\Gamma}\bar{\gamma}}(r) \psi_{\Gamma\gamma_2} d\tau \tag{3.34}$$

where $K_\Gamma(\bar{\Gamma})$ is a constant which depends on the vibronic properties of the Γ state and the symmetry $\bar{\Gamma}$ of the operator X, independent of its nature. The constant $K_\Gamma(\bar{\Gamma})$ is called the vibronic reduction factor (not to be confused with the curvature or force constants denoted by K_Γ).

From this vibronic reduction theory it follows that *if the vibronic reduction factors $K_{\bar{\Gamma}}(\bar{\Gamma})$ are known, there is no necessity to solve the vibronic problem in order to obtain electronic properties of the ground state; they can be calculated by wave functions of the initial electronic term.* In particular, the $K_\Gamma(\bar{\Gamma})$ constants can be determined from certain experimental data, and then used to predict the vibronic contribution to all the other observables.

Approximate analytical expressions for the vibronic reduction factors in the linear E–e problem $K_E(A_2)$ and $K_E(E)$, often denoted by p and q, respectively, can be derived immediately from the above approximate solutions to the problem (Sections 2.1, 3.2, and 3.3). It can be shown that for the ideal linear case the following relationship is valid:

$$2q - p = 1 \tag{3.35}$$

In the case of weak vibronic coupling $E_{JT} \ll \hbar\omega_E$,

$$p \approx \exp(-4E_{JT}/\hbar\omega_E) \tag{3.36}$$

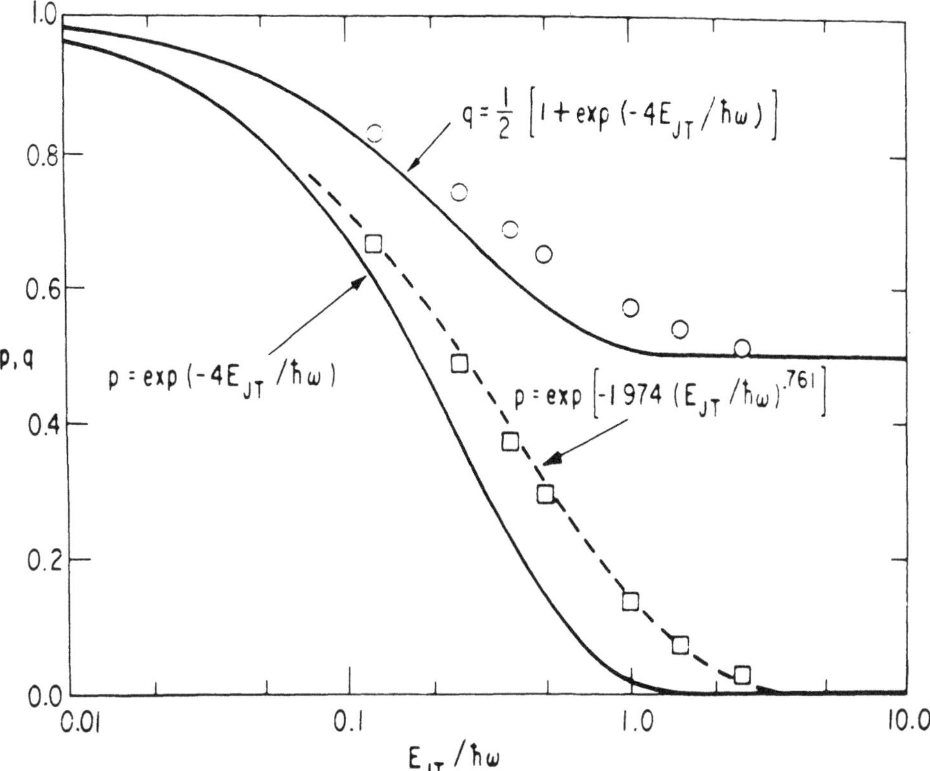

Figure 3.6. Vibronic reduction factors $p = K_E(A_2)$ and $q = K_E(E)$ as functions of dimensionless parameter $\lambda_E = E_{JT}/\hbar\omega_E$ for the linear E–e problem after equations (3.36) and (3.37). Exact numerical values are shown by points; the dashed line is chosen to fit the numerical data in the region $0.1 \lesssim \lambda_E \lesssim 3.0$ (after Ham[6]).

For arbitrary coupling the value of p can be derived from the numerical solutions[77]; for $0.1 \lesssim E_{JT}/\hbar\omega_E \lesssim 3.0$, they obey the function[6]

$$p = \exp[-1.974(E_{JT}/\hbar\omega_E)^{0.761}] \tag{3.37}$$

These data are illustrated in Figure 3.6. For sufficiently strong vibronic coupling, $p = 0$ and $q = \frac{1}{2}$.

Note that the relation (3.35) is, strictly speaking, invalid in the general case. In particular, it fails in the multimode problem, as well as under quadratic VI. In general $2q - p \leqslant 1$, and the deviation of the value $2q - p$ from unity may be regarded as an indicator of multimodal or quadratic effects. It can be shown[73] that in the multimode E–e problem with strong vibronic coupling, $p = 0$ and $q = \frac{1}{2}\exp(-M/2)$, where $M \geqslant 0$ and depends on the dimensionless VC value.

For the T–$(e + t_2)$ problem in the limiting case of weak vibronic coupling,

$$K_T(E) = 1 - (9E_{JT}^T/4\hbar\omega_T) \tag{3.38}$$

$$K_T(T_1) = 1 - (3E_{JT}^E/2\hbar\omega_E) - (9E_{JT}^T/4\hbar\omega_T) \tag{3.39}$$

$$K_T(T_2) = 1 - (3E_{JT}^E/2\hbar\omega_E) - (3E_{JT}^T/4\hbar\omega_T) \tag{3.40}$$

It follows that

$$K_T(E) + \tfrac{3}{2}[K_T(T_2) - K_T(T_1)] = 1 \tag{3.41}$$

and this relation remains valid also for the strong vibronic coupling limit.[78] More precise values of these constants have been obtained by numerical calculation.[41,63]

In the absence of VI with T_2 displacements the linear T–e problem can be solved exactly, the corresponding reduction factors being

$$\begin{aligned} K_T(E) &= 1 \\ K_T(T_1) &= K_T(T_2) = \exp(-3E_{JT}^E/2\hbar\omega_E) \end{aligned} \tag{3.42}$$

For the linear T–t_2 problem the following approximate expressions

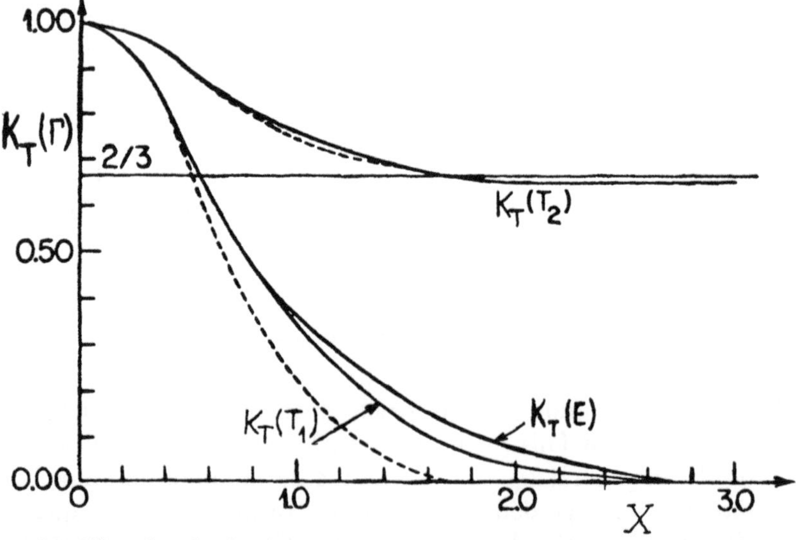

Figure 3.7. Vibronic reduction factors for the linear T–t_2 problem obtained from numerical solutions[6] as functions of dimensionless parameter $X = [3E_{JT}^T/2\hbar\omega_T]^{1/2}$. The dashed lines correspond to the approximate expressions (3.43).

are known[6] (Figure 3.7):

$$K_T(E) \approx K_T(T_1) = \exp(-9E_{JT}^T/4\hbar\omega_T)$$
$$K_T(T_2) \approx \tfrac{1}{3}[2 + \exp(-9E_{JT}^T/4\hbar\omega_T)] \qquad (3.43)$$

These expressions satisfy relation (3.41) and agree qualitatively with the results of numerical calculations.[78] It is seen that, in the limit of strong vibronic coupling, $K_T(E) = K_T(T_1) = 0$, $K_T(T_2) = \frac{2}{3}$.

In the case of orthorhombic minima of the AP in the quadratic T–$(e + t_2)$ problem (Section 2.2) subject to equal tetragonal and trigonal vibration frequencies $\omega_T = \omega_E = \omega$ and hence $K_E = K_T = K$ (where K is the force constant), one can obtain[43,13]

$$K_T(E) = (1 + 8S + 6\tilde{S})/(4 + 8S)$$
$$K_T(T_1) = (3S + \tilde{S})/(1 + 2S)$$
$$K_T(T_2) = (1 + 6S)/(2 + 4S) \qquad (3.44)$$
$$S = \tfrac{1}{2}\exp[-K(3Q_\theta^{(0)2} + 2Q_\zeta^{(0)2})/4\hbar\omega]$$
$$\tilde{S} = \tfrac{1}{2}\exp(-KQ_\zeta^{(0)2}/\hbar\omega)$$

where the coordinates of the minimum points $Q_\gamma^{(0)}$ are given by equation (2.31).

The above expressions show explicitly that the reduction of electronic properties due to VI depends exponentially on the relation $E_{JT}/\hbar\omega$ (the dimensionless VC), which is a measure of the JTE. The stronger the JTE, the larger the reduction. Especially, in every case the T_1 type operators, to which the orbital angular moment of the electrons and hence the spin–orbit interaction pertain, are heavily reduced. The essential reduction of spin–orbit splitting of the ground states (sometimes by 1–2 orders of magnitude; see e.g., Bersuker and Vekhter[75]) is one of the clearest manifestations of the JTE for systems with T terms (the E term of cubic systems is not split by spin–orbit interaction in the first order).

Examples of other vibronically reduced quantities are: the orbital (anisotropic) part of the Zeeman interaction, all kinds of interactions between the electronic shell and the nucleus (dipole–dipole, quadrupole, etc.), Coulomb and exchange interactions between the electronic shells of JT centers in crystals.[79]

It must be emphasized that the idea of vibronic reduction loses its simplicity and attractiveness as soon as the physical quantity in question ceases to be determined by the ground state alone. In particular, in the case of strong vibronic coupling the first excited vibronic level Γ_2 is spaced quite close to the ground multiplet Γ_1 (Figures 3.3 and 3.4), and their mixing may essentially influence the physical properties under consideration. In these cases, similar to equation (3.34), the vibronic mixing factor $r = K(\Gamma_1|\bar{\Gamma}|\Gamma_2)$, equal to the ratio of the two matrix elements calculated by total and pure electronic wave functions, respectively, can be introduced. In the E–e problem, mixing the ground E vibronic (tunneling) level with the first excited A level by E type operators is described by the mixing factor $r = K(E|E|A)$, which in the limiting case of strong vibronic coupling[6] is $r = \pm 1/\sqrt{2}$. If $K_{\Gamma'}(\bar{\Gamma}) \ll 1$, second-order perturbation theory corrections to the vibronic reduction may be important.[6]

As indicated above, the use of vibronic reduction factors is limited by operators of physical quantities which depend on electronic variables only, and not on nuclear coordinates. If the operator depends on electronic variables alone, but in the calculation of its physical manifestations nuclear coordinates are also involved, the above relations for the vibronic reduction factors are also invalid. For instance, when the electronic operator is calculated by second-order approximation of perturbation theory, in which excited electronic states are involved (such as in the calculation of Zeeman splitting of an E term in octahedral systems), taking into account the VI in general makes this operator dependent upon nuclear coordinates, and hence the vibronic reduction factors, obtained above, become inapplicable here.[80] The solution to the vibronic problem in these cases may be simplified by introducing second-order vibronic reduction factors $K^{(2)}(\Gamma)$.[80,13] Fortunately, in many cases $K^{(2)}_{\bar{\Gamma}}(\Gamma) \approx K_{\bar{\Gamma}}(\Gamma)$.

3.4. Numerical Solutions

The above analytical solutions of the vibronic problems determine the energy levels and wave functions of the system in the limiting cases of weak and strong vibronic coupling, but only for a limited number of the lowest energy states. A more complete solution of the problem for arbitrary values of VC can be obtained by means of numerical methods. The well-known method in quantum chemistry, namely, expansion of the unknown function as a finite number of zero-order basis functions (of the initial undistorted configuration), is widely used. However, this standard method is not always acceptable, since it involves laborious computation and computer time.

For instance, in the T–$(e + t_2)$ problem, if the vibrational functions of the lowest unperturbed E and T_2 vibrations up to $n = 8$ are included in the basis of calculations, the secular equation to be solved has order 3861×3861, and if the spin–orbit interaction is included the matrix becomes of order 15444×15444 (the group-theoretical classification by irreducible representations of the symmetry group reduces these dimensions by a factor of 3–4). On the other hand, in order to obtain accurate results many basis functions are needed. The greater the VC, the larger the basis required. Therefore, more economical methods of calculation, such as the Lanczos method and the method of coordinate relaxation[81] (see also Bersuker and Polinker[13]), are increasingly used in vibronic state calculations.

The first numerical results for the vibronic energy levels and wave functions of the linear $E-e$ problem as functions of the dimensionless VC were obtained by Longuet-Higgins, Öpik, Pryce, and Sack.[82,83] As expected, for small values of λ ($\lambda \leqslant 0.25$) the positions of the vibronic levels resulting from the numerical solutions agree well with that obtained from

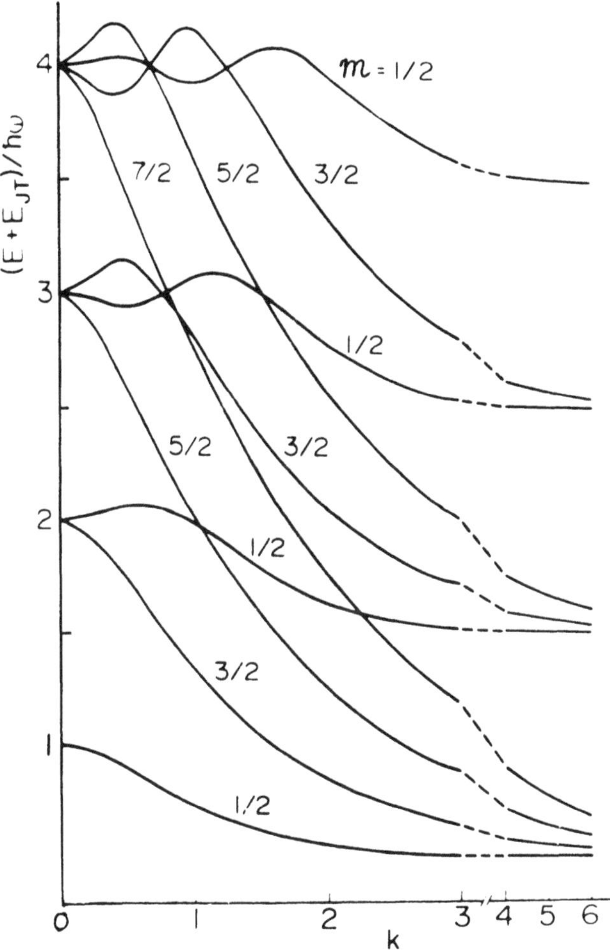

Figure 3.8. Vibronic energy levels for the linear $E-e$ problem obtained by numerical solutions without separation of the radial and angular motion.[84] $k = (2E_{\mathrm{JT}}/\hbar\omega_E)^{1/2}$ is a dimensionless coupling parameter.

the analytic expression (3.1) for the weak coupling case, while for large values of λ ($\lambda \gtrsim 4$) they follow expression (3.4). With the energies and wave functions obtained the authors calculated the probabilities of transition to these states from the vibrational states of a nondegenerate electronic A term, and vice versa ($A \to E$ and $E \to A$ transitions). The resulting spectra will be discussed in Chapter 4. Calculations of matrix elements of physical quantities by means of such numerical data are given in more detail elsewhere[77] (Section 4.3).

The numerical solution of the E–$(b_1 + b_2)$ problem (Section 2.1) was obtained by Muramatsu and Sakamoto[84] by means of diagonalization of the 231×231 matrix (equivalent to mixing the first $n_1 + n_2 \leqslant 20$ vibrational levels of the unperturbed system). The results also include data for the linear E–e problem as a particular case when $K_1 = K_2$ and $F_1 = F_2$ (Figure 3.8). An interesting feature in the behavior of the excited vibronic levels occurs for large VC: with increasing VC their energies oscillate near the mean value $\hbar\omega_E(n+1) - E_{JT}$ and then one of these levels descends smoothly, ultimately taking its place among the levels of the rotational structure of the strongly coupled spectrum. (See also Figure 3.9.)

Another important feature is the strong dependence of the vibronic level positions on the VC even in the case of weak coupling (see Figures 3.8 and 3.9). It follows that vibronic effects in the cases under consideration

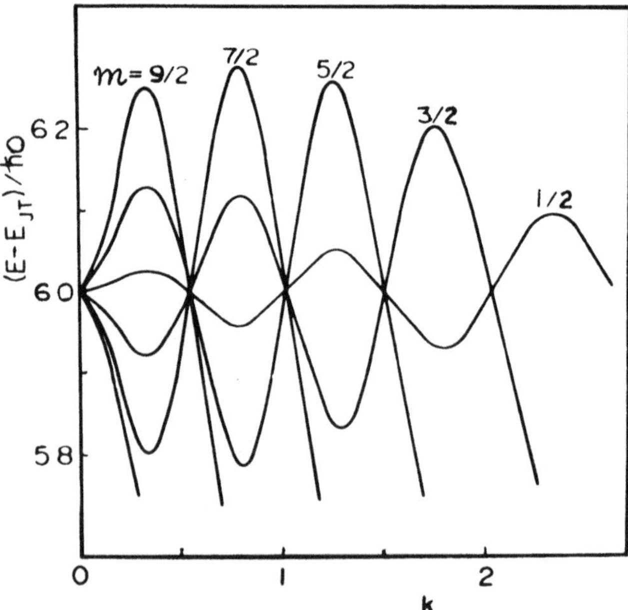

Figure 3.9. Energy levels for high excited states of the linear E–e problem[84] (the same as shown in Figure 3.8 for low-lying states). Accidental multiple degeneracy is a special feature of these states.

cannot be neglected even when vibronic coupling is weak. Other numerical calculations for E terms were conducted by Struck and Herzfeld[85] and Uehara.[86]

The first numerical calculations for the T–t_2 problem were carried out by Caner and Englman.[78] The vibronic energy levels as functions of the ratio $E_{JT}/\hbar\omega_T$, illustrated in Figure 3.10, were obtained by expanding the unknown wave functions in terms of the corresponding linear combinations

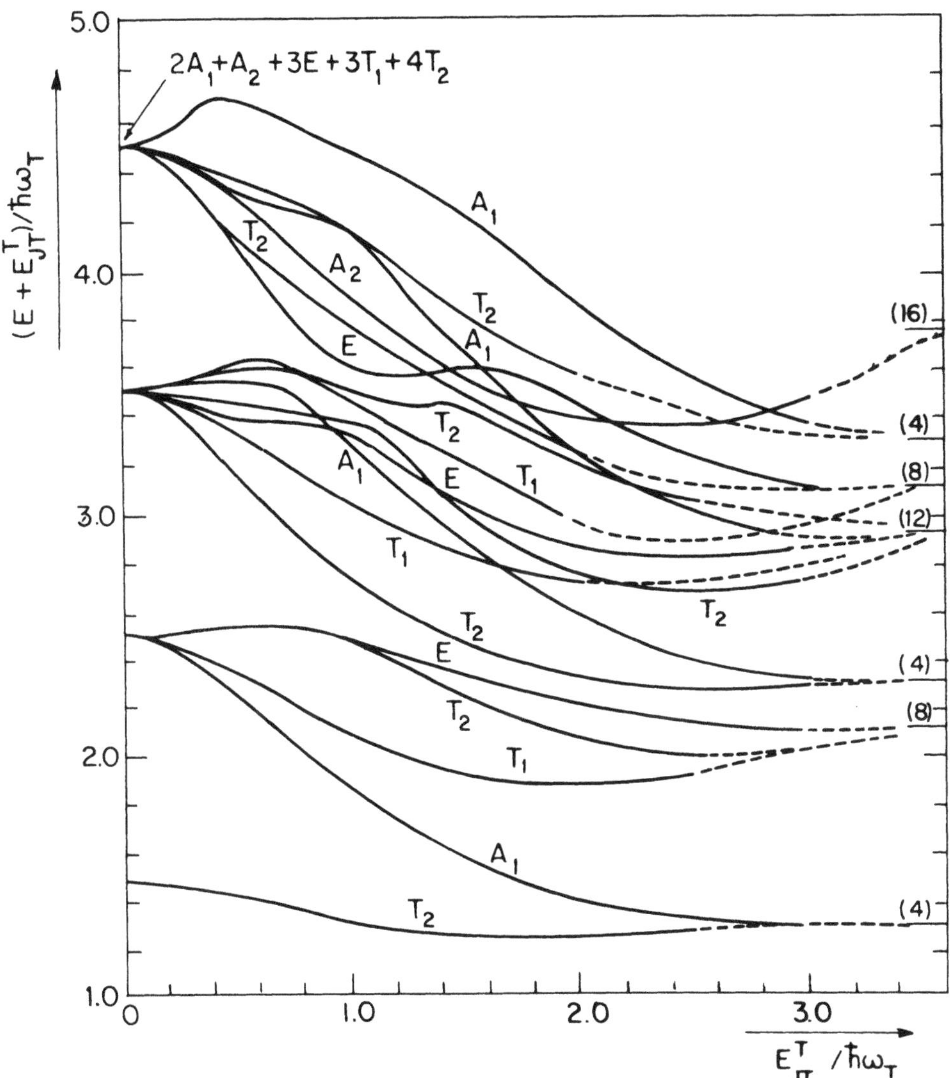

Figure 3.10. Vibronic energy levels for the T_1–t_2 problem (after Caner and Englman[78]). For the T_2–t_2 problem, subscripts 1 and 2 in the term labels must be interchanged.

of functions governing the unperturbed three-dimensional oscillator. It is seen that without VI the spectrum contains equidistant levels spaced according to those of the three-dimensional oscillator $E_n = \hbar\omega_T(n + \frac{3}{2})$. In the other limiting case of very strong vibronic coupling, all the levels are fourfold degenerate (in accordance with four minima of the AP) and join into two groups of equidistant levels with steps $\Delta E = \hbar\omega_T = \hbar\omega_A$ and $\Delta E = \hbar\omega_T(2/3)^{1/2} = \hbar\omega_E$, respectively; these two frequency distances are

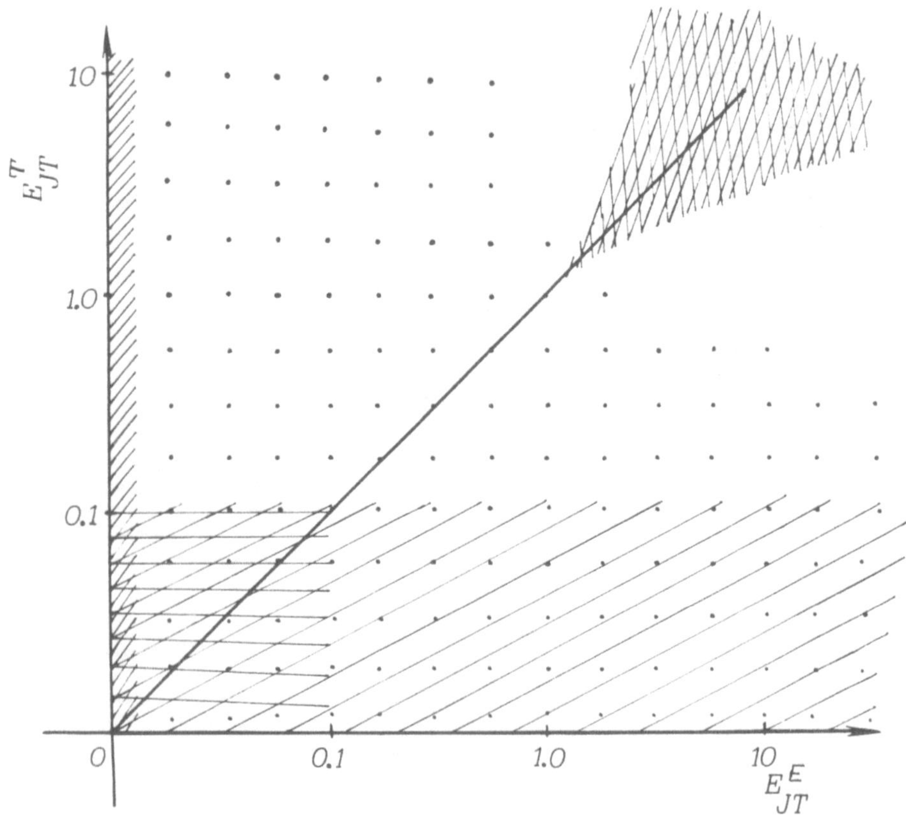

Figure 3.11. Illustration of the present-day state of calculations of energy levels and wave functions for the T–$(e + t_2)$ problem. Dimensionless values of JT stabilization energies E^E_{JT} and E^T_{JT} in units $\hbar\omega_E = \hbar\omega_T = \hbar\omega$ are plotted on the axes in a logarithmic scale. In the area with horizontal shading the perturbation theory of Moffitt and Thorson[62] is valid. The thick oblique shading shows the area of applicability of the numerical results of Caner and Englman.[78] The light oblique shading corresponds to the validity of the perturbation theory of Bersuker and Polinger.[63] Along the line $E^T_{JT} = E^E_{JT}$ the results of O'Brien's d-model are valid.[93] Double shading at large E^T_{JT} and E^E_{JT} values corresponds to the qualitative results of O'Brien.[41,87] The area of applicability of the numerical results of Sakamoto and Muramatsu[89] and Boldyrev, Polinger, and Bersuker[90] is shown by points.

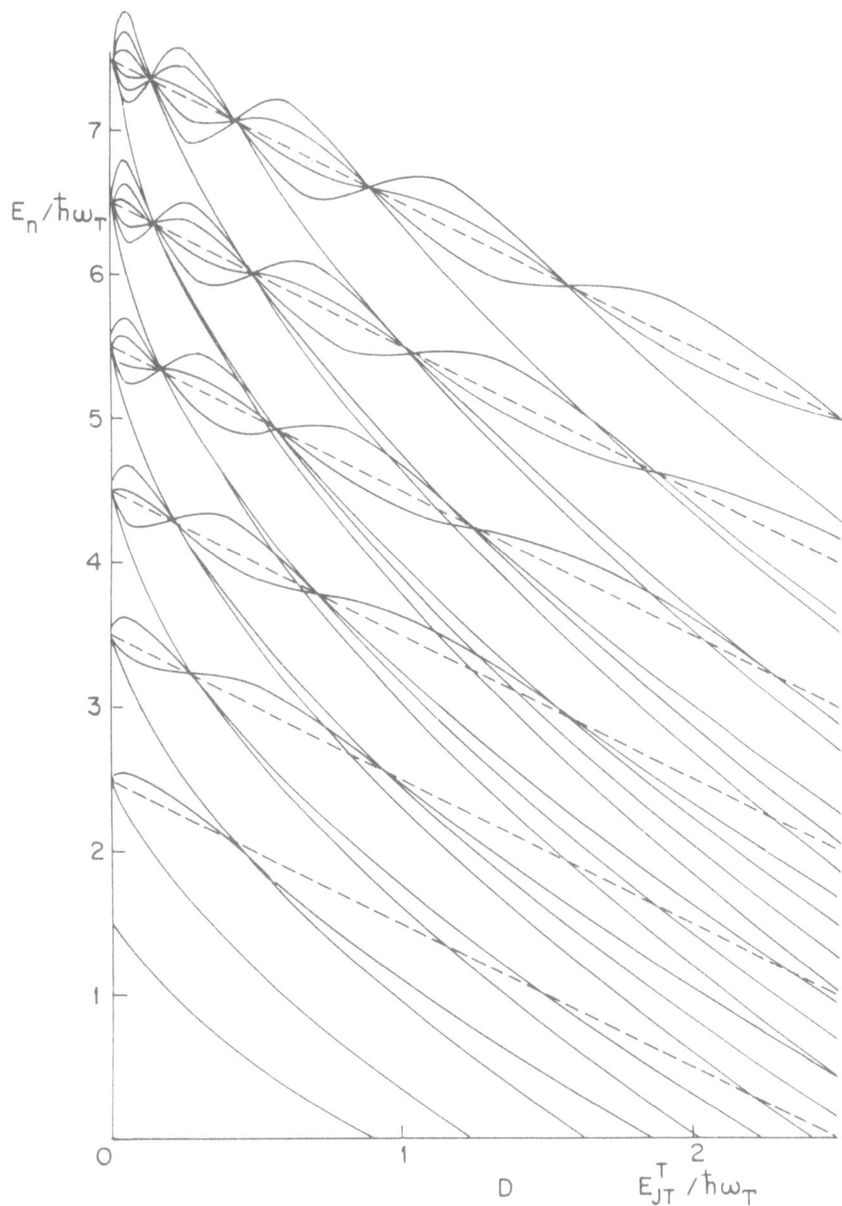

Figure 3.12. Vibronic energy levels for the linear Γ_8–t_2 problem.[93] Dashed lines show mean values around which the excited state energy levels oscillate (cf. Figures 3.8 and 3.9).

due to splitting of the ω_T frequency into two components $\omega_{A1} = \omega_T$ and $\omega_E = (2/3)^{1/2}\omega_T$ in the trigonal minima (Section 2.2). In the area where $E_{JT}/\hbar\omega_T \gtrsim 2$, the levels correspond to those predicted by the theory of tunneling splitting.

Calculations for the T–$(e + t_2)$ problem are the most difficult. Numerical results[41,62,78,93,63,87–90] for different regions of VC values are illustrated in Figure 3.11 and cover the main areas of possible VC. Recently, the problem was solved[90] taking into account spin–orbit interaction and additional external perturbations (trigonal distortions). Numerical solutions for the Γ_8–t_2 problem (Figure 3.12) can be found elsewhere.[83,91–93]

One limitation of the numerical solutions is the impossibility of presenting them in a visual way, especially when the number of vibronic parameters is large. In such cases the value of an economical computer program depends on the possibility of obtaining independently some of the vibronic parameters from experimental data or from calculations (at least approximately).

4

Applications to Spectroscopy

Vibronic effects directly influence resonance interactions of a system with external perturbations and play a role in spectroscopic methods of research over the whole range of electromagnetic waves from radio frequencies through infrared, visible and ultraviolet to X-ray and hard γ quanta, as well as ultrasonics.

4.1. Electronic Spin Resonance and Related Methods

The Method of ESR. The method of electronic spin resonance (ESR) has played a most important role in the development of research on the Jahn–Teller effect and vibronic interactions in the post-war years. In the early fifties Bleaney and his coworkers[94,95] discovered unusual magnetic resonance properties of divalent copper compounds, in particular, a distinctive temperature dependence of the fine and hyperfine structure of the spectrum. A qualitatively correct interpretation of the high-temperature component of these spectra as based on the Jahn–Teller dynamics of the nuclear configuration was first given by Abraham and Pryce.[96] Subsequently, the origin of the main features of the more complicated low-temperature spectrum was clarified on the basis of the theory of tunneling splitting.[97,9] These and other related results have had a strong impact on the development of this area of research.

ESR theory which takes into account vibronic JTE and PJTE comprises a large proportion of the work on, and applications of the ESR method, often occupying whole chapters in relevant monographs and textbooks.[5,6,7,13,98] The achievements in this area are well presented also in reviews[8,9,11,99] (see also the Bibliographic Review[14]). In this section, applications of vibronic interaction theory to ESR and related techniques are presented from the viewpoint of recent achievements.

It is known that ESR spectra may be described satisfactorily by means of the spin Hamiltonian H, its matrix elements determining the positions and intensities of the absorption lines in the radio-frequency range of electromagnetic radiation. The electronic (S) and nuclear (I) spin, and the external magnetic field intensity $\mathcal{H}$ (and their combinations) in the spin Hamiltonian are operators with coefficients depending on the structure and properties of the system. It is convenient to group the terms of the spin Hamiltonian in such a way that they pertain to (transform as) the irreducible representations of the point group symmetry of the problem. This allows us to obtain the selection rules for the matrix elements directly. For instance, in a system with cubic symmetry (tetrahedral, octahedral, cubic) the corresponding symmetry adapted parts of the spin Hamiltonian are

$$H_{A_1} = g_1\beta(\mathcal{H}, S) + \tilde{A}_1(I, S) + g_{N_1}(\mathcal{H}, I) \tag{4.1}$$

$$\begin{aligned} H_{E\theta} &= \tfrac{1}{2}g_2\beta[3\mathcal{H}_z S_z - (\mathcal{H}, S)] + \tfrac{1}{2}\tilde{A}_2[3I_z S_z - (I, S)] \\ &\quad + \tfrac{1}{2}g_{N_2}\beta_N[3\mathcal{H}_z I_z - (\mathcal{H}, I)] + \tfrac{1}{2}P_2(3I_z^2 - I^2) \end{aligned} \tag{4.2}$$

(and a similar component $H_{E\varepsilon}$);

$$\begin{aligned} H_{T_2\xi} &= g_3\beta(\mathcal{H}_y S_z + \mathcal{H}_z S_y) + \tilde{A}_3(I_y S_z + I_z S_y) \\ &\quad + g_{N_3}\beta_N(\mathcal{H}_y I_z + \mathcal{H}_z I_y) + P_3(I_y I_z + I_z I_y) \end{aligned} \tag{4.3}$$

(and similar components $H_{T_2\eta}$ and $H_{T_2\zeta}$);

$$H_{T_1X} = g_L\beta\mathcal{H}_X + \lambda S_X + aI_x \tag{4.4}$$

(and similar components H_{T_1Y} and H_{T_1Z}).

In these expressions g_i is the electronic g-factor, g_L is its pure orbital part, g_{N_i} is the nuclear g-factor, β and β_N are the electronic and nuclear Bohr magnetons, respectively, $\tilde{A}_i$ is the hyperfine structure constant, P_i is the quadrupole interaction constant, λ is the electronic spin–orbit interaction parameter, and a is a constant determining the magnitude of the nuclear magnetic moment interaction with the magnetic field of the electrons.

The physical meaning of each component of the spin Hamiltonian, describing the corresponding magnetic interaction within the system and with the external field, is well known.[7] As a result of averaging with electronic functions, there are no electronic coordinates in equations (4.1)–(4.4). The matrix elements — parameters of the spin Hamiltonian — can be calculated approximately. For example, in the E state formed by d electrons in cubic-symmetry systems in the crystal field theory approximation of second-order perturbation theory,

$$\begin{aligned} g_1 &= g_s - (4\lambda/\Delta E), \qquad g_2 = -(4\lambda/\Delta E) \\ \tilde{A}_1 &= -\mathcal{P}(\kappa - 4\lambda/\Delta E), \qquad \tilde{A}_2 = -\mathcal{P}(6\xi + 4\lambda/\Delta E + 4\lambda\xi/\Delta E) \end{aligned} \tag{4.5}$$

where $\Delta E = 10Dq$ is the parameter of the splitting of the d states in cubic crystal fields, $g_s = 2.0023$ is the pure spin g-factor, $P = 2g_N\beta\beta_N\langle r^{-3}\rangle$, $\langle r^{-3}\rangle$ is the averaged value of r^{-3}, κ is the constant of the contact Fermi interaction of the electronic shell with the nucleus, and ξ is a numerical factor (for the ^2D term $\xi = 2/21$).

If the energy levels and wave functions of the system are known, equations (4.1)–(4.5) can be used to estimate the perturbation of the states by the magnetic field (Zeeman effect), transitions between Zeeman levels, and resonance absorption of electromagnetic radiation in the radio-frequency range which, together with relaxations and other temperature effects, determine the ESR spectrum.

Note that the parameters (4.5) are calculated on the assumption that the nuclei are fixed, and hence the vibronic effects in the ESR spectrum in this approach are ignored completely.

Vibronic Reduction in ESR Spectra. If vibronic effects are taken into account, the solution of the problem becomes complicated and the resulting ESR spectrum changes considerably. In the case of isolated systems *the main effects are vibronic reduction and tunneling splitting. In actual cases these effects are modified by relaxations and small random deformations.* We shall consider each of these effects in turn.

In accord with the results obtained in Section 3.3, the matrix elements of physical properties which depend only on electronic coordinates as determined by the ground state Γ, are, as a result of VI, multiplied by the vibronic reduction factor $K_{\Gamma}(\bar{\Gamma})$, where $\bar{\Gamma}$ is the irreducible representation to which the physical quantity in question corresponds. Therefore, if the ESR spectrum is determined only by the ground vibronic state, the influence of the VI can be evaluated immediately by means of direct multiplication of the matrix elements — the parameters of the spin Hamiltonian — by the respective reduction factors $K_{\Gamma}(\bar{\Gamma})$. Since the totally symmetric operators are not reduced, $K_{\Gamma}(A_1) = 1$ (provided second-order vibronic reduction factors may be neglected; see Section 3.3), all the parameters of the spin Hamiltonian H_{A_1} remain unchanged. However, the parameters of H_E, H_{T_2}, and so on (see equations (4.1)–(4.4)) become multiplied by $K_{\Gamma}(E)$, $K_{\Gamma}(T_2)$, etc., respectively. The resulting reduced parameters of the ESR spectra with vibronic interactions, in the case under consideration, are

$$\begin{gathered}\tilde{g}_1 = g_1, \quad \tilde{g}_2 = K_{\Gamma}(E)g_2, \quad \tilde{g}_3 = K_{\Gamma}(T_2)g_3, \quad \tilde{g}_L = K_{\Gamma}(T_1)g_L \\ A_1' = \tilde{A}_1, \quad A_2' = K_{\Gamma}(E)\tilde{A}_2, \quad A_3' = K_{\Gamma}(T_2)\tilde{A}_3\end{gathered} \tag{4.6}$$

Taking into account vibronic parameters, the frequencies of the two ESR transitions between the components of the two Kramers doublets of the ground vibronic ^2E level in the E–e problem for a cubic symmetry (O_h, O, T_d, T) system are

$$\hbar\Omega_{\pm} = (g_1\beta\mathcal{H} + \tilde{A}_1\nu_I) \pm q(g_2\beta\mathcal{H} + \tilde{A}_2\nu_I)f$$
$$f = [1 - 3(l^2m^2 + l^2n^2 + m^2n^2)]^{1/2} \tag{4.7}$$

where $q = K_E(E)$, ν_I is the quantum number of the hyperfine interaction (I is the nuclear spin), and l, m, and n are direction cosines of the magnetic field vector $\mathcal{H}$. In particular, for the angular dependence of the corresponding two g-factors, we have

$$g_{1,2} = g_s - g_2 \pm qg_2f \tag{4.8}$$

Note that, without vibronic interaction, $q = 1$, whereas in the limiting case of strong vibronic coupling $q = \frac{1}{2}$. These two limiting cases of angular dependence of the ESR lines are presented in Figure 4.1 together with experimental data for Cu^{2+} : MgO obtained at temperature $T = 1.2$ K.[100,101] It is seen that the experimental measurements clearly confirm the results of vibronic theory with a reduction factor $q \approx \frac{1}{2}$. At present, there are quite a number of such clear experimental illustrations of the importance of VI theory in the ESR spectra of systems with degenerate E terms (see reviews[6,11] and the Bibliographic Review[14]). For trigonal systems with an E term the vibronic effects in the ESR spectra are different.[102]

The threefold orbitally degenerate T term behavior differs from the E term case in cubic systems by (1) a strong spin–orbit splitting in first-order

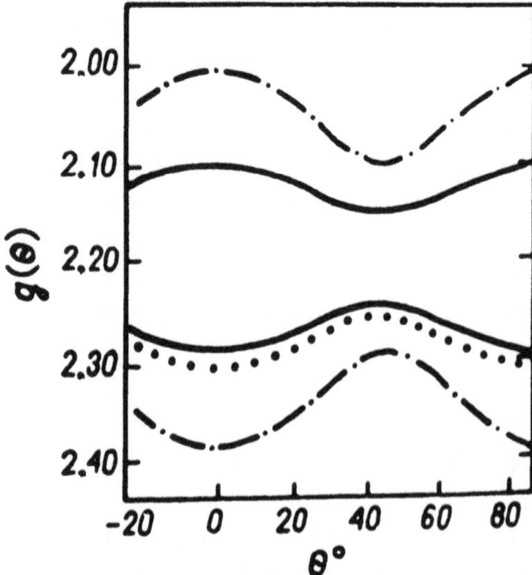

Figure 4.1. Two limiting cases of angular dependence of the g-factor for the linear E–e problem with strong ($q = \frac{1}{2}$, solid line) and without ($q = 1$, dashed line) vibronic coupling. Experimental data are shown by points.

perturbation theory (the E term in trigonal systems is also split in the first order), and (2) a larger orbital contribution to the Zeeman energy. As a result, the ground vibronic level ^{2s+1}T of the electronic ^{2s+1}T state (Section 3.2) is split by the spin–orbit interaction. If allowance is made for Zeeman interaction, the spin Hamiltonian for this ground vibronic level gives, using first-order perturbation theory,[7]

$$H = K_T(T_1)g_L\beta(\mathcal{H}, \boldsymbol{L}) + g_s\beta(\mathcal{H}, \boldsymbol{S}) + \lambda K_T(T_1)(\boldsymbol{L}, \boldsymbol{S}) \tag{4.9}$$

where $K_T(T_1)$ is the vibronic reduction factor which, according to equation (3.43) (Figure 3.7), varies from zero in the limiting case of strong vibronic coupling to unity in the limiting case of weak coupling.

Assuming that Zeeman splitting is smaller than reduced spin–orbit and tunneling splittings, one obtains, with vibronic interactions, the following expression for the g-factor of the ground state in the crystal field approximation (the Landé formula with vibronic reduction):

$$\begin{aligned} g(J) = \{\alpha K_T(T_1)g_L\,[J(J+1)+2-S(S+1)] \\ + g_s\,[J(J+1)-2+S(S+1)]\}/2J(J+1) \end{aligned} \tag{4.10}$$

where J is the quantum number of the total angular momentum and assumes all values between $S-1$ and $S+1$ via unity, and α is a known numerical factor.

For example, consider, for the $^5T_2(3d^6)$ term $(S=2, J=1, \alpha=-1)$, $g=\frac{1}{2}[3g_s + K_{T_2}(T_1)g_L]$, and for the $^4T_1(3d^7)$ term $(S=\frac{3}{2},\ J=\frac{1}{2},\ \alpha=-\frac{3}{2})$, $g=\frac{5}{3}g_s + K_{T_1}(T_1)g_L$. Comparison of these results with experimental data shows that, say, for Fe^{2+} : CaO (the $^5T_2(3d^6)$ term) the experimental value is $g=3.30$, and hence the nonvibronic value $g_L=1$ is reduced by $K_{T_2}(T_1)=0.6$ (a part of this reduction may be due to covalency). For another case Cr : Si with the same term, $g=2.97$, and hence $K_{T_2}(T_1)\approx 0$. This complete reduction of the orbital contribution is inherent to the presence of strong vibronic coupling (see equation (3.43) and Figure 3.7). Similarly, complete reduction of the orbital contribution takes place in the case of Mn : Si (term $^4T_1(3d^7)$), whereas for the same term in Fe^+ : MgO, $K_{T_1}(T_1)=0.2$ (other examples are given elsewhere[5,6,11,14]).

The angular dependence of the ESR spectrum can be used to reveal the nature of the T term AP minima. In particular, if the VC for coupling with E and T_2 displacements are of the same order of magnitude, orthorhombic minima may occur, as shown in Section 2.2. The observed angular spectra for Ni^- : Ge, Pd^- : Si, and Pt^- : Si[103] may be assumed due to orthorhombic minima.

In the case of a strong spin–orbit splitting of the T term, the vibronic effect in the Γ_8 state must be considered (Section 2.2). Such a term occurs commonly with the $^2T_2(3d^1)$ term, which is split into Γ_8 and Γ_6. It can be shown that without vibronic interactions, the Γ_8 term is nonmagnetic, i.e., it does not split under the influence of the external magnetic field, its g-factor vanishing due to complete compensation of the orbital and spin contributions of opposite sign.[7] When vibronic interactions are taken into account, the orbital contribution to the g-factor becomes reduced, but the spin contribution does not; the Γ_8 term becomes magnetic.[43] The resulting g-factor is given by $g = \frac{2}{3}[1 - K_{\Gamma_8}(T_1)]$ and it is seen that $g = 0$ only when the vibronic reduction is neglected, in which case $K_{\Gamma_8}(T_1) = 1$.

Influence of Tunneling Splitting. Transition from Dynamic to Static JTE. In the case of strong vibronic coupling, when the conditions of the tunneling splitting approximation are valid, there is a low-lying excited vibronic level (Section 3.2, Figure 3.4) which influences the ground state properties. In this case, the concept of vibronic reduction of the corresponding orbital contribution to the g-factor is insufficient, since the two vibronic levels mix under the magnetic field. For the ^2E term, Zeeman mixing of the tunneling ground (^2E) and excited (^2A) levels can be determined by perturbation theory. Taking into account that the two spin states with $s_z = \pm\frac{1}{2}$ are independent of the orbital state, the corresponding sixth-order secular equation may be reduced to two equations of third order[97]:

$$\begin{vmatrix} 3\Gamma \pm \frac{1}{2}g_1\beta\mathcal{H} - \varepsilon & \pm\frac{1}{4}rg_2\beta\mathcal{H}(3n^2-1) & \pm\frac{\sqrt{3}}{4}rg_2\beta\mathcal{H}(l^2-m^2) \\ \pm\frac{1}{4}rg_2\beta\mathcal{H}(3n^2-1) & \pm\frac{1}{2}\beta\mathcal{H}[g_1 - \frac{1}{2}qg_2(3n^2-1)] - \varepsilon & \pm\frac{\sqrt{3}}{4}qg_2\beta\mathcal{H}(l^2-m^2) \\ \pm\frac{\sqrt{3}}{4}rg_2\beta\mathcal{H}(l^2-m^2) & \pm\frac{\sqrt{3}}{4}qg_2\beta\mathcal{H}(l^2-m^2) & \pm\frac{1}{2}\beta\mathcal{H}[g_1 + \frac{1}{2}qg_2(3n^2-1)] - \varepsilon \end{vmatrix} = 0 \tag{4.11}$$

Here, the notation $\delta \equiv 3\Gamma$ for the tunneling splitting as assumed in ESR spectroscopy is used, and r is the second-order vibronic reduction factor, $r = K_E(A_1|E|E)$; for strong vibronic coupling $r = -q\sqrt{2} = -1/\sqrt{2}$ (see Section 3.3).

For the magnetic field along the $0z$ axis (the fourfold axis of the octahedron), $l = 1$, $m = n = 0$, equation (4.11) is simplified and its roots can be derived directly:

$$\begin{aligned} \varepsilon_1^{\pm} &= \tfrac{1}{2}\{3\Gamma \pm \beta\mathcal{H}(g_1 - qg_2) + [(3\Gamma \pm \tfrac{1}{2}qg_2\beta\mathcal{H})^2 + r^2g_2^2\beta^2\mathcal{H}^2]^{1/2}\} \\ \varepsilon_2^{\pm} &= \tfrac{1}{2}\{3\Gamma \pm \beta\mathcal{H}(g_1 - qg_2) - [(3\Gamma \pm \tfrac{1}{2}qg_2\beta\mathcal{H})^2 + r^2g_2^2\beta^2\mathcal{H}^2]^{1/2}\} \\ \varepsilon_3^{\pm} &= \tfrac{1}{2}\beta\mathcal{H}(g_1 + qg_2) \end{aligned} \tag{4.12}$$

The behavior of these levels with respect to the magnetic field intensity is shown in Figure 4.2; the arrows indicate the allowed ESR transitions as determined by the wave functions of the states (4.12) — solutions of equation (4.11). In order to analyze the expected ESR spectrum, it is convenient to distinguish three ranges, depending on the relation between the tunneling splitting magnitude 3Γ and the anisotropic part of the Zeeman interaction $g_2\beta\mathscr{H}$: (I) the low-frequency range, $g_2\beta\mathscr{H} \ll 3\Gamma$; (II) the intermediate range; and (III) the high-frequency range, $g_2\beta\mathscr{H} \gg 3\Gamma$. In range I where mixing of the ground vibronic state with the excited one is relatively weak, the ESR spectrum consists of three lines, two of which have the g-factor (4.8) of the isolated ground orbital doublet. The third line arises from the excited orbital singlet and is isotropic with $g = g_1$. The intensity of this line is proportional to the population of the excited level, which is zero at $T = 0$ and increases with temperature.

In high-frequency range III the other three transitions are allowed with g-factors which, in the limiting case of strong vibronic coupling when $q = \frac{1}{2}$ and $3\Gamma \approx 0$, are $g_{\parallel} = g_1 + g_2$ and $g_{\perp} = g_1 - \frac{1}{2}g_2$ if the excited tunneling

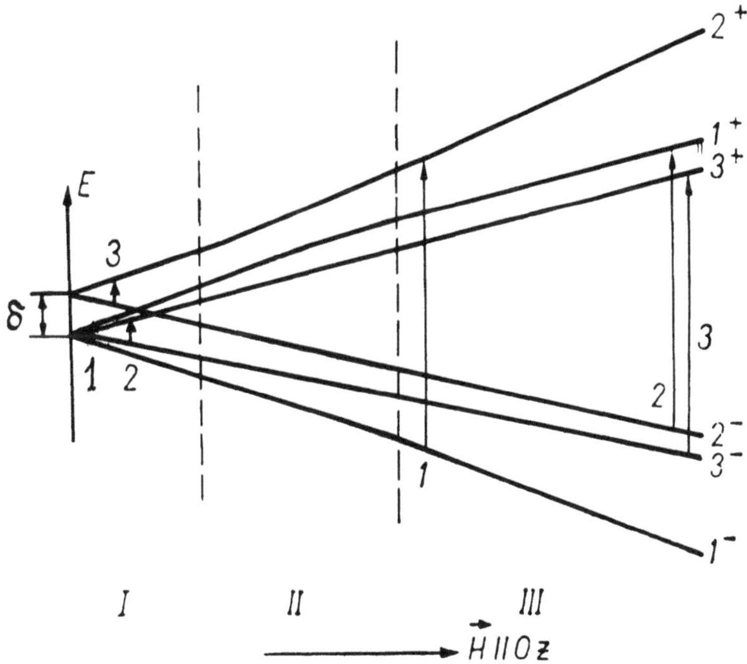

Figure 4.2. Energy levels of the ground state $E-e$ problem in magnetic fields $\mathscr{H}\parallel Oz$ taking into account tunneling splitting. Magnetic-dipole allowed transitions in regions I and III are shown by arrows.

level is A_2, and $g_\| = g_1 - g_2$ and $g_\perp = g_1 + \frac{1}{2}g_2$ if this level is A_1. Similar relationships for the constants of the hyperfine structure are $\tilde{A}_\| = \tilde{A}_1 + \tilde{A}_2$ and $\tilde{A}_\perp = \tilde{A}_1 - \frac{1}{2}\tilde{A}_2$ if A_2 is the nearest excited level, and $\tilde{A}_\| = \tilde{A}_1 - \tilde{A}_2$ and $\tilde{A}_\perp = \tilde{A}_1 + \frac{1}{2}\tilde{A}_2$ if A_1 is the nearest level. In the intermediate range II the two spectra coexist, the low-frequency spectrum fading slowly and the high-frequency spectrum growing uniformly when moving from range I to range III.

The spectrum in range III is seen to be that of the statically distorted system frozen at the AP minimum, whereas the spectrum in range I corresponds to its free or hindered rotations (Sections 3.1). Therefore, the transition from the spectrum in range I to the spectrum in range III is often called the transition from the static to the dynamic JTE. Taking into account the relativity rule concerning the means of observation (Section 5.1), the static JTE in the ESR spectra of real systems (for which $3\Gamma \neq 0$) can be observed if the lifetime of the distorted configuration τ in the pulsating motion (Section 3.2) is larger than the characteristic time of measurement interaction τ'. In the case under consideration, τ' is determined by the anisotropic part of the Zeeman interaction $g_2\beta\mathcal{H}$. For typical situations this means that the tunneling (pulsation) frequency is slower than 10^{-9} sec^{-1}. In the above case of $\mathcal{H} \parallel Oz$ *the strong external field* in range III *depresses the pulsations, locking the system in the minimum* along the Oz axis.

When the direction of the magnetic field $\mathcal{H}$ does not coincide with Oz (with the tetragonal axes of an octahedron) the energy levels (4.12) approach each other and the differences in the g-factors of different lines decrease. For the trigonal axes of a cubic system, all the g-factors become identically equal to g_1.

Low-Symmetry External Perturbations. The vibronic ESR spectrum described above relates to "ideal" systems with electron degeneracy or pseudodegeneracy, when each molecule may be regarded as isolated, unaffected by the environment. In real systems, as mentioned above, there are two main effects which influence the spectra in question: random strain and relaxations. The latter are inherent in any system with energy dissipation, whereas random strain has special importance for a vibronic system. This is due to the fact that, since either Zeeman or tunneling splitting is relatively small, even small external perturbations split the degenerate ground vibronic state, and this splitting may suffice to cause changes in the expected ESR spectra.

If the low-symmetry perturbation is "regular," i.e., it is the same for all paramagnetic centers, the resulting spectrum is not affected very much since the system, in principle, remains "ideal." Such is the case of a polyatomic molecule (or a crystal), for example, when the paramagnetic center is weakly affected by the second coordination sphere having lower symmetry than the first coordination sphere. However, there are cases when these low-symmetry perturbations are of random magnitude and direction. For instance, if in the above example the molecules or paramagnetic centers in crystals are arbitrarily oriented with respect to the magnetic field, the changes in the ESR spectrum (compared with that for the unperturbed system) will be as if the perturbation distortions are random.

The cases of paramagnetic centers in crystals, for which the most precise ESR measurements are performed, are of special interest. Here, random distortions occur due to imperfections of the host crystal lattice, i.e., due to small local distortions of the regular symmetry caused by dislocations, mosaic structure, impurities, locally uncompensated charges, and other defects.

Qualitatively, the role of random strain is obvious. For example, in the $E-e$ problem with three equivalent AP minima, external distortions, when strong enough, make these minima nonequivalent. As a result the pulsating motion and tunneling cease, and the system is locked in one (the deepest) minimum (similar to the above case when the system is locked in the minimum by the magnetic field). The random nature of the distortion direction is not very important, since it determines only at which of the three equivalent minima the system is locked. Comparing this result with the above transition from the dynamic to the static JTE, we conclude that random deformations promote a transition to the static JTE.

Quantitatively, the influence of random strain is determined by the magnitude of the vibronic level splitting Δ produced by this strain. In general, the expected ESR spectrum depends on the relationship between the mean value of the strain effect $\bar{\Delta}$, the tunneling splitting 3Γ, and the anisotropic part of the Zeeman interaction $g_2\beta\mathcal{H}$. Note that the stronger the vibronic coupling, the larger the effect of the strain influence. Indeed, the external distorting influence displaces the nuclear coordinates and the latter influences the electronic states through vibronic coupling, resulting in a corresponding energy level splitting. Therefore the random strain is most important in cases of strong vibronic coupling.

Role of Random Strain in the ESR $E-e$ Problem. The effect of random strain on the vibronic ESR spectra is best studied for the $E-e$ problem. Consider first the isolated ground vibronic doublet ^2E when the first excited vibronic singlet ^2A is well removed, $3\Gamma \gg g\beta\mathcal{H}$.[104,6] In order to describe the low-symmetry strain perturbation $\tilde{W}$, we introduce the strain tensor and the electron–strain interaction constants in the form of respective components $e_{\Gamma\gamma}$ and P_Γ transforming according to the corresponding irreducible representations ($\tilde{W} \sim \Sigma_{\Gamma\gamma} C_{\Gamma\gamma} e_{\Gamma\gamma} P_\Gamma$). Only one constant P_E and two components, $e_{E\theta} \equiv e_\theta$ and $e_{E\varepsilon} \equiv e_\varepsilon$, are essential to the E term splitting. Instead of these two components, their modulus e and orientation φ, given by

$$e = [e_\theta^2 + e_\varepsilon^2]^{1/2}, \qquad \tan\varphi = e_\varepsilon / e_\theta \tag{4.13}$$

are more convenient for practical use. Under this strain the E level splits:

$$\begin{aligned} E_1 &= qeP_E, \qquad \psi_1 = \frac{1}{\sqrt{2}}[(1-\cos\varphi)^{1/2}|\theta\rangle + (1+\cos\varphi)^{1/2}|\varepsilon\rangle] \\ E_2 &= -qeP_E, \qquad \psi_2 = \frac{1}{\sqrt{2}}[-(1+\cos\varphi)^{1/2}|\theta\rangle + (1-\cos\varphi)^{1/2}|\varepsilon\rangle] \end{aligned} \tag{4.14}$$

It follows that the orientation φ of the strain [in the $(Q_\theta, Q_\varepsilon)$ plane] does not affect the splitting, but considerably influences the wave functions.

The energy levels (4.14) are Kramers spin doublets. If the strain splitting $\Delta = 2qeP_E$ is large compared with the Zeeman interaction (but still small compared with the tunneling splitting, $3\Gamma \gg \Delta \gg g\beta\mathscr{H}$), the splitting of each of the levels (4.14) in the magnetic field may be treated

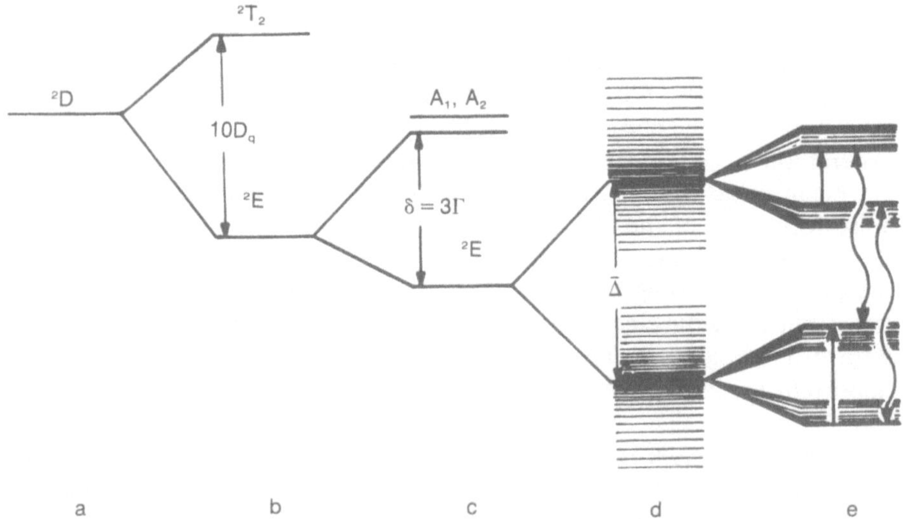

Figure 4.3. Lowest energy level scheme for an octahedral system in the ^2E state: (a) free ion term; (b) crystal field splitting; (c) tunneling splitting; (d) splitting of the ground vibronic ^2E level into two Kramers doublets by random strain; (e) Zeeman splitting by magnetic fields. Allowed transitions in the radio-frequency field are shown by straight arrows, whereas relaxation transitions without spin reversal are demonstrated by wavy arrows (after Reynolds and Boatner[110]).

independently by means of perturbation theory. Then (Figure 4.3)

$$E_1^{\pm} = qeP_E \pm \tfrac{1}{2}[g_1 + qg_2 \cos(\varphi - \alpha) f]\beta\mathcal{H} \tag{4.15}$$

$$E_2^{\pm} = -qeP_E \pm \tfrac{1}{2}[g_1 - qg_2 \cos(\varphi - \alpha) f]\beta\mathcal{H} \tag{4.16}$$

where the magnetic field direction α satisfies

$$\tan\alpha = \sqrt{3}(l^2 - m^2)/(2n^2 - l^2 - m^2) \tag{4.17}$$

The two pairs of levels (4.15) and (4.16) give rise to two allowed ESR transitions with the following frequencies (the expressions in brackets equal the g-factors):

$$\hbar\Omega_{1,2} = [g_1 \pm qg_2 \cos(\varphi - \alpha) f]\beta\mathcal{H} \tag{4.18}$$

Thus, with the strong strain influence in question, the frequencies of the ESR transitions at a given paramagnetic center depend on the angle $\varphi - \alpha$ of the strain orientation with respect to the direction of the magnetic field. In particular, for $\varphi - \alpha = \pm\pi/2$ the ESR spectrum is isotropic with $g = g_1$, and for $\varphi - \alpha = 0$ or π, the spectrum is the same as in the absence of strain. With intermediate values of $\varphi - \alpha$ (between 0 and $\pi/2$) the anisotropic part of the g-factor varies from 0 to $\pm qg_2f$. It can be shown[13] that the transition probabilities (intensities) at frequencies given by equation (4.18) are independent of the strain and magnetic field directions.

Therefore, in the presence of strain with random orientations with respect to the direction of the magnetic field $\mathcal{H}$ (originating both from random orientations of the strain with regard to different paramagnetic centers, and constant orientation of the strain with regard to the centers but random orientation of the centers with respect to the magnetic field), the ESR frequencies (or the corresponding resonance values of $\mathcal{H}$) vary from one center to another within $\pm qg_2f$, which is equivalent to inhomogeneous broadening of the ESR lines. Assuming that the ESR line of each paramagnetic center has a δ character ($\delta(x) = 0$ if $x \neq 0$, and $\delta(x) = \infty$ if $x = 0$), and averaging over the random strain orientations φ, one can derive the absorption coefficient K as a function of the resonance magnetic field intensity in the form[6,13]

$$K(\mathcal{H}) = (\pi^2 Z)^{-1/2}\Theta(Z), \qquad Z = (qg_2 f\beta\mathcal{H})^2 - (\hbar\Omega - g_1\beta\mathcal{H})^2 \tag{4.19}$$

where Ω is the radio frequency and $\Theta(Z)$ is a step function: $\Theta(Z) = 1$ if $Z > 0$, and $\Theta(Z) = 0$ if $Z < 0$.

The line shape determined by equation (4.19) is illustrated in Figure 4.4a. Its unusual, abrupt form is due to the assumed δ-character of the absorption at each center. If a Gaussian form for this absorption (with a

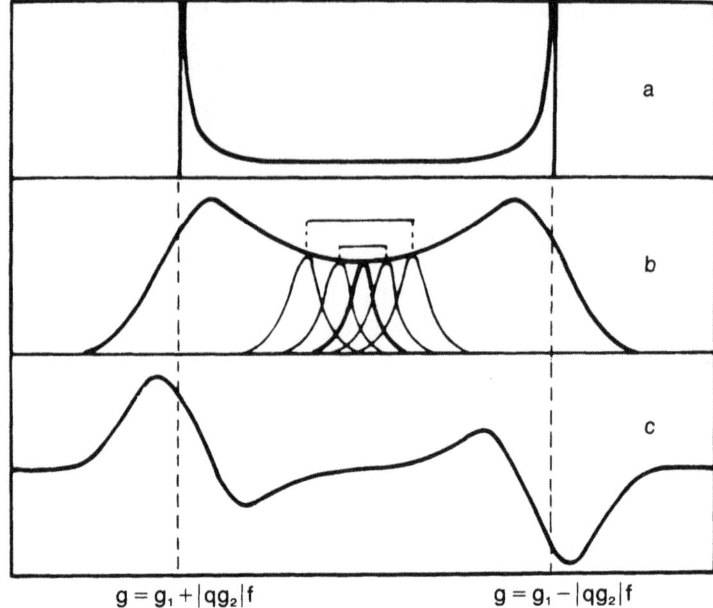

Figure 4.4. ESR line shape for the ground ^2E state taking into account random strain averaged over all orientations: (a) under the assumption that individual transitions contributing to the envelope satisfy a δ-function; (b) the individual transitions are Gaussian bands with a half-width equal to 0.2Ω; (c) the first derivative of the function given in (b).

certain half-width) is assumed instead of the δ one, the line shape becomes more natural (Figure 4.4b). Another feature of the line shape (4.19) is its temperature independence, although a Boltzmann temperature population of the strain split levels was assumed when deriving equation (4.19); this is a consequence of averaging over random orientations of the strain.

With hyperfine splitting, each of the lines (4.19) contains $2I+1$ components having the following frequencies:

$$\hbar\Omega_{1,2}(\nu_I) = (g_1\beta\mathcal{H} + \tilde{A}_1\nu_I) \pm \tfrac{1}{2}q(g_2\beta\mathcal{H} + \tilde{A}_2\nu_I)\cos(\varphi - \alpha)f \quad (4.20)$$

(ν_I is the appropriate quantum number, introduced above.)

The coefficient of absorption (line shape) is

$$\begin{aligned} K(\mathcal{H}, \nu_I) &= (\pi^2 Z)^{-1/2}\Theta(Z), \\ Z &= (qg_2\beta\mathcal{H} + \tilde{A}_1\nu_I)^2 f^2 - (\hbar\Omega - g_1\beta\mathcal{H} - \tilde{A}_2\nu_I)^2 \end{aligned} \quad (4.21)$$

Relaxation Effects. The above picture of the ESR spectrum cannot be regarded as complete, since the influence of the dissipative subsystem (lattice vibrations, collisions, etc.) has been disregarded. Under the influence of the dissipative subsystems (thermostat) there are continuous transitions between the strain split energy levels. These transitions lower the lifetime of the corresponding states and broaden their energy levels (Figure 4.3). Besides, the usual relaxation transitions take place between the energy levels of Kramers doublets, being split in the magnetic field. However, these transitions lead to spin reversal which is affected by the thermostat indirectly through rather small spin–orbit interaction; therefore they are much less probable than direct nonspin transitions and can be neglected.

Let the probability of relaxation transition without spin reversal be denoted by W. The general theory of relaxation[105,106] yields, instead of equation (4.19), the following expression for the line shape[107]:

$$K(\mathscr{H}) \sim \frac{2W\gamma^2}{(E^2 - \gamma^2) + 4E^2W^2} \tag{4.22}$$

where $E(\mathscr{H}) = \hbar\Omega - g_1\beta\mathscr{H}$ and $\gamma(\mathscr{H}) = qg_2\beta\mathscr{H}\cos(\varphi - \alpha)f$.

The relaxation probability depends strongly on temperature. For low temperatures $W \ll \gamma$, and equation (4.22) describes two Lorentzian lines (with centers at frequencies given by equation (4.18)):

$$K(\mathscr{H}) \sim \frac{W}{(E-\gamma)^2 + W^2} + \frac{W}{(E+\gamma)^2 + W^2} \tag{4.23}$$

For high temperatures $W \gg \gamma$, and equation (4.22) describes one Lorentzian line centered at frequency $\hbar\Omega = g_1\beta\mathscr{H}$:

$$K(\mathscr{H}) \sim \frac{W_{\text{eff}}^2}{E^2 + W_{\text{eff}}^2}, \qquad W_{\text{eff}} = \gamma^2/W \tag{4.24}$$

As a result the vibronic ESR spectrum, which at low temperatures has a well-resolved doublet structure, narrows and transforms into one isotropic line when the temperature increases. Visually, this result is due to the fact that at low temperatures relaxation transitions seldom occur, and the lifetime of the energy level states is larger than the characteristic time of the act of measurement, so that the latter finds the system in one of the two Kramers doublets given by equation (4.15). As a result, two lines are seen in the ESR spectrum at the same frequencies Ω_1 and Ω_2, as in the absence of relaxation. At high temperatures the relaxation is so fast, that during the measurement the system performs multiple transitions from one state to another, resulting in an averaged isotropic line at the mean frequency $(\Omega_1 + \Omega_2)/2$. Such a temperature narrowing of the ESR band is, in principle, similar to the well-known exchange narrowing in concentrated magnetic systems, and to motional narrowing in liquids.[105,106] Similar temperature effects are observed in the hyperfine structure of Mössbauer spectra.

The anisotropic part of the Zeeman interaction

$$\gamma = qg_2\beta\mathcal{H}\cos(\varphi - \alpha)f$$

and hence the line shape $K(\mathcal{H})$, depend on the angle φ of strain orientation. If the latter has a random character, averaging must be performed. At low temperatures, for most orientations φ the inequality $W \ll \gamma(\varphi)$ is valid, and the vibronic ESR spectrum has the form illustrated in Figure 4.5a. Compared with the above spectrum without relaxation (Figure 4.4b), the edges here are smoother due to relaxation broadening. A

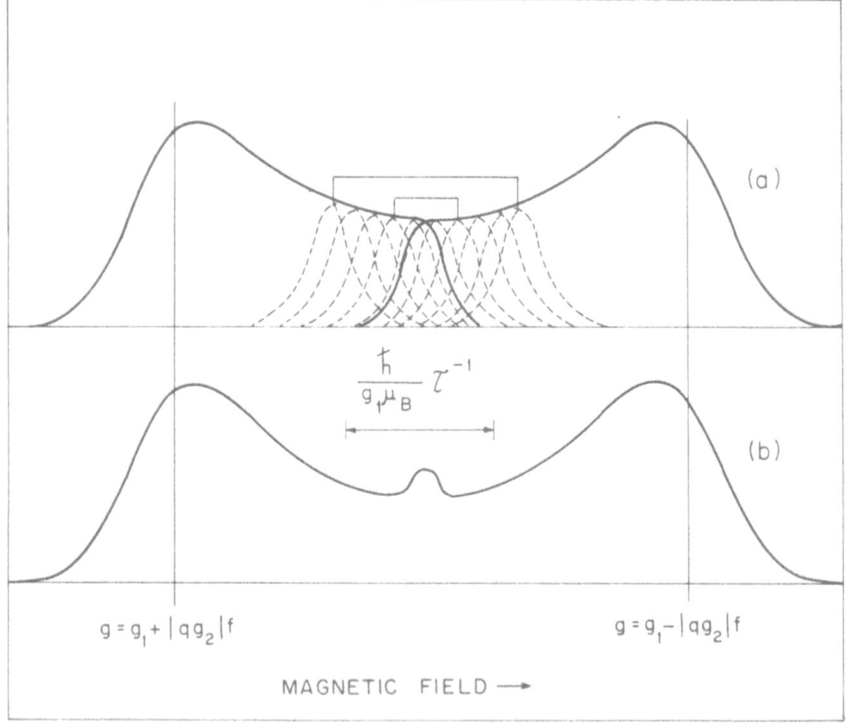

Figure 4.5. Influence of relaxation transitions without spin reversal on ESR line shape–temperature narrowing: (a) line shape at low temperatures when the number of pairs of transitions for which the condition of narrowing is fulfilled, is small; (b) at higher temperatures an isotropic line occurs in the center of the band at the expense of the intensity of the wings.

small absorption in the central range at $\hbar\Omega = g_1\beta\mathcal{H}$ is due to the presence of paramagnetic centers, for which $\varphi - \alpha \approx \pm\pi/2$ and the inequality $W \ll \gamma(\varphi)$ is invalid. This absorption is isotropic in contrast to the edges of the band, for which the angular dependence of the absorption follows function f in equation (4.7).

With increasing temperature, the number of centers increases for which $W \gg \gamma(\varphi)$ and the corresponding pairs of lines converge into one averaged isotropic line at $\hbar\Omega = g_1\beta\mathcal{H}$ (Figure 4.5b). This leads to enlarging of the isotropic band in the center of the spectrum at the expense of the intensity of the anisotropic wings. The temperature dependence of this spectrum, as seen from equations (4.22)–(4.24), is determined by the temperature dependence of the relaxation probabilities W, their calculation being an independent, complicated task[107–108a] based on the solutions of the multimode JT problem.

Tunneling Splitting Plus Random Strain and Relaxation in ESR. The above vibronic ESR spectrum becomes more complicated when the vibronic singlet (Kramers doublet) ^2A ($^2A_1$ or $^2A_2$) approaches the ground state ^2E and the tunneling splitting 3Γ becomes comparable to Zeeman or strain splitting.[109–111] Consider first the case when 3Γ is still large, approaching the $\bar{\Delta}$ value, and $\bar{\Delta} \gg g_2\beta\mathcal{H}$. In this case, mixing of vibronic levels A_1 and E under the magnetic field can be neglected, while their mixing by the strain $e_\theta P_E$ may be taken into account by means of perturbation theory. In the zero approximation the spectrum is as that obtained above in the absence of the A level; at low temperatures it has the form given in Figure 4.6 for small $\bar{\Delta}/3\Gamma$ values. Assume that the magnetic field lies in the plane (110), i.e., $l = m$. Then $\alpha = 0$ (see equation (4.17)) and for the perturbation of the two peaks for which $\varphi - \alpha = 0$ or π, we have $\varphi = 0$ or π. On substituting these values into equation (4.13), we obtain $e_\varepsilon = 0$, i.e., the peaks of the spectrum are due to the e_θ component of the strain tensor. The resulting g-factors corresponding to the two peaks of the spectrum perturbed by the approaching A_1 level are[112]:

$$
\begin{aligned}
g_\theta &= g_1 - \tfrac{1}{2}[q + (2r^2 eP_E/3\Gamma)]g_2(3n^2 - 1) \\
g_\varepsilon &= g_1 + \tfrac{1}{2}qg_2(3n^2 - 1)
\end{aligned}
\tag{4.25}
$$

It follows that only one of the two peaks of the spectrum is shifted; which of them and in what direction depends on the signs of e_θ, g_2, and $(3n^2 - 1)$. For example, if $g_2 > 0$, $e_\theta > 0$, and $3n^2 - 1 > 0$, the right-side peak corresponding to larger resonance fields $\mathcal{H}$ is shifted to the right. If instead of the vibronic A_1 level the A_2 level approaches the ground E state, the sign of the effect changes, and the left-side peak is shifted to the left. All

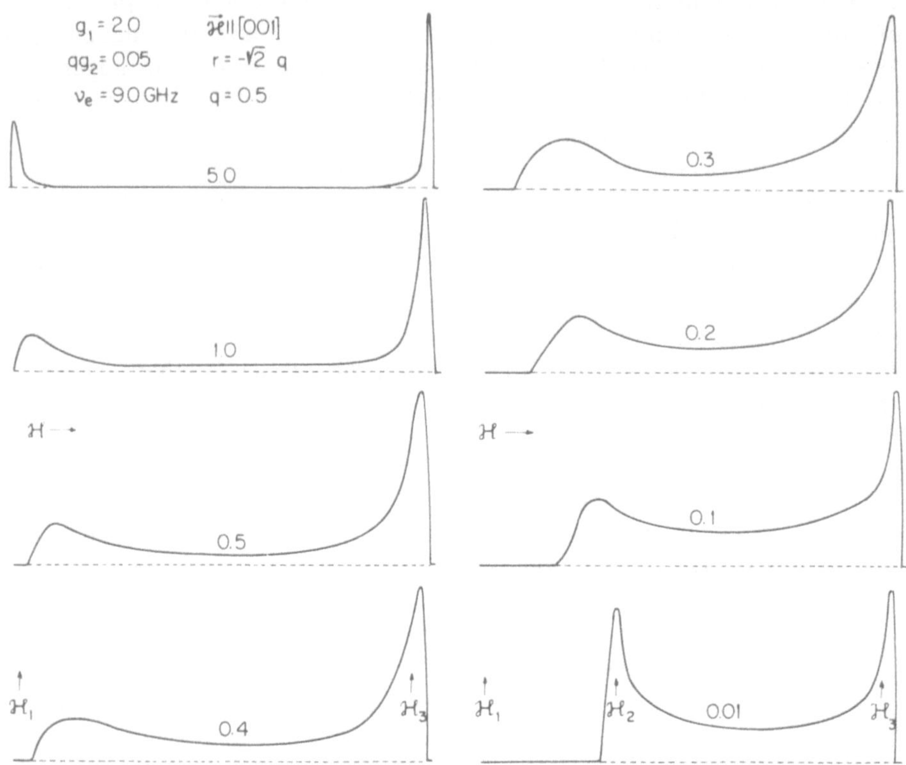

Figure 4.6. ESR line shape calculation for a ^2E term in $\mathscr{H} \parallel [001]$ at low temperatures and different values of $\bar{\Delta}/3\Gamma$ (indicated on the curves). The lowest vibronic singlet is assumed to be A_2. $g_1 = 2.0$, $qg_2 = 0.05$, $q = \frac{1}{2}$, $r = -\sqrt{2}q$, $\nu_e = \Omega/2\pi = 9.0$ GHz. $\beta\mathscr{H}_1 = \hbar\Omega/(g_1 + 2qg_2)$, $\beta\mathscr{H}_2 = \hbar\Omega/(g_1 + qg_2)$, $\beta\mathscr{H}_3 = \hbar\Omega/(g_1 - qg_2)$ (after Reynolds and Boatner[110]).

these possible cases are summarized in Table 4.1. Also, with the dispersion of the absolute values of the strain, the perturbed peak is not only shifted, but in addition becomes somewhat flattened (i.e., wider and lower in its intensity).

When tunneling splitting becomes comparable to strain splitting, $3\Gamma \sim \bar{\Delta}$, perturbation theory becomes invalid, and the problem must be solved by numerical solution of the sixth-order secular equation, in which splittings of the three vibronic doublets by the magnetic field and their mixing by random strain are taken into account simultaneously. The results obtained by Boatner, Reynolds, Abraham, and Chen[109–111] are illustrated in Figures 4.6–4.14.

The dependence of the spectrum on the ratio $\bar{\Delta}/3\Gamma$ is illustrated in Figure 4.6 for low temperatures, when only the lowest Kramers doublet is

Table 4.1. Shift of One of the Two Peaks (Right and Left) in the Vibronic ESR Spectrum of a ^2E Term under the Influence of Mixing with the ^2A Level. $\mathscr{H} \parallel (110)$, $l = m$, $3n^2 > 1$, and $\mathscr{H}$ Increases from Left to Right. In Case $3n^2 < 1$ all the Results Change to the Opposite Sign[110]

Sign of g_2	Nearest excited singlet	Shifted peak	Direction of shift
	Ground Kramers doublet		
+	A_1	right	right
+	A_2	left	left
−	A_1	left	left
−	A_2	right	right
	Excited Kramers doublet		
+	A_1	right	left
+	A_2	left	right
−	A_1	left	right
−	A_2	right	left

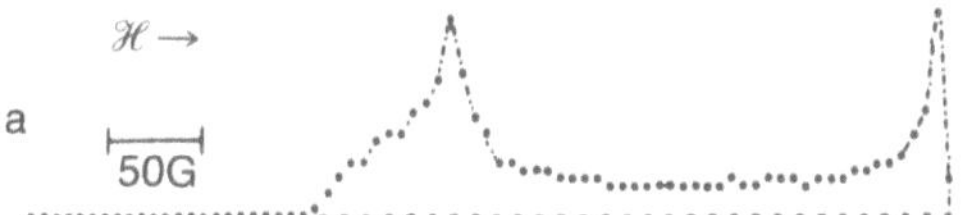

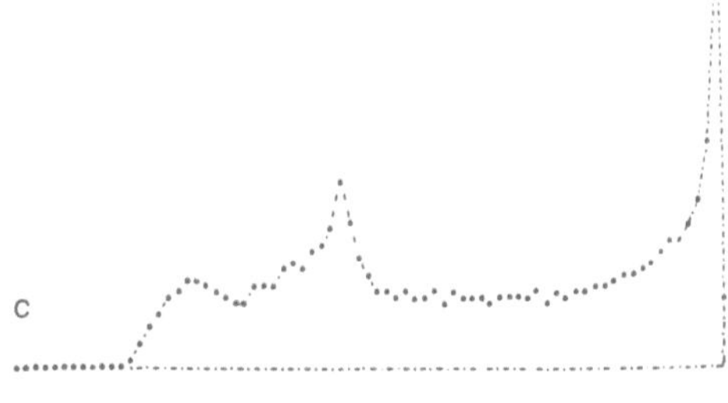

Figure 4.7. Change in ESR line shape due to population of the first excited Kramers doublet in the E–e problem with random strain: (a) line shape of the excited doublet; (b) line shape of the ground doublet; (c) summarized line (T = 1.5 K, $\bar{\Delta}/3\Gamma = 0.13$, $\Gamma = 1.0$, $qg_2 = 0.1$, and other data as in Figure 4.6) (after Reynolds and Boatner[110]).

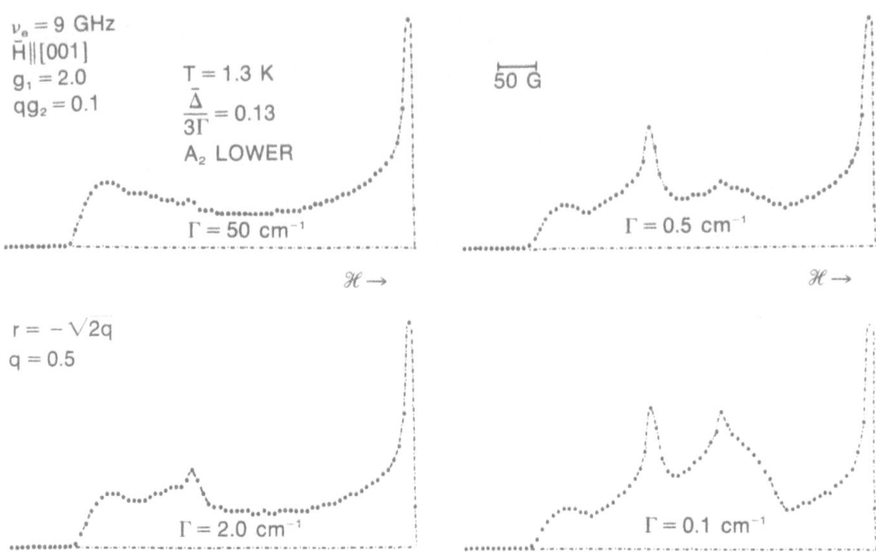

Figure 4.8. ESR line shape taking into account the contribution of all three Kramers doublets at $T = 1.3$ K and different values of tunneling splitting 3Γ (other data as in Figure 4.7) (after Reynolds and Boatner[110]).

populated. It is seen that a small broadening begins even at $\bar{\Delta}/3\Gamma = 0.01$ and rapidly increases, simultaneously shifting to the left, when the $\bar{\Delta}/3\Gamma$ value increases. At $\bar{\Delta}/3\Gamma = 1$ the rate of shifting decreases, and at $\bar{\Delta}/3\Gamma = 5$ the extreme value of the static JTE corresponding to $\beta\mathscr{H} = \hbar\Omega/(g_1 + 2qg_2)$ is attained. As the temperature increases, first the next Kramers doublet becomes populated (Figure 4.7), and then the third one, resulting in a complicated spectrum (Figure 4.8).

In Figure 4.9, the angular dependence of the spectrum is presented with respect to the magnetic field $\mathscr{H}$ lying in the plane $(1\bar{1}0)$ (containing the main directions $\mathscr{H} \parallel [001]$, [110], and [111]) at low temperatures, when only the lowest Kramers doublet is populated. At $\bar{\Delta}/3\Gamma = 0.01$, the angular dependence is symmetric with respect to the $\mathscr{H} \parallel [111]$ direction of the magnetic field for which, according to equation (4.18) with $\cos(\varphi - \alpha) = 1$, the two peaks coincide. When $\bar{\Delta}/3\Gamma$ increases, the anisotropy of the perturbed peak (Table 4.1) increases, and at $(\bar{\Delta}/3\Gamma) \gtrsim 5$ the angular dependence corresponds to that of an axially symmetric spectrum of the type [100], [010], or [001] of the static JTE.

Using the data of Figure 4.9, the ratio $\bar{\Delta}/3\Gamma$ can be derived from the ESR spectrum. For this purpose the angular dependence of the unperturbed peak of the observed spectrum must first be matched with the single curve in Figure 4.9, and the values of g_1 and qg_2 (as well as the hyperfine constants $\tilde{A}_1$ and $\tilde{A}_2$) need to be estimated. Then, matching the angular dependence of the perturbed peak with one of the curves in the figure, we get the $\bar{\Delta}/3\Gamma$ value indicated at the left- and right-hand sides of this curve. Similar considerations can be applied to the case of angular dependence of the spectrum when rotating the magnetic field in the plane (001) (Figures 4.10 and 4.11).

The two cases of $\mathscr{H}$ rotating in the planes (110) and (001), respectively, cover the main resonances of the spectrum but, in principle, there may be more complicated spectra in other directions. Figure 4.12 shows the low-temperature spectra of the ${}^2E$ term systems in question with the direction of the magnetic field $\mathscr{H}$ in the (100) plane at an angle of 30° to the [001] axis. It is seen that, besides the usual dynamic and static JTE spectra at $\bar{\Delta}/3\Gamma = 0.0033$ and at $\bar{\Delta}/3\Gamma = 5$, respectively, there is a rather compli-

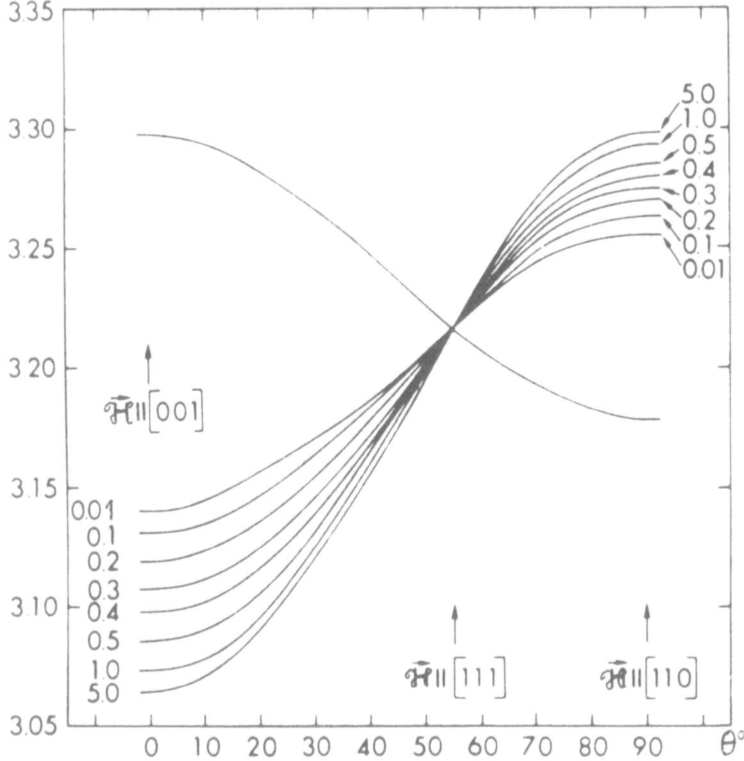

Figure 4.9. Angular dependence of the low-temperature ESR spectrum of a JT system in the ${}^2E$ state by rotating the magnetic field $\mathscr{H}$ in the plane (110). The absolute values of $\mathscr{H}$ in kG are plotted on the ordinate. The lowest vibronic singlet is A_2. In case of A_1 the figure must be reflected with respect to the axis $g - g_1 = 0$. The $\bar{\Delta}/3\Gamma$ values for different curves are shown on the left and right sides (after Reynolds and Boatner[110]).

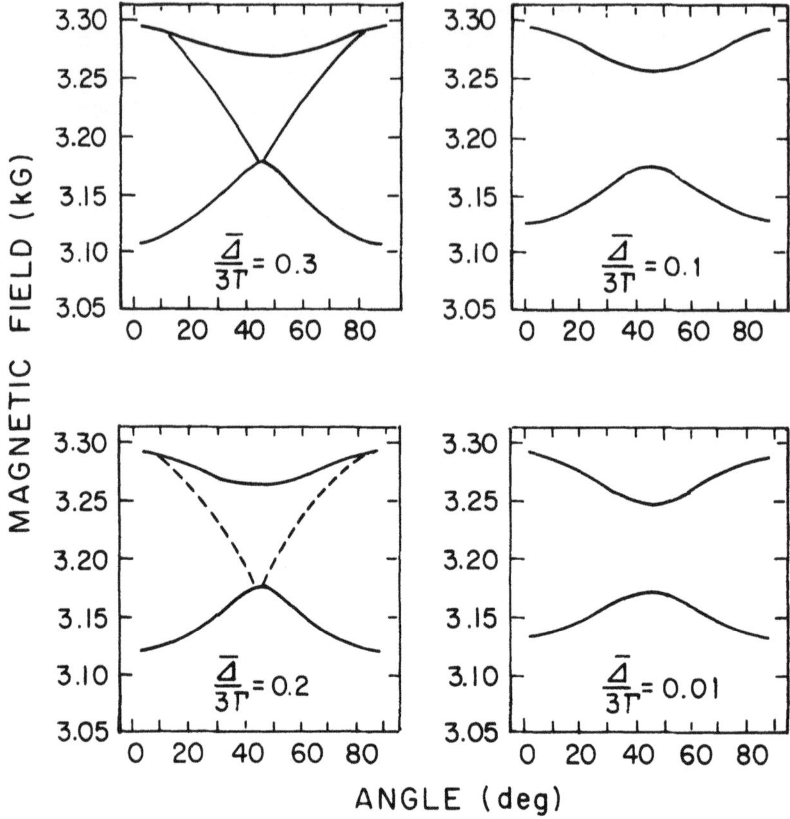

Figure 4.10. Angular dependence of the low-temperature ESR spectrum of a system in the ^2E state by rotating the magnetic field $\mathscr{H}$ in the plane (001) at small values of $\bar{\Delta}/3\Gamma$ (other data as in Figures 4.6–4.9). Angles 0, 45°, and 90° correspond to the field directions $\mathscr{H}$ along 100, [110], and [010], respectively. At values for which $\bar{\Delta}/3\Gamma \gtrsim 0.2$, a third line appears in the spectrum (after Reynolds and Boatner[110]).

cated spectrum at intermediate values of $\bar{\Delta}/3\Gamma$. Also complicated is the spectrum at $\mathscr{H} \parallel [111]$ and intermediate JT effects (Figure 4.13); in the limit cases of dynamic and static JTE it converts into one single line with g-factors

$$g_{\text{dyn}} = [g_1^2 + \tfrac{1}{2}(qg_2)^2]^{1/2}$$
$$g_{\text{stat}} = [g_1^2 + 2(qg_2)^2]^{1/2} \tag{4.26}$$

Comparison of the spectra described above with experimental spectra enables the vibronic parameters of the system to be determined. Such a comparison is illustrated in Figure 4.14 for the Cu^{2+} ion in the CaO lattice.[109] Since, in the above expressions, the orders of magnitude are estimated with respect to the radio-frequency quantum $\hbar\Omega = g\beta\mathscr{H}$ of the ESR measuring instrument, by changing this $\hbar\Omega$ value we can obtain different spectra for the same specimen (cf. the relativity rule concerning means of observation, Section 5.1). Therefore, as mentioned earlier, at small $\hbar\Omega$ the case $|g_2\beta\mathscr{H}| \ll \bar{\Delta}$, 3Γ is realized, for which the value of $\bar{\Delta}/3\Gamma$ can be derived from comparison of experimental and theoretical spectra. At sufficiently large values of $\hbar\Omega$ the ratio $g_2\beta\mathscr{H}/3\Gamma$ can be determined in much the same way. The values of $\bar{\Delta}$ and 3Γ can be estimated by combining the data obtained via these two methods of measurements (by means of low and high frequencies). Some of these data are listed in Table 4.2 as examples.

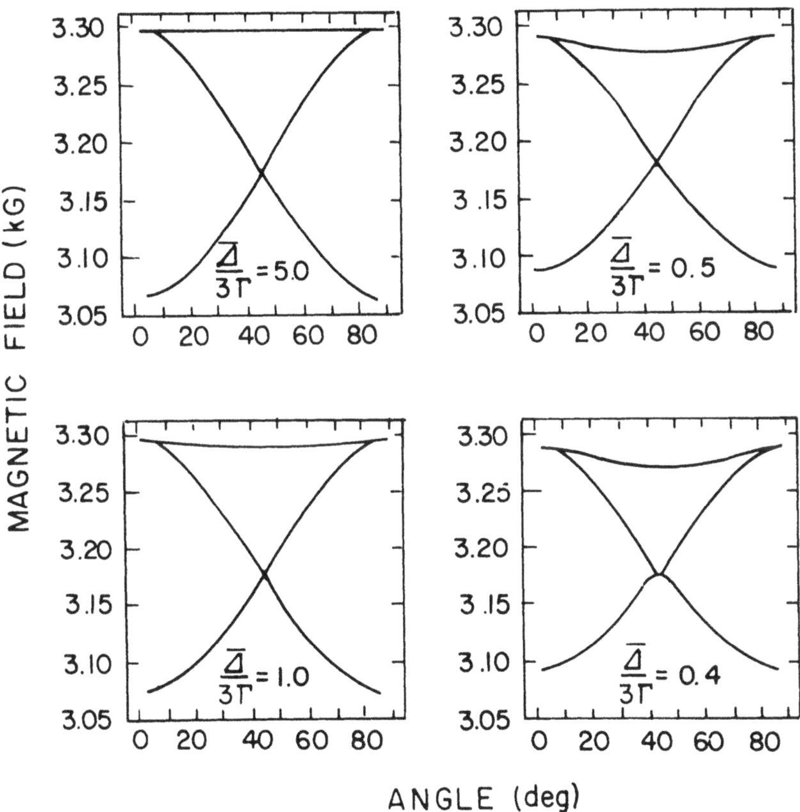

Figure 4.11. Same as Figure 4.10 for larger values of $\bar{\Delta}/3\Gamma$. At $\bar{\Delta}/3\Gamma \sim 5.0$ one of the lines becomes isotropic.

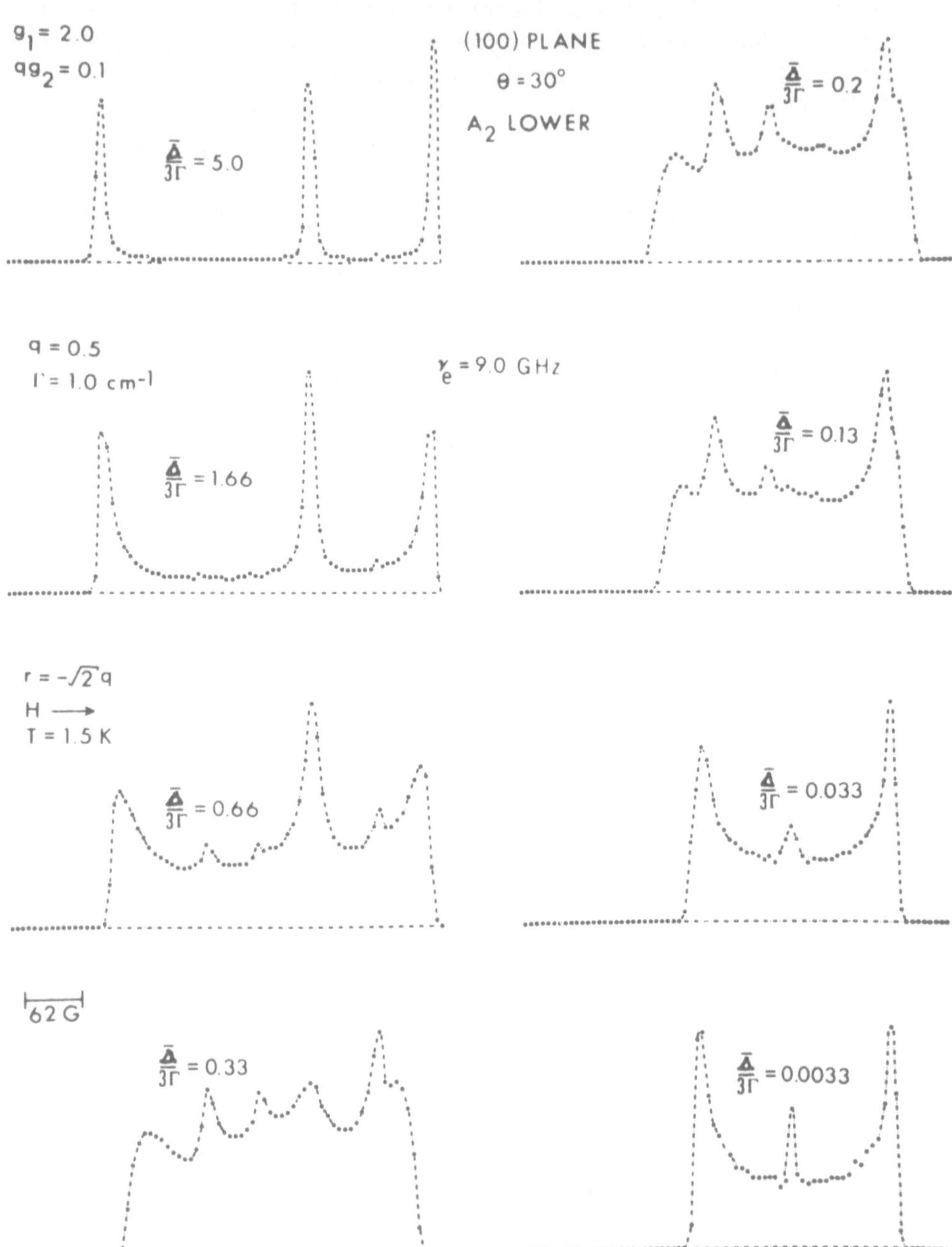

Figure 4.12. Low-temperature ESR line shape for a JT system in the ^2E state and different values of $\bar{\Delta}/3\Gamma$, when the $\mathscr{H}$ direction lies in the plane (100) at 30° to the [001] axis (other conditions remaining unchanged) (after Reynolds and Boatner[110]).

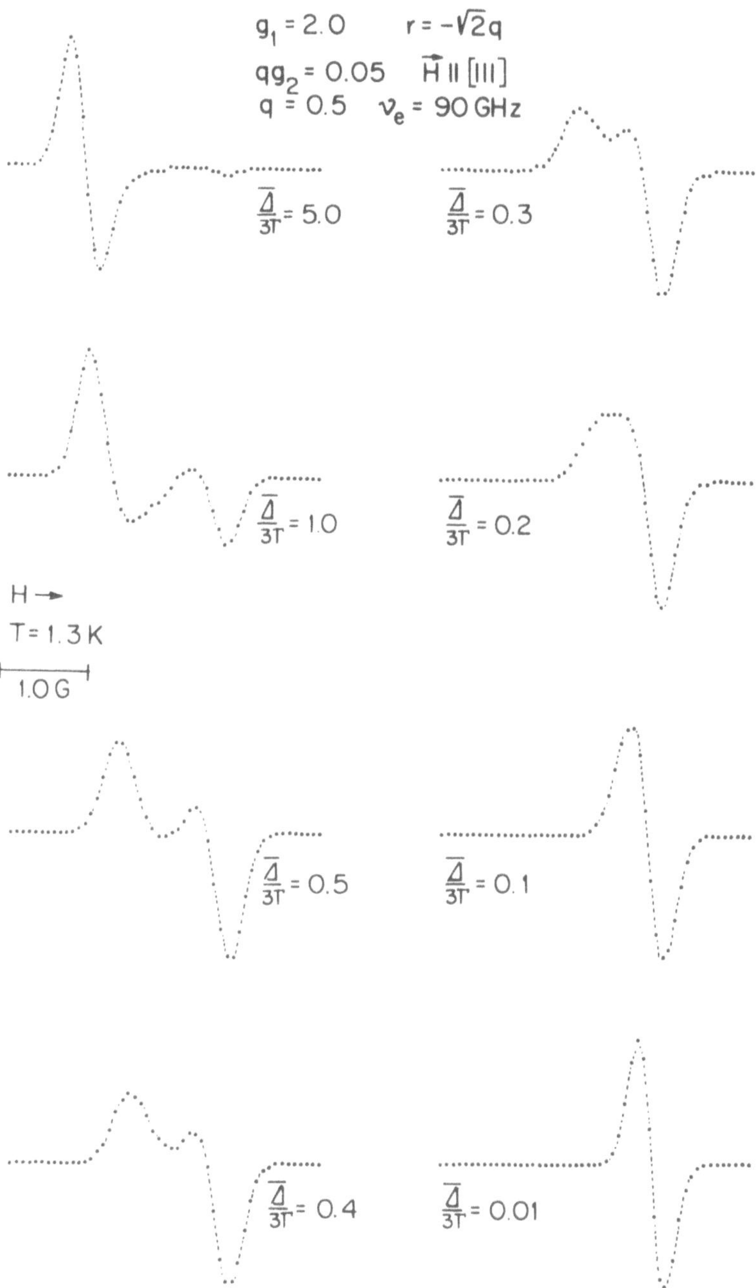

Figure 4.13. Same as Figure 4.12 for $\mathscr{H} \parallel [111]$, the first derivative of the line shape being shown (after Reynolds and Boatner[110]).

Table 4.2. Tunneling Splitting 3Γ and Averaged Splitting by Random Strain $\bar{\Delta}$ Obtained from ESR Data for Several JT Systems with ^2E Ground State (after Boatner et al.[111])

System	Lowest vibronic singlet	3Γ (cm^{-1})	$\bar{\Delta}$ (cm^{-1})	$\bar{\Delta}/3\Gamma$	g_1	qg_2	q
$SrO:Ag^{2+}$	A_2	—	—	—	2.049	0.032	0.682
$CaO:Ag^{2+}$	A_2	3.9	4.7	1.2	2.076	0.045	0.610
$AgO:Ag^{2+}$	A_2	4.8	0.62	0.13	2.0998	0.0563	0.577
$CaO:Cu^{2+}$	A_1	$\simeq 4.0$	2.7	0.67	2.2211	0.122	0.557
$MgO:Cu^{2+}$	A_2	—	—	—	—	—	—

With increasing temperature the vibronic ESR spectrum changes for two reasons: relaxation broadening and population of higher Kramers doublets. Relaxation transitions in the static JTE, similar to the dynamic JTE discussed above, result in the occurrence of a single isotropic line at the average frequency

$$\begin{aligned}(\overline{\hbar\Omega}) &= (\tfrac{1}{3}g_{\parallel} + \tfrac{2}{3}g_{\perp})\beta\mathscr{H} + (\tfrac{1}{3}\tilde{A}_{\parallel} + \tfrac{2}{3}\tilde{A}_{\perp})\nu_I \\ &= g_1\beta\mathscr{H} + \tilde{A}_1\nu_I \end{aligned} \tag{4.27}$$

On the other hand, the same isotropic line may occur as a result of the population of the Kramers doublet originating from the nearest vibronic singlet (A_1 or A_2). Therefore, it is often impossible to distinguish between these two reasons for the origin of the averaged spectrum.

There may be a third reason for the temperature dependence of the ESR spectrum, and that is the strong temperature dependence of the tunneling splitting itself caused by the temperature dependence of the barrier height between the minima of the AP which, in turn, is due to the temperature dependence of the population numbers for the vibrational modes in the multimode problem (see Section 3.2). Visually, with an increase in temperature the frequency of pulsations increases due to the decrease of the "viscosity" of the environment.[97] More rigorously, the temperature dependence of the tunneling splitting 3Γ occurs in a solution of the multimode problem when taking into account anharmonicity. But the problem has not yet been solved quantitatively.

Taking into account the temperature dependence of the tunneling splitting 3Γ, both the static JTE at low temperatures when 3Γ is small (but $\bar{\Delta}/3\Gamma$ and $g_2\beta\mathscr{H}/3\Gamma$ are large), and the dynamic JTE at higher temperatures can be obtained for the same values of $\bar{\Delta}$ and $g_2\beta\mathscr{H}$. All the above temperature effects may be present in combined form in experimentally

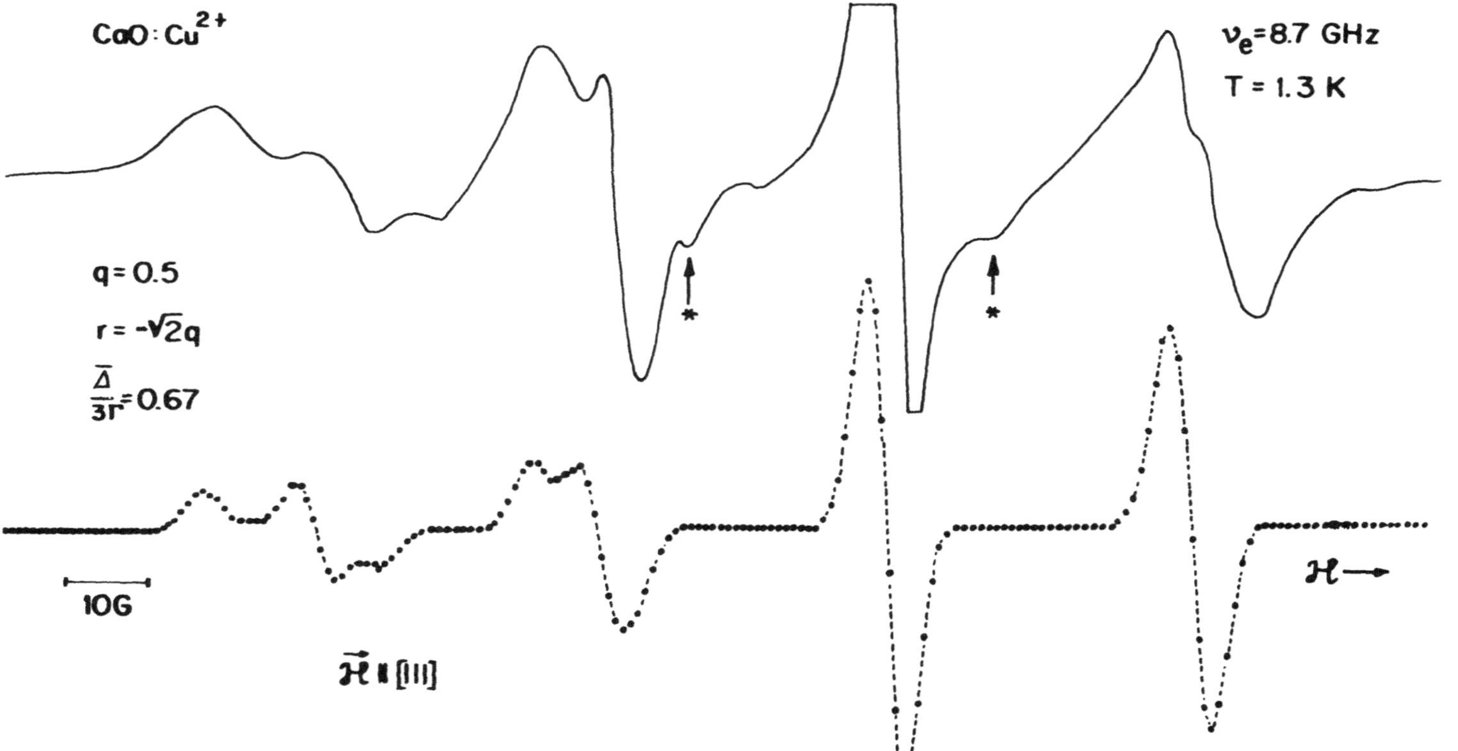

Figure 4.14. Experimental (upper) and calculated (lower) ESR spectrum of the Jahn–Teller ^2E system CaO : Cu^{2+} at $T = 1.3$ K ($\nu_e = \Omega/2\pi = 8.7$ GHz). Vibronic parameters obtained by comparing experimental and theoretical angular dependence of the spectrum in the plane (110) (Figure 4.9) are given on the left. Arrows indicate forbidden transitions induced by quadrupole interactions (after Reynolds et al.[109]).

observed temperature transitions from one ESR spectrum to another, illustrated in Table 4.3.

Low-symmetry environment, cooperative phenomena, and structural phase transitions (Section 5.2), which stabilize the low-symmetry configurations of the JT system, influence the expected ESR spectra.[125–130]

Let us summarize briefly the above results obtained for the vibronic ESR spectra of $E-e$ problem JT systems with cubic type symmetry. The dynamic JTE spectrum is observed at low temperatures with $g_2\beta\mathscr{H}/3\Gamma < 0.1$ and $\bar{\Delta}/3\Gamma < 0.1$. When $g_2\beta\mathscr{H}/3\Gamma \gtrsim 5$ or $\bar{\Delta}/3\Gamma \gtrsim 5$ the spectrum becomes that of the static JTE. In the intermediate range, when $0.1 < g_2\beta\mathscr{H}/3\Gamma \lesssim 5$ or $0.1 < \bar{\Delta}/3\Gamma \lesssim 5$, the following special features of the spectrum are expected: (1) selective broadening and shift of one of the two peaks of the dynamic spectrum; (2) special angular dependence of the spectrum different from both the dynamic and static JTE; (3) complicated fine structure of the spectrum for arbitrary orientations of the magnetic field.

With a rise in temperature, the increase of 3Γ in crystals results in a shift of the spectrum from a static to a dynamic one, while the population of the excited tunneling energy level and the relaxation gradually reduce the whole complicated spectrum to one isotropic line.

Tunneling Splitting in Mössbauer Spectra. Similar vibronic effects should be inherent in nuclear magnetic resonance spectra, nuclear quadrupole resonance, and nuclear γ-resonance (NGR) (*Mössbauer*) *spectra.* By way of illustration consider the hyperfine structure of the Mössbauer NGR spectrum for an $E-e$ system with strong vibronic coupling. The Mössbauer nucleus is assumed to be at the center of a high-symmetry coordination system which gives no quadrupole splitting, if vibronic interactions are disregarded. Using the wave functions of the tunneling energy levels given by equation (3.7), one can show[131] that for the nuclear transition $(I = \frac{1}{2}) \rightarrow (I = \frac{3}{2})$ (e.g., in the case of the iron nucleus), taking into account vibronic interactions, six lines should be observed, their frequencies (read off the undisplaced line) and intensity ratios being given by

$$\hbar\Omega_{1,4} = -\frac{3\Gamma}{2}[1 \pm (1+x)^{1/2}], \qquad \hbar\Omega_{2,5} = \mp 2qa$$
$$\hbar\Omega_{3,6} = \frac{3\Gamma}{2}[1 \mp (1+x)^{1/2}] \tag{4.28}$$

$$I_{1,6} : I_{2,5} : I_{3,4} = \frac{2x}{x + [1 + (1+x)^{1/2}]^2} : 1 : \frac{2[1 + (1+x)^{1/2}]^2}{x + [1 + (1+x)^{1/2}]^2} \tag{4.29}$$

Here, a is the usual constant of quadrupole coupling of the electrons with

Table 4.3. Examples of Temperature Transitions in ESR Spectra of JT Systems with Tunneling Splitting (^2E term)

Electronic configuration	Ion	Host system	Range of transition temperature	Low-temperature spectrum		High-temperature spectrum	References
				$g_\parallel$	$g_\perp$	g	
$3d^9$	Cu^{2+}	$ZnSiF_6 \cdot 6H_2O$	12–50	2.46 ± 0.01	2.10 ± 0.01	2.23 ± 0.01	94, 95, 113, 114
$3d^9$	Cu^{2+}	$Zn(BrO_3)_2 \cdot 6H_2O$	7–35	2.46 ± 0.01	2.10 ± 0.01	2.22 ± 0.01	94, 95, 113, 114
$3d^9$	Cu^{2+}	$La_2Mg_3(NO_3)_{12} \cdot 24H_2O$	44–42	2.470 ± 0.002	2.097 ± 0.002	2.218 ± 0.003	94, 95, 113, 114
$3d^9$	Cu^{2+}	$Bi_2Mg_3(NO_3)_{12} \cdot 24H_2O$	> 20–< 90	2.454 ± 0.003	2.096 ± 0.003	2.218 ± 0.003	94, 95, 113, 114
$3d^9$	Cu^{2+}	$Cu_3La_2(NO_3)_{12} \cdot 24H_2O$	170–270	2.41	2.10	2.22	
$3d^9$	Cu^{2+}	AgCl	90–300	2.30	2.07	2.15	115
$3d^9$	Cu^{2+}	MgO	> 1.2–< 77	2.303	2.087	2.190	100, 101
$3d^9$	Cu^{2+}	CaO	> 1.2–< 77	2.331	2.110	2.221	116
$3d^9$	Ni^{1+}	NaF	130–230	2.067 ± 0.004	2.464 ± 0.004	2.332 ± 0.004	117
$4d^1$	Y^{2+}	CaF_2	4–10	2.00	1.968	1.971	118
$5d^1$	La^{2+}	CaF_2	10–12	2.00 ± 0.01	1.904 ± 0.002	1.937 ± 0.02	119
$3d^1$	Sc^{2+}	CaF_2	> 1.5–< 10	1.995	1.951	1.967	120, 121
	Sc^{2+}	SrF_2	> 1.5–< 10	1.991	1.935	1.963	120, 121
$5d^7$	Pt^{3+}	Al_2O_3	> 4.2–< 79	2.011 ± 0.006	2.328 ± 0.004	2.220 ± 0.001	122
$3d^7$	Ni^{3+}	CaO	62–68	2.0672 ± 0.0006	2.3828 ± 0.0006	2.2814 ± 0.0006	123, 124

the nucleus, $q = K_E(E)$ is the vibronic reduction factor, $x = 2(B/3\Gamma)^2$, $B = bqa$, b is a numerical factor of the order of unity depending on the quadratic barrier height, and 3Γ is the tunneling splitting, as above.

It is seen that in this case, for arbitrary values of 3Γ, the NGR spectrum has six lines symmetric about their positions and intensities (Figure 4.15a).

In the limiting case of very deep minima when $3\Gamma \ll a$, the six lines are arranged in two groups of three lines positioned at $\pm a$ with near values of frequencies and intensities, the spectrum as a whole being similar to that expected for a statically distorted system, but with a quadrupole splitting reduced by one half, since in this case $q = \frac{1}{2}$ (static JTE). For larger 3Γ, two lines from each of the two groups of three lines split off, one shifting toward

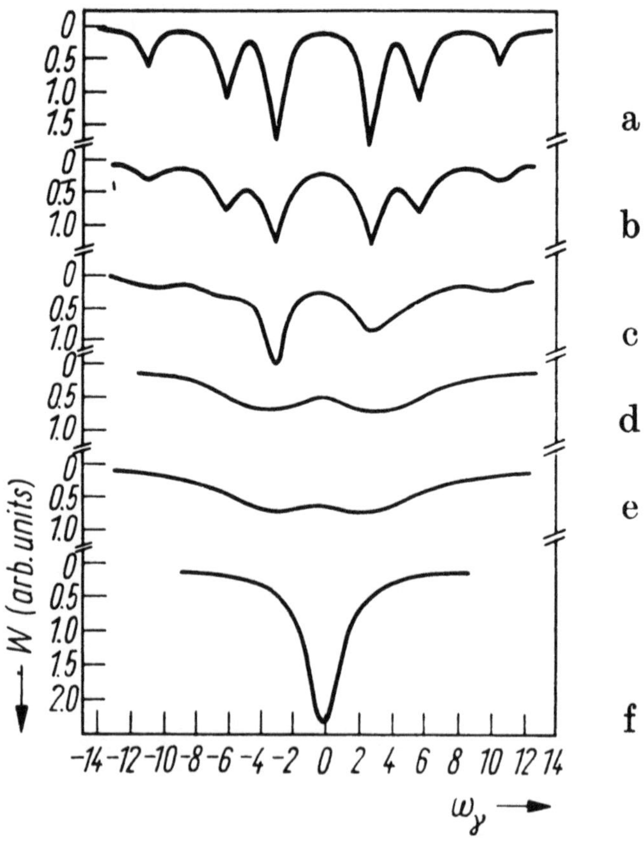

Figure 4.15. Mössbauer absorption line shapes for a JT system in an E state calculated with VI and relaxation effects taking $qa = 3$, $b = 4$, $3\Gamma = 8$, and: (a) $\lambda_1 = \lambda_2 = 0$; (b) $\lambda_1 = \lambda_2 = 0.1$; (c) $\lambda_1 = 1.0$, $\lambda_2 = 0.01$; (d) $\lambda_1 = 0.01$, $\lambda_2 = 1$; (e) $\lambda_1 = \lambda_2 = 1.0$; (f) $\lambda_1 = \lambda_2 = 10$ (all in units of the line width).

the center of the spectrum and increasing its intensity, the other shifting in the opposite direction and lowering its intensity. If the resulting spectrum is, as usual, not well resolved, it appears as if the quadrupole splitting should decrease when 3Γ is increased. Taking into account the above statement about the growth of the tunneling splitting 3Γ as the temperature increases, we reach the general qualitative conclusion that in the JT systems under consideration, a reduction of quadrupole splitting should be observed as the temperature increases. Such a temperature reduction of quadrupole splitting in the NGR spectra has been observed experimentally in many cases (e.g., see the spectra of $FeCl_4^{2-}$ [(132)] and $FeCr_2O_4$[(133)]).

As for ESR, relaxation processes are of great importance here. In the E–e problem under consideration, with the lowest E and A tunneling energy levels, one may introduce two relaxation transition probabilities, for the transitions within the two E states λ_1 and between the E and A states λ_2, respectively. Then, the influence of relaxation transitions on the NGR spectrum can be taken into account by means of the stochastic approach.[(131)] The results for some of the important values of the relaxation constants λ_1 and λ_2 are given in Figure 4.15. With an increase in constants λ_1 and λ_2 (for a given system — with increase in temperature) the six line spectrum changes first to a doublet and then to the usual singlet inherent in an undistorted system, the two constants λ_1 and λ_2 affecting these changes differently.

The case of the T term may be even more important in the NGR spectra, since the most studied high spin Fe^{2+} and low spin Fe^{3+} octahedral complexes pertain to this case. The problem has been considered with allowance for magnetic hyperfine interaction with electron spin $S = 0, \frac{1}{2}$, and 2, but disregarding relaxation.[(134)] Random strain may also modify the results.[(135)]

Absorption of Microwaves and Ultrasonics. Vibronic effects in ultrasonic and electromagnetic microwave (nonmagnetic) absorption are closely related to the effects in ESR and NGR spectra discussed above. To begin with, resonance absorption of microwaves due to direct transitions between the tunneling energy levels under electromagnetic radiation perturbation is to be expected.[(136)] Transitions between the lowest vibronic levels, in general only allowed as magnetic dipolar transitions in the absence of an inversion center, may be allowed also as electric dipolar transitions. Taking into account relaxations and crystal imperfections, which are especially important in the cases under consideration when tunneling splitting is small, the expected absorption lines may be rather broad and sometimes unresolved. Experimentally observed[(137)] additional microwave loss due to Mn^{3+} ions in yttrium–iron garnets (E term) seems to be of the nature of the tunneling under consideration. Indeed, this absorption has a clear-cut frequency dependence with a temperature-dependent maximum position ($\nu_{max} = 15$ kMc at 37 K and $\nu_{max} = 56$ kMc at 58 K) in accord with the strong dependence of tunneling splitting on crystal temperature, as described above.

Similarly, transitions between tunneling levels under acoustic perturbations which lead to ultrasonic absorption can be considered.[138,139] As distinct from acoustic spin resonance concerned with acoustic transitions between spin levels in magnetic fields, tunneling transitions are much more intense, since they are not concerned with the low-probability processes of spin reversal, mentioned earlier. For the $E \to A$ transition between the vibronic (tunneling) levels of the E–e problem for a cubic type symmetry system under the assumption that $\hbar\Omega \ll kT$, we obtain the following expression for the sound absorption coefficient as a function of frequency Ω[138]:

$$\sigma_{E\to A} = \frac{\pi}{3}\frac{NR^2F_E\Omega^2}{kTv^3d}\, g(\Omega)L \tag{4.30}$$

where N is the number of absorption JT centers per unit volume in the crystal, R is the minimal metal–ligand distance, F_E is the linear vibronic constant, v is the velocity of sound, $g(\Omega)$ is the form factor of the line determining its shape and depending on the relaxation processes in the crystal,

$$\int_0^\infty g(\Omega)d\Omega = 1$$

and L is the factor which takes into account the sound propagation direction $\boldsymbol{n}$ and polarization $\boldsymbol{m}$

$$L = \sum_i m_i^2 n_i^2 - \tfrac{1}{2}\sum_{i\neq j}{}' m_i n_i m_j n_j, \qquad i, j = x, y, z \tag{4.31}$$

(m_i and n_i are the direction cosines).

Estimates for hydrated copper complexes (taking $F_E = 2.5 \cdot 10^{-4}$ dyn), say, give $\sigma \approx 3.5 \cdot 10^{-11}\ \Omega^2(4\pi T)^{-1} \cdot L$ (in cgs units), which (neglecting random strain, see below) is several orders of magnitude greater than the magnetic acoustic resonance absorption. Note that, under the same conditions, the probability of relaxation of direct transitions under thermal vibrations is given by[140]: $W = 4 \cdot 10^{-13}\ \Omega^2 T/4\pi^2$. It follows that, at least for low temperatures, the width of the tunneling levels is small compared with the transition frequency. The absorption frequency can be regulated by means of magnetic fields.[139]

Introducing the relaxation time $\tau \sim W^{-1}$, the absorption line shape can be expressed as

$$g(\Omega)=\frac{2}{\pi}\frac{\tau}{1+(\Omega-\Omega_0)^2\tau^2} \tag{4.32}$$

where $\hbar\Omega_0=3\Gamma$ is the tunneling splitting. As τ is strongly temperature-dependent, $g(\Omega)$ has a maximum as a function of temperature which can be derived from the expression $(\Omega-\Omega_0)\tau(T_{max})=1$, provided the temperature dependence of 3Γ is neglected.

Acoustic loss due to the JTE was observed experimentally in $Ni^{3+}:Al_3O_3$ by Sturge et al.(141) (see also Sturge(142)). However, Sturge et al.(141) claim that this loss is due to relaxation processes involving tunneling effects, not to resonance transitions between tunneling levels, the motivation being that the random strain causes much greater splitting ($\Delta \sim 2$ cm^{-1}) than the tunneling (estimated as $3\Gamma \sim 10^{-1}$ cm^{-1}). These objections are unconvincing since it can be shown(9) that the sound absorption intensity depends strongly on the influence of the strain and decreases with strain increase as the ratio $(3\Gamma)^2/[(3\Gamma)^2+\Delta^2]$ (for a two-minimum system). It follows that the centers, which are strongly affected by random strain, do not actually absorb the sound, and only those centers which are not distorted (or are only slightly distorted) by the strain absorb the sound. The number of resonance-absorbing centers, and hence the total absorption intensity, is thereby strongly decreased by random strain. However, this has little influence on the possibility of experimentally observing the sound absorption since, as shown above, the effect itself is very strong. A more complete treatment of the influence of random strain on sound absorption has yet to be carried out.

Acoustic paramagnetic resonance in JT systems was considered elsewhere.(11,143–145) Experimental data allow one to estimate the tunneling splitting 3Γ which, for $Cr^{2+}:MgO$ and $Cr^{2+}:KMgF_3$, lie in the range 7.6–32 cm^{-1}.

4.2. Vibronic Infrared and Raman Spectra

Vibronic IR Spectra. Selection Rules. Infrared (IR) and Raman spectra that allow for vibronic JTE, PJTE, and the Renner effect (vibronic IR and Raman spectra) differ fundamentally from those without vibronic effects. This is seen from the AP shapes considered in Chapter 2. In the case of weak vibronic coupling, the energy gap between different sheets of the AP can be of the order of the vibration quantum $\hbar\omega$. Therefore, transitions between states of different sheets (accompanied by a change of electronic state) fall within the same frequency region as pure vibrational transitions.

In other words, in contrast to nonvibronic systems (such as diatomics), in vibronic systems there may be IR spectra which are classified in nonvibronic systems as electronic spectra. Those IR spectra in which vibronic effects are important, are termed vibronic IR spectra.

As for any resonance interaction, the probability of transitions between two stationary states Γ_1 and Γ_2 accompanied by absorption (or emission) of electromagnetic radiation in the IR region is determined by the transition moment $\boldsymbol{M}_{12}$:

$$\boldsymbol{M}_{12} = \langle \Gamma_1 | \boldsymbol{M} | \Gamma_2 \rangle \tag{4.33}$$

where M is the dipole (electric or magnetic), or quadrupole, etc., moment of the system.

Consider electric dipole transitions, which are most important in IR absorption (in the absence of electric dipole transitions, the magnetic dipole, quadrupole, etc., transitions are essential, but they are less intensive by several orders of magnitude). For such transitions, $\boldsymbol{M}$ is the dipole moment of the system and can be expressed in terms of two components:

$$\boldsymbol{M} = \boldsymbol{M}_c(Q) + \boldsymbol{M}_e(r) \tag{4.34}$$

where $\boldsymbol{M}_e(r) = -\Sigma_{i=1}^{n} e\boldsymbol{r}_i$ is the dipole moment of the electrons, and $\boldsymbol{M}_c = \Sigma_{\alpha=1}^{N} Z_\alpha \boldsymbol{R}_\alpha$ is the dipole moment of the nuclei (or of the core; see Section 1.2).

Using symmetrized coordinates (Section 1.2), the core (nuclear) dipole moment can be written in the form

$$\boldsymbol{M}_c(Q) = \boldsymbol{M}_c^{(0)} + \sum_{\bar{\Gamma}\bar{\gamma}} Z_{\bar{\Gamma}} Q_{\bar{\Gamma}\bar{\gamma}} \boldsymbol{n}_{\bar{\Gamma}\bar{\gamma}} \tag{4.35}$$

where the summation is over the irreducible representations to which the components of a vector pertain, $Z_{\bar{\Gamma}}$ is the effective charge of the core corresponding to the symmetrized $Q_{\bar{\Gamma}}$ mode ($Z_{\bar{\Gamma}}$ is a linear combination of the Z_α values occurring as a result of the transformation of R_α to $Q_{\bar{\Gamma}\bar{\gamma}}$ coordinates), $\boldsymbol{n}_{\bar{\Gamma}\bar{\gamma}}$ is a combination of the unit vectors of the coordinate system transforming as $\bar{\Gamma}\bar{\gamma}$, and $\boldsymbol{M}_c^{(0)}$ is the dipole moment of the core in the fixed initial configuration $Q_{\bar{\Gamma}\bar{\gamma}}^{(0)} = 0$.

It is evident that $\boldsymbol{M}_c^{(0)} \neq 0$ only for rigid dipole molecules. If $\boldsymbol{M}_c^{(0)} = 0$, the nuclear dipole moment is formed only by the symmetrized displacements $Q_{\bar{\Gamma}\bar{\gamma}}$ transforming as vector components. These displacements are called dipole-active displacements. The electronic dipole moment $\boldsymbol{M}_e(r)$, as any other vector, transforms according to the same representations. Note that in the case under consideration the dipole-active displacements cannot be totally symmetric.

Expression (4.33) allows for a relatively easy deduction of the selection rules for the vibronic IR spectra. Some general features can be revealed first. In the IR absorption we deal with transitions between states within a given electronic term Γ, or between near terms Γ and Γ'.

In the former case, vibronic effects are possible if and only if the term Γ is degenerate, and the JT active displacements include dipole-active displacements. Indeed, consider first the matrix elements of the electronic dipole moment $\boldsymbol{M}_e$. Since the latter depends only on electronic coordinates, its matrix element with electronic functions $\langle\Gamma|\boldsymbol{M}_e|\Gamma\rangle$ is nonzero only if $[\Gamma\times\Gamma]$ contains dipole-active representations. However, the nontotally symmetric components of $[\Gamma\times\Gamma]$ are just the JT-active modes (Section 1.6).

For the nuclear part of the dipole moment $\boldsymbol{M}_c(Q)$, which depends on the dipole-active displacements according to equation (4.35), the matrix element for transitions between vibronic states is nonzero only if their wave functions contain dipole-active coordinates, which means that the latter are JT-active coordinates.

The requirement that the JT active coordinates should also be dipole-active restricts the number of systems to which the vibronic IR spectrum pertains. For example, for an E term in a triangular molecule of type X_3, the E vibrations are JT-active and transform as vector components, i.e., they are also dipole-active. It follows that in such a molecule a vibronic IR spectrum is possible. For the same E term in an octahedral system, the vibronic IR spectrum is impossible, since the JT-active E vibrations do not transform as vectors, i.e., they are not dipole-active.

For transitions between the states of two electronic terms of different symmetry Γ and Γ', vibronic effects in the IR spectra occur if these terms mix by dipole-active displacements $\bar{\Gamma}\in\Gamma\times\Gamma'$ (see Section 5.1 for dipole instability in the PJTE). In principle, such a situation may occur in any system. In particular, in the above-mentioned octahedral system the vibronic IR spectrum may occur as a result of mixing of the A_{1g} and T_{1u} terms ($A_{1g}\times T_{1u}=T_{1u}$ and the vector components transform as T_{1u}).

It should be emphasized that, although the vibronic IR spectrum may occur only when the JT-active modes include dipole-active modes, all the JT-active vibrations will appear in the spectrum itself.[146] This is because all the above-mentioned modes are interdependent (mixed) in the JTE. For instance, in a tetrahedral system only the T_2 modes (from the JT-active E and T_2 modes) are dipole-active, but in the vibronic IR spectrum both the T_2 and E modes are evident (transitions to the vibronic E states being allowed due to the admixture of T_2 states in the T–$(e+t_2)$ problem).

Another special feature of the vibronic IR spectrum is its strong temperature dependence, due to the nonequidistant positioning of the vibronic levels (see illustrations in Section 3.4). As the temperature increases, new states, from which transitions with essentially different frequencies are possible, become populated (new spectral lines appear). This effect is not inherent in the usual IR spectrum, since in the absence of VI the equidistant vibrational levels are only slightly perturbed by anharmonicity (which is much smaller than vibronic anharmonicity; see Section 1.4).

More specific cases must be examined in order to derive selection rules for vibronic IR spectra. Consider the E term in a triangular molecule of type X_3, for which there may be a vibronic IR spectrum of dipole-active E vibrations. The energy levels and wave functions of the vibronic E–e problem are given in Section 3.1. In the case of weak linear vibronic coupling they are given by equation (3.1). It can be shown[13,77] that under the influence of the nuclear part of the dipole moment (4.35), vibronic transitions within a given sheet of the AP are allowed with a change of vibrational quantum number $\Delta n = \pm 1$, and a change of the quantum number of the total angular momentum of the electrons and nuclei $\Delta m = \pm 1$: $(\pm, n, m) \rightarrow (\pm, n \pm 1, m \pm 1)$.

In particular, two types of transition are possible from the ground state: $(-, 0, \pm\frac{1}{2}) \rightarrow (-, 1, \pm\frac{1}{2})$ and $(-, 0, \pm\frac{1}{2}) \rightarrow (-, 1, \pm\frac{3}{2})$, their frequency difference being $\hbar\Delta\Omega = 4E_{JT}$. It follows that in the vibronic IR spectrum the fundamental tone is split into two components (Figure 3.1). The electronic dipole moment (the first term in equation (4.34)) does not cause new transitions in the weak coupling approximation in question.

In the case of strong vibronic coupling, both intersheet and intrasheet transitions are possible. The nuclear dipole moment (4.35) causes transitions only within a given AP sheet $(\pm, n, m) \rightarrow (\pm, n', m \pm 1)$. In particular, four types of such transitions are allowed from the ground state:

$$(-, 0, \pm\tfrac{1}{2}) \rightarrow (-, 0, \mp\tfrac{1}{2}) \tag{4.36}$$

$$(-, 0, \pm\tfrac{1}{2}) \rightarrow (-, 0, \pm\tfrac{3}{2}) \tag{4.37}$$

$$(-, 0, \pm\tfrac{1}{2}) \rightarrow (-, 1, \pm\tfrac{1}{2}) \tag{4.38}$$

$$(-, 0, \pm\tfrac{1}{2}) \rightarrow (-, 1, \pm\tfrac{3}{2}) \tag{4.39}$$

The frequencies of transitions (4.36) and (4.37) with $\Delta n = 0$ are $\Omega = 0$ and $\hbar\Omega = (\hbar\omega_E)^2/2E_{JT}$, respectively. Although the line with $\Omega = 0$ cannot be observed directly, this transition is essential as it allows appropriate pure rotational transitions of a free molecule to take place (see below). The frequencies of vibrational transitions (4.38) and (4.39) with $\Delta n = 1$ are $\hbar\omega_E$ and $\hbar\omega_E[1 + \hbar\omega_E/2E_{JT}]$, respectively. The anharmonicity of the AP surface sheet $\varepsilon_-(\rho, \phi)$ allow overtone transitions with $n' > n + 1$ to exist, too, their intensity increasing with n for large values of n.

The electronic dipole moment in the case of strong vibronic coupling allows both intrasheet and intersheet transitions; for the former, $\Delta n = 0$, $\Delta m = \pm 1$: $(\pm, n, m) \rightarrow (\pm, n, m \pm 1)$, while for the latter, $\Delta m = \pm 1$ and Δn is arbitrary: $(\pm, n, m) \rightarrow (\mp, n', m \pm 1)$. The intensities of the intersheet transitions are determined by an additional factor $\langle \chi_n (\rho - \rho_0) | \chi_{n'}(\rho) \rangle$ (the Franck–Condon integral), which is less than unity, and due to which these lines may be much weaker than those lines of intrasheet transitions. Two types of intersheet transition are possible from the ground state: $(-, 0, \pm\frac{1}{2}) \rightarrow (+, n, \mp\frac{1}{2})$ and $(-, 0, \pm\frac{1}{2}) \rightarrow (+, n, \pm\frac{3}{2})$. Each of these leads to an almost equidistant spacing of spectral lines with a maximum of the envelope at $\hbar\Omega_{\max} = \varepsilon_+(\rho_0) - \varepsilon_-(\rho_0) = 4E_{\mathrm{JT}}$ (see Figure 4.16).

In the case of strong vibronic coupling, taking into account quadratic terms of the VI, there are three minima on the lowest sheet of the AP of the E–e problem. At each minimum, the triangle molecule X_3, for

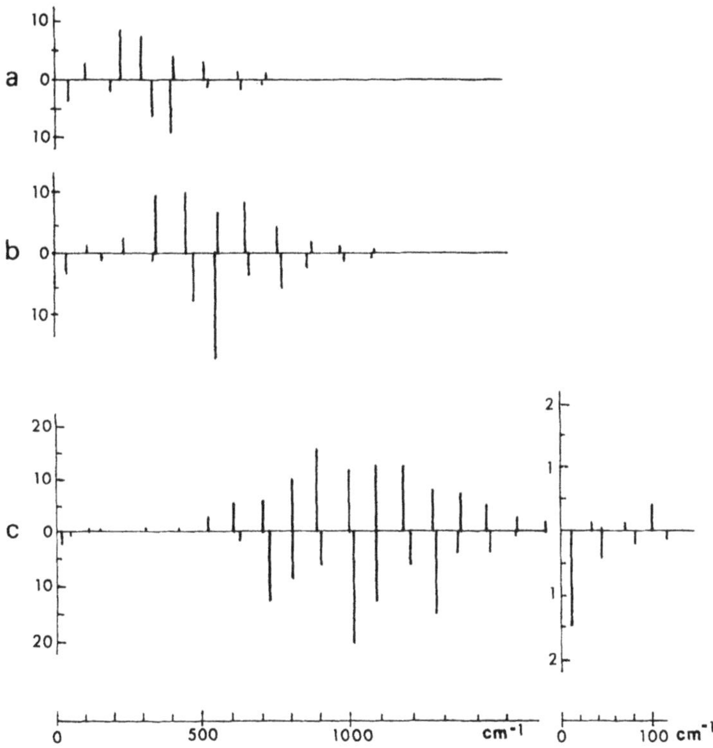

Figure 4.16. Vibronic IR spectrum for a JT system with a linear E–e problem calculated for the following dimensionless VC: (a) $\lambda_E = E_{\mathrm{JT}}/\hbar\omega_E = 0.5$; (b) $\lambda_E = 1.0$; (c) $\lambda_E = 2.5$. Transitions $|m| = \frac{1}{2} \rightarrow |m| = \frac{1}{2}$ and $|m| = \frac{1}{2} \rightarrow |m| = \frac{3}{2}$ are plotted above and below the abscissa, respectively (after Child and Longuet-Higgins[77]).

instance, becomes isosceles (with C_{2v} symmetry, Figure 2.4), while an octahedral MX_6 system of O_h symmetry distorts to D_{4h} symmetry (Figure 2.3). In these minimum states the E vibration splits. When the pulsating motion is taken into account, an additional transition may occur at the tunneling frequency $\delta = 3\Gamma$. If the quadratic barriers between the minima are large enough, this frequency falls within the microwave region (Section 4.1).

For arbitrary values of the VC, the vibronic IR spectrum can be obtained by numerical solution of the vibronic equations (Section 3.4). In Figure 4.16 some such results are illustrated for the E–e problem of an X_3 system.[77] In the limiting cases of weak and strong coupling, the numerical results confirm the features of the vibronic IR spectrum obtained above by means of analytical expressions.

The general features of the vibronic IR spectra of a triangular molecule in an E term state are, in principle, inherent to all JT systems, although the actual appearance of these spectra may differ for different types of systems. Vibronic IR spectra for E–$(b_1 + b_2)$, T–$(e + t_2)$, and Γ_8–$(e + t_2)$ problems with weak vibronic coupling have been considered by Child.[147] Strong coupling to one JT active vibration and weak coupling to another in the Γ_8–$(e + t_2)$ problem were also investigated by Child.[148] Numerical solutions for arbitrary linear coupling to E and T_2 vibrations with allowance for spin–orbit interactions of first and second order for a ^3T term have been obtained recently.[90] The temperature dependence of polarizations in the vibronic IR spectrum owing to the PJTE caused by mixing two singlet electronic states were studied by Fulton and Gouterman.[149]

The Pure Rotational Vibronic Spectrum. Transitions which occur without change in the vibrational quantum number $\Delta n = 0$, as allowed in the vibronic IR spectrum, are of prime importance for a possible new type of pure rotational spectra. In the absence of VI, pure rotational transitions without change of the vibrational state are possible only if the proper dipole moment $\boldsymbol{M}_c^{(0)}$ is nonzero (rigid dipole molecules). In this case, as a result of transformation from the molecular coordinate system to the fixed laboratory system, the constant dipole moment $\boldsymbol{M}_c^{(0)}$ becomes dependent on nuclear coordinates describing the rotations of the molecule as a whole, and since $\boldsymbol{M}_c^{(0)}$ does not contain vibration coordinates, its matrix elements are nonzero for pure rotational transitions (the electronic dipole moment has no such zeroth-order terms). If $\boldsymbol{M}_c^{(0)} = 0$, the corresponding (nondipolar) molecules have no pure rotational absorption spectrum. The same conclusions can be drawn with pure rotational Raman spectra.

If vibronic interactions are taken into account, transitions with $\Delta n = 0$ become allowed due to the variable part of the nuclear dipole moment (the second term in equation (4.35)), and hence, in the vibronic IR spectrum, rotational transitions are possible for nondipolar molecules having $\boldsymbol{M}_c^{(0)} = 0$. (Without vibronic interactions, the variable part of the nuclear dipole moment allows rotational transitions only if there are simultaneous vibrational transitions with $\Delta n = \pm 1$.) We may conclude that *because of vibronic effects a new type of spectrum becomes possible: a pure rotational IR and Raman spectrum for molecules having no average dipole moment — the vibronic rotational spectrum.*[150-152,77]

This result may be given a visual physical treatment. If there are dipole-active JT distortions, the nondipolar system (having no proper dipole moment in the high-symmetry configuration) possesses a dipole moment in the AP minima configurations. For example, the triangular molecule in the E term state with strong vibronic coupling (and quadratic barriers sufficiently high) pulsates between the three equivalent configurations, in each of which the isosceles triangle has a dipole moment (Figure 2.4). Therefore, although in the high-symmetry configuration of the perfect triangle the molecule has no dipole moment, this configuration is unstable due to the JTE, and the system is mostly in the AP minima configurations with a dipole moment. Similarly, a tetrahedral molecule with a T term and without a proper dipole moment (CH_4^+, say) under certain conditions pulsates between the configurations of the four AP minima at each of which the molecule has a dipole moment directed along the third-order axis.

Dipolar distortions of high-symmetry configurations may also occur in the absence of electronic degeneracy, provided there is vibronic mixing of a given state with the near states by dipole-active displacements (PJTE in dipole-active directions). The JT distortions are not dipole-active in systems having a center of inversion. Therefore, the new type of purely rotational vibronic IR spectrum for such systems may occur only as a result of the PJTE. In systems without a center of inversion, the new spectra may occur due to both electronic degeneracy (JTE) and pseudodegeneracy (PJTE).

It follows also that molecules having no proper dipole moment may behave as rigid dipole molecules if there is dipole instability of vibronic origin (see Section 5.1).

The position and intensities of the vibronic pure rotational lines can be found by solving the rotation problem for the vibronic system. The X_3 molecule is a symmetric top molecule, described (i) by the wave function φ_{JKM}, where J is the quantum number of the total rotational momentum, and K, $M = 0, \pm 1, \ldots, \pm J$ are the quantum numbers of its projections, and (ii) by the energy levels

$$E_{\mathrm{rot}} = BJ(J+1) + (A-B)K^2 - 2AK\zeta \qquad (4.40)$$

where A and B are rotational constants, and ζ is the constant of Coriolis interaction. In the E term state with only linear vibronic coupling[77,153] (see also solutions (3.7)),

$$\zeta = m[(p/|m|)(\zeta_e - \tfrac{1}{2}) + 1] \tag{4.41}$$

where ζ_e is the electronic orbital momentum of the degenerate term ($\zeta_e - \frac{1}{2}$ is the total momentum of "internal" rotation; see Section 3.1), and $p = K_E(A_2)$ is the vibronic reduction factor (equations (3.36) and (3.37)). In particular, with strong vibronic coupling $p \approx 0$ and $\zeta \approx m$. From equation (4.41) it follows that, by taking into account vibronic effects, the Coriolis constant in the usual formula of rotational energy levels becomes dependent on the energy level m.

Neglecting rotational-vibronic interactions, we can present the total energy of the system as a sum of energies (3.7) and (4.40), and the total wave function as the product $\Psi_{nm} \cdot \varphi_{JKM}$, and then determine the selection rules for pure rotational transitions. For transitions $(-,0,\frac{1}{2}) \rightarrow (-,0,-\frac{1}{2})$ (cf. equation (4.36)), we have $\Delta K = \pm 1$, which leads to IR absorption with three characteristic branches P, Q, R. For example, the Q branch (for which $J = J'$) consists of a set of lines with intervals $\Delta E = [A(1+2\zeta) - B](2K+1)$.[77] In particular, it is seen that if A, B, and ζ are known, the vibronic reduction factor p, and hence the VC, can be obtained.

Consider now the rotational-vibronic spectrum of a tetrahedral system with T_d symmetry (a spherical top molecule) in a T term state, when in the T–$(e + t_2)$ problem the linear interaction with the T_2 displacements is predominant and the AP has four equivalent minima (Section 2.2). The simplest of such systems is CH_4^+ in which the JTE with trigonal minima is apparently confirmed experimentally[154] (see also Section 5.1). The lowest are the tunneling T_2 and A_1 levels ($E_{A_1} > E_{T_2}$) with wave functions given by equation (2.27) and Table 3.1. Taking account of rotations (but disregarding vibronic-rotational interaction), each vibronic level is accompanied by a series of rotational levels of a spherical top molecule with energies (Figure 4.17)

$$E_{T_2J} = BJ(J+1), \qquad E_{A_1J} = BJ(J+1) + \delta \tag{4.42}$$

where B is the rotational constant, and δ is the tunneling splitting. As in the above case of the E term, the wave function for each of these levels may be presented in the form of a product of the wave function of the vibronic state $|\Gamma\gamma\rangle$ and the rotational function of the spherical top φ_{JKM}:

$$\begin{gathered} |\Gamma\gamma JKM\rangle = |\Gamma\gamma\rangle\varphi_{JKM} \\ \Gamma = A_1, T_2; \qquad K, M = 0, \pm 1, \ldots, \pm J \end{gathered} \tag{4.43}$$

The known wave functions can be employed to calculate the intensities of the induced dipolar transitions $\Gamma J \to \Gamma' J'$ which, for each unit of incident radiation density and N absorbing centers in a unit volume, are[151]

$$\alpha_{\Gamma J \to \Gamma' J'} = \frac{8\pi^3 p_0^2}{27\hbar^2 c^2}\frac{N}{Z}(E_{\Gamma' J'} - E_{\Gamma J})[\exp(-E_{\Gamma J}/kT - \exp(-E_{\Gamma' J'}/kT)]g^I_{\Gamma J}C_{JJ'} \tag{4.44}$$

where Z is the statistical sum given by (B in cm^{-1})

$$Z = \sum_J (2J+1)^2 \left[3g^I_{T_2 J} + g^I_{A_1 J}\exp\left(-\frac{\delta}{kT}\right)\right]\exp\left[-\frac{B\hbar c J(J+1)}{kT}\right] \tag{4.45}$$

$g^I_{\Gamma J}$ being the statistical weight owing to nuclear spin I,

$$C_{JJ'} = \begin{cases} (2J+1)(2J+3), & J' = J+1 \\ (2J+1)^2, & J' = J \\ (2J+1)(2J-1), & J' = J-1 \end{cases}$$

and p_0 is the absolute value of the dipole moment of the system at the dipolar minimum.

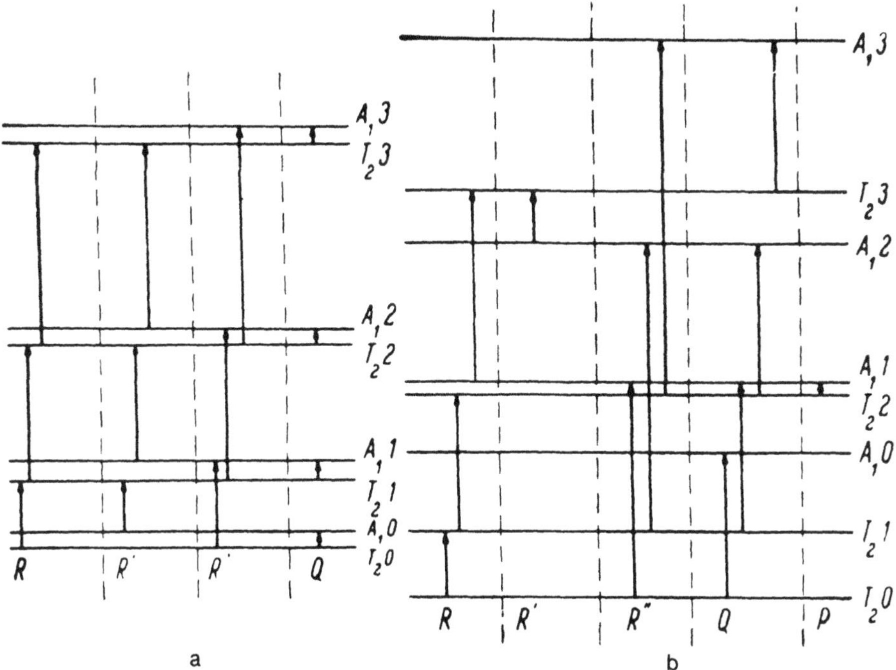

Figure 4.17. Tunneling–rotational energy levels and allowed transitions at: (a) $\delta < 2B$; (b) $\delta > 2B$. The dashed line separates transitions pertaining to different branches.

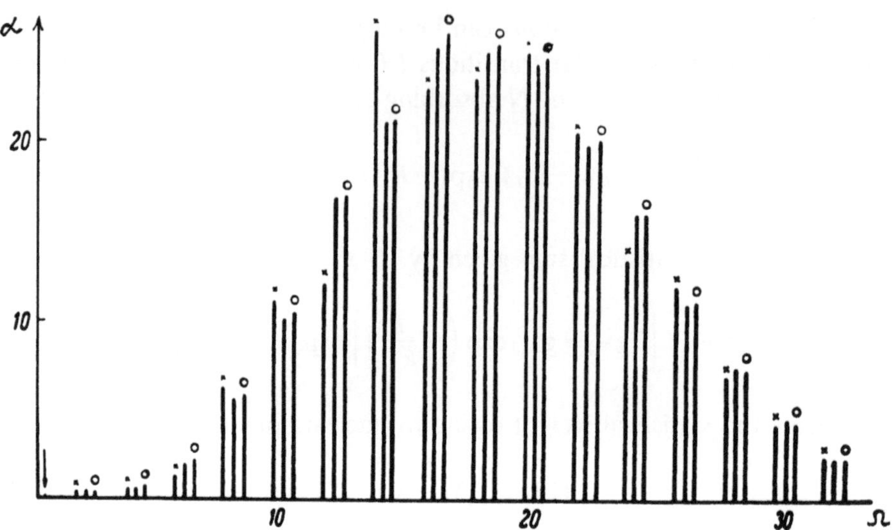

Figure 4.18. Calculated line positions and intensities of the rotational vibronic spectrum of dipolar unstable tetrahedral systems at $\delta = 1\ \text{cm}^{-1}$, $B = 5.24\ \text{cm}^{-1}$, and $kT = 200\ \text{cm}^{-1}$. Branches R'', R', and R are marked by rings, crosses, and not marked, respectively. Frequency Ω is given in units of B. The Q transition is shown by an arrow (after Bersuker et al.[151]).

It is seen from expression (4.44) that the transition with $J' = J + 1$ (R branches) can be of three types:

$$T_2J \to T_2(J+1) \quad (R), \quad A_1J \to T_2(J+1) \quad (R'),$$

$$\text{and} \quad T_2J \to A_1(J+1) \quad (R'')$$

while the Q transitions ($J' = J$) and the P transitions ($J' = J - 1$) are allowed only as $T_2J \to A_1J$ (Q branch) and $T_2J \to A_1(J-1)$ (P branch). If $\delta \ll B$ the frequencies in the three R branches (R, R', R'') coincide, while the P and Q branches vanish, resulting in a series of equidistant lines spaced $2B$ apart. As δ increases, each of these lines splits (Figure 4.18) and a Q transition occurs at frequency δ, its intensity being small due to the almost equal population of the two combining levels.

If $\delta > B$ the spectrum changes (Figure 4.19). Along with an increase in the difference of line frequencies in the three R branches, their intensity changes, increasing for $T_2 \to A_1$ transitions (R'') and decreasing for $A_1 \to T_2$ transitions (R'). Simultaneously, the Q transition becomes stronger and the lines of the P branch appear. The number and intensity of the latter increase with the degree of the inequality $\delta > B$. Note that in the common, pure rotational spectra for systems with a proper dipole moment, only one R branch occurs, three branches R, P, and Q being incidental only for the rotational structure of the vibrational spectrum. In the vibronically pure rotational spectrum inherent in nondipolar molecules with dipolar vibronic instability, all the branches, R, R', R'', P, and Q, may be present, the latter two being required for large values of δ.

When the temperature drops, the intensity of the Q transition increases faster than the others, and therefore the Q line begins to stand out against the background formed by the band of the other lines (Figure 4.20). This line corresponds to transition without change of rotational quantum number; its origin, appearance, and temperature are quite similar to the zero-phonon line in optical spectra (Section 4.3), and it can therefore be called a "zero-rotation" line. It is discussed above in Section 4.1 as a transition between tunneling levels.

The intensity of this new spectrum is proportional to p_0^2, and hence it is $(p_0/p_M)^2$ times weaker than the pure rotational spectra, where p_M is the dipole moment of a usual rigid dipole molecule. Although quantitative numerical estimates of the value of p_0 are still unavailable, it seems to be of the order of 0.1 debye for strong dipolar unstable systems, judging by the known orders of magnitude of JT distortions. This leads to a value of

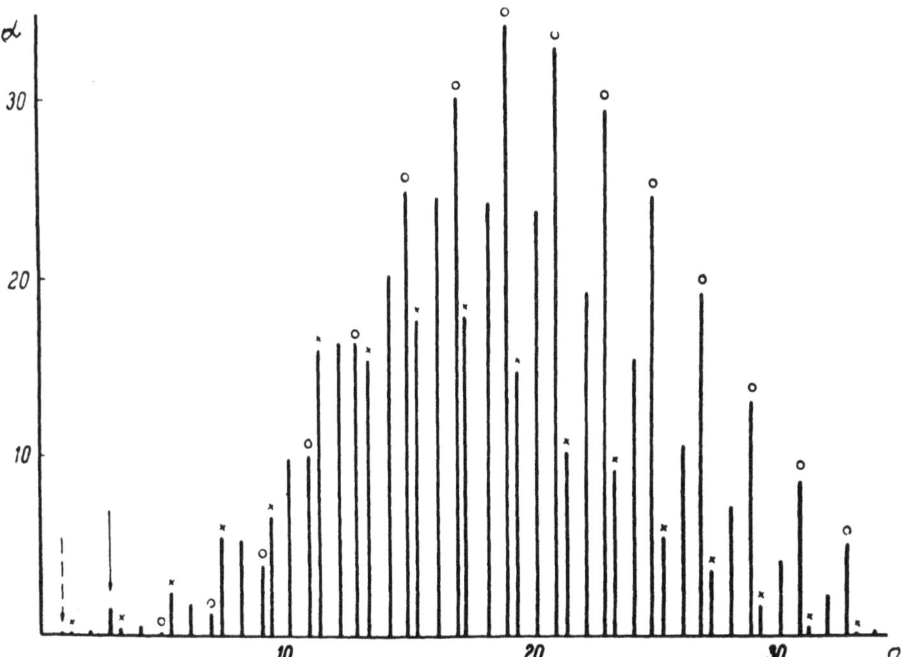

Figure 4.19. Same as Figure 4.18 at $\delta = 15\ \text{cm}^{-1}$. The dashed arrow indicates the P transition (after Bersuker et al.[151]).

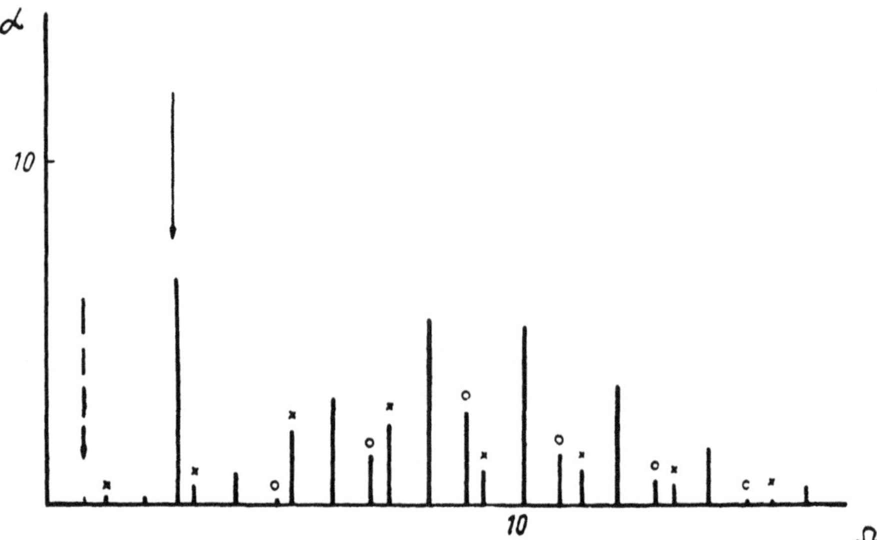

Figure 4.20. Same as Figure 4.19 at $kT = 50\ \text{cm}^{-1}$, the intensity scale being reduced by a factor of 100 (after Bersuker et al.[151]).

$(p_0/p_M)^2 \sim 10^{-2}$. It is certainly possible to observe such an absorption and even much weaker ones. (The observed Coriolis interaction-induced absorption is of the order of 10^{10} times weaker than the common rotational absorption.[155])

Vibronic Raman Spectra. In order to analyze possible vibronic Raman spectra, we introduce the cross-section for light scattering $\sigma_{12}(\Omega)$ due to the transition of the system from state 1 to state 2 in terms of its component cross-sections[156]:

$$\sigma_{12}(\Omega) = (4\Omega^4/c^4)[G_{12}^{(0)} + G_{12}^{(s)} + G_{12}^{(a)}] \tag{4.46}$$

where $G_{12}^{(0)}$, $G_{12}^{(s)}$, and $G_{12}^{(a)}$ are the scalar, symmetric, and antisymmetric components of the scattering, respectively, and Ω is the frequency of the incident light. The scalar part of the scattering is due to transitions leaving unchanged the state of the system or the frequency of the incident light (Rayleigh scattering). The symmetric and antisymmetric Raman scatterings are due to quadrupole and magnetic dipole transitions, respectively.

The component cross-sections $G_{12}^{(i)}$ are determined in simple cases as matrix elements of the relevant components of the tensor operator of the molecular polarizability. These components transform as vector products. With the known vector representations Γ of the symmetry group of the system, the representations of the tensor components, as well as of their symmetric ($G^{(s)}$) and antisymmetric ($G^{(a)}$) parts, can be determined. To do this, the representation $\bar{\Gamma} \times \bar{\Gamma}$ must be decomposed into the symmetric $[\bar{\Gamma} \times \bar{\Gamma}]$ and antisymmetric $\{\bar{\Gamma} \times \bar{\Gamma}\}$ parts. The nontotally symmetric components of $[\bar{\Gamma} \times \bar{\Gamma}]$ represent the symmetric part of the polarizability tensor $G^{(s)}$ (the totally symmetric representation corresponding to its scalar component), whereas the antisymmetric part of this tensor refers to the irreducible representations in the $\{\bar{\Gamma} \times \bar{\Gamma}\}$ part. For instance, the vector components of an octahedral system with O_h symmetry transform according to $\bar{\Gamma} = T_{1u}$. Since $[T_{1u} \times T_{1u}] = A_{1g} + E_g + T_g$ and $\{T_{1u} \times T_{1u}\} = T_{1g}$, the polarizability tensor has scalar (A_{1g} type), symmetric (E_g and T_{2g} type), and antisymmetric (T_{1g} type) parts. For an X_3 molecule with C_{3v} symmetry, the vector components transform according to the representations A_1 and E. Since $[A_1 \times A_1] = A_1$, $\{A_1 \times A_1\} = 0$, $[E \times E] = A_1 + E$, and $\{E \times E\} = A_2$, the polarizability tensor transforms according to A_1 (scalar part), E (symmetric part), and A_2 (antisymmetric part).

Consider pure rotational transitions in which the vibronic state Γ does not change.[150,152,77] If Γ is nondegenerate, $[\Gamma \times \Gamma] = A_1$, $\{\Gamma \times \Gamma\} = 0$, and therefore only Rayleigh scattering is allowed. For a doubly degenerate E term, $[E \times E] = A_1 + E$, $\{E \times E\} = A_2$, and hence pure rotational Raman scattering may occur alongside Rayleigh scattering provided the polarizability tensor contains symmetric E or antisymmetric A_2 representations. As shown above, the polarizability tensor of octahedral O_h systems contains E components in the symmetric part and no A_2 components. It follows that in these systems the pure rotational Raman scattering is allowed as symmetric (quadrupole) in origin. For C_{3v} molecules both the E and A_2 components are present in the polarizability tensor, and hence both symmetric (quadrupole) and antisymmetric (magnetic dipole) types of Raman scattering are allowed.

For threefold degenerate T terms $[T \times T] = A_1 + E + T_2$ and $\{T \times T\} = T_1$, and both types of pure rotational Raman scattering (symmetric (quadrupole) scattering by the E and T_2 components of the polarizability tensor and antisymmetric (magnetic dipole) scattering by the T_1 component) are possible for the octahedral system (the C_{3v} system has no T terms).

For JT systems in an E term state (symmetric or spherical top molecules) the vibronic Raman spectrum consists of quadrupole transitions of type $A \rightarrow E$, $E \rightarrow A$, and $E \rightarrow E$.[77,152] Accordingly, say for a spherical top molecule,[152] with strong vibronic coupling and tunneling splitting, three types of O, P, Q, R, and S lines (for which $\Delta J = -2, -1, 0, 1, 2$, respectively, where J is the rotational quantum number) are possible: $E \rightarrow E(O, P, Q, R, S)$, $E \rightarrow A(O', P', Q', R', S')$, and $A \rightarrow E(O'', P'', Q'', R'', S'')$. The lines labeled by single or double primes are shifted relative to the nonlabeled lines by an amount δ and $-\delta$, respectively, where δ is the tunneling splitting. It follows that δ equals the frequency of the Raman-shifted Q' lines, for which $\Delta J = 0$ (the "no-rotation" line of the $E \to A$ transition from the ground state), or of the Q'' line, shifted in the opposite (short-wave) direction. The Q line of the "no-rotation" $E \to E$ transition corresponds to the nonshifted Rayleigh line.

The degree of depolarization of scattered light in the vibronic Raman spectrum can be determined using the general relationships

$$\rho = \frac{3G_{12}^{(s)} + 5G_{12}^{(a)}}{10G_{12}^{(0)} + 4G_{12}^{(s)}}, \qquad \rho_n = \frac{2\rho}{1+\rho} \tag{4.47}$$

where ρ is the degree of depolarization of the incident light polarized normal to the plane of the incident and scattered beam, and ρ_n is the degree of depolarization of natural incident light. Inserting appropriate values into equation (4.47), we find that for all the above lines, except Q lines, $G_{12}^{(0)} = G_{12}^{(a)} = 0$ and $\rho = \frac{3}{4}$, $\rho_n = \frac{6}{7}$. For the Q lines, $G^{(0)} \neq 0$ and $\rho \leqslant \frac{3}{4}$, $\rho_n \leqslant \frac{6}{7}$.

In the case of a T term with strong vibronic coupling to T_2 vibrations (T–t_2 problem), there may be also three types of, formally the same, O, P, Q, R, and S transitions: $T \to T(O, P, Q, R, S)$, $T \to A(O', P', Q', R', S')$, and $A \to T(O'', P'', Q'', R'', S'')$. As distinct from the case of the E term, magnetic dipole (antisymmetric) transitions with $G_{12}^{(a)} \neq 0$ (as indicated above) are possible, leading to other expressions for the degree of depolarization: for the Q lines[152] $\rho \geqslant 0$, $2 \geqslant \rho_n \geqslant 0$, while for the P and R lines $\rho \geqslant \frac{3}{4}$, $2 \geqslant \rho_n \geqslant \frac{6}{7}$.

The intensities of the vibronic Raman lines were calculated for an X_3 molecule using the data obtained by numerically solving the vibronic E–e problem.[77] The features of this spectrum in the analogous multimode problem were considered elsewhere.[72,73a,157] For selection rules in the vibronic Raman spectra see also Mulazzi and Terzi.[158,159]

The analysis of experimental data on Raman spectra with the Jahn–Teller effect began before the theory was developed (see, for example, the review of Weinstock and Goodman[160] on the Raman spectra of hexafluorides of transition metals). Recent work[161,162] describes Raman scattering by Cu^{2+} impurities in CaO crystals at 4 K. Part of the spectrum is shown in Figure 4.21. The line at $\delta \sim 4\ \text{cm}^{-1}$ has been interpreted as the Q' line due to the $E \to A$ transition between tunneling levels in the E–e problem of the Cu^{2+} ion in the octahedral environment (cf. this δ value with almost the same value obtained from ESR measurements; see Table 4.2).

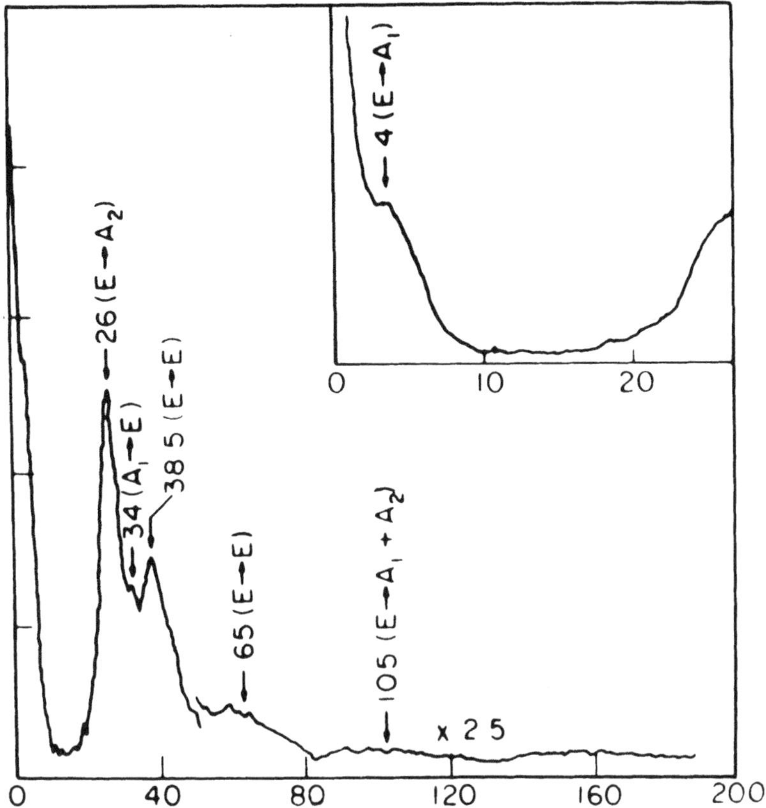

Figure 4.21. Vibronic E type Raman spectrum for the CaO : Cu^{2+} system at $T = 4.2$ K (frequencies in cm^{-1}). The top right corner shows that part of the spectrum containing the line of the $E \rightarrow A$ transition between tunneling levels at a frequency of 4 cm^{-1} on an enlarged scale (after Guha and Chase[161]).

Of special interest is the vibronic spectrum of XeF_6 (see Section 5.1).

Research has recently begun on the resonance Raman scattering of light (when the incident light falls in resonance with one possible transition of the system), taking into account vibronic effects. The spectra here depend not only on the initial and final states, but also on the intermediate state to which the system transforms upon absorbing a quantum of light, and from which it passes to the final state irradiating another quantum of light. In particular, if the initial and final states are nondegenerate but the intermediate state is degenerate, information about this state may be obtained from the resonance vibronic Raman spectrum. Computation of the expected spectrum in the case of an intermediate E term has been carried out on the basis of numerical solutions of the E–e problem with allowance for the linear and quadratic terms of vibronic interactions.[163,163a] Some of the results are illustrated in Figure 4.22. Similar results have been obtained for an intermediate T term[164] and for intermediate vibronically coupled states.[164a]

Vibronic effects in electron scattering have been considered.[165] For multimode aspects of the resonance vibronic Raman spectrum see Muramatsu et al.[166] For other publications on the subject see the Bibliographic Review.[14]

4.3. Electronic Spectra

This heading refers to spectra due to transitions between different electronic terms, as opposed to electronic transitions between different sheets of the same degenerate term, considered in Section 4.2. It is clear that if one or both of the combining terms are degenerate or pseudo-degenerate (in the sense that a third term is present, with which one of the combining terms is vibronically mixed), corresponding vibronic effects appear in the spectra. These effects may be conventionally divided into two kinds: (1) the fine structure of the electronic transition is determined by the vibronic structure of the terms (Chapter 3); (2) the band shape (the evelope of the fine structure lines) is determined by special vibronic effects via the frequency dependence of the transition intensities. We shall examine these two aspects of the problem in turn.

Vibronic Fine Structure of the Spectra in Gas Phase or in Matrixes. The fine structure of electronic absorption can be observed either in the gas phase (where relaxation broadening is small even at relatively high temperatures), or in crystals at low temperatures. Systematic research on vibronic effects in the gas phase was started recently. Some aspects of the problem were considered earlier[167,168] (see also Herzberg[169]). The first extensive investigations of this kind dealt with the fine structure of the absorption spectrum of s-triazine in gas and crystalline phases over the range of the forbidden transition ${}^1E'' \leftarrow \tilde{X}\,{}^1A_1''$.[170]

A study of the vibronic electronic emission spectra of cations $C_6F_6^+$, 1,3,5-$C_6F_3H_3^+$, 1,3,5-$C_6F_3D_3^+$, and 1,3,5-$C_6Cl_3H_3^+$ in the gas phase was carried out in a series of papers by Leach and coworkers.[171–176] Similar investigations of the fluorescence and emission spectra of these ions in the neon matrix were performed by Miller, Bondybey, and English.[177–181] These ions have a doubly degenerate electronic E ground term, and the fine structure of the allowed band ${}^2E \leftarrow {}^2A_2$ ($\tilde{X}\,{}^2E_1g \leftarrow \tilde{B}\,{}^2A_{2u}$ for $C_6F_6^+$, and

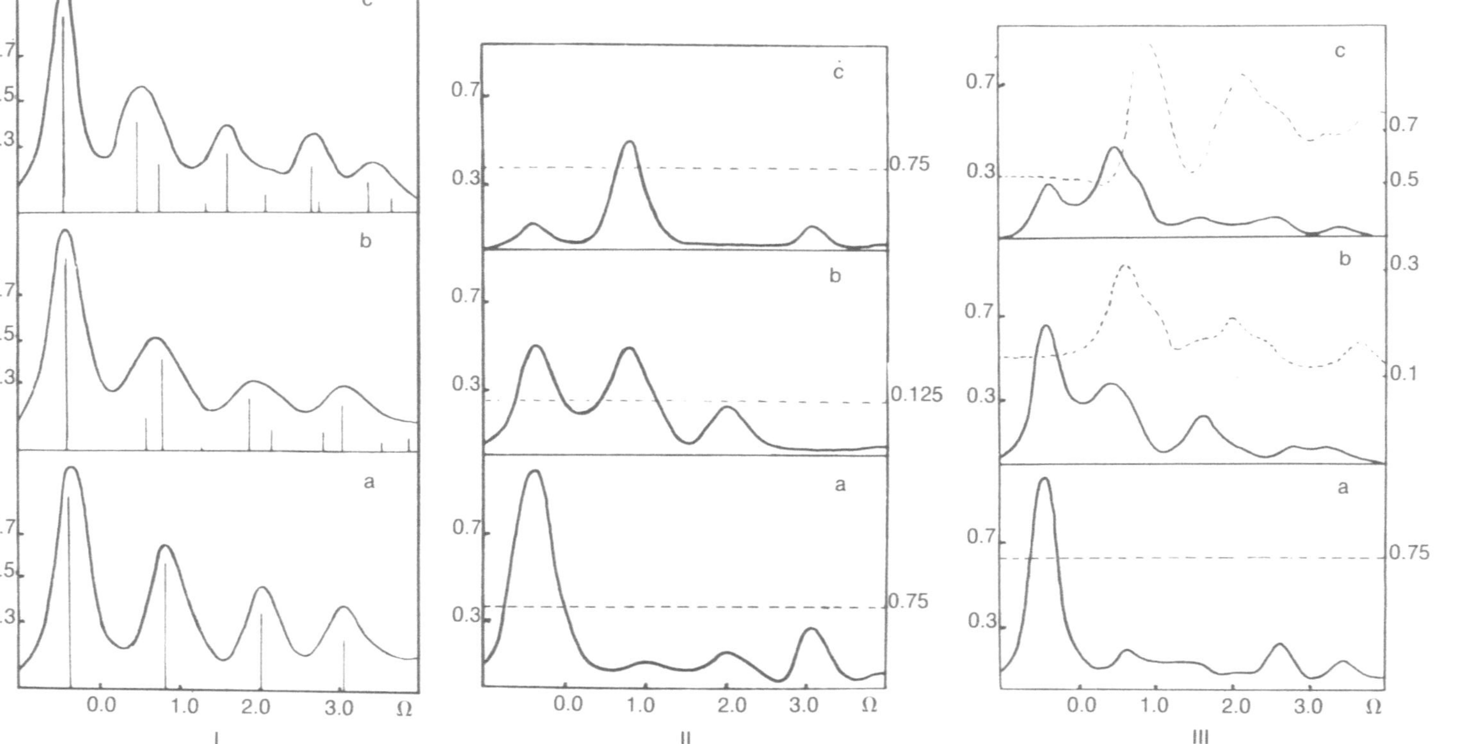

Figure 4.22. I. Calculated absorption spectra of a $A \to E$ transition with $\lambda_E = 1$ and $x = \lambda_E G_E / K_E = 0$ (a); $x = 0.05$ (b); $x = 0.10$ (c). II. Excitation profiles (solid lines) and depolarization ratios (broken lines, right-hand side figures) for the resonance Raman scattering via the excited E term (case Ia) as functions of the incident light frequency Ω (in ω_E units; the zero energy corresponds with the position of the degenerate electronic term): (a) fundamental tone; (b) first overtone; (c) second overtone. III. The same for the case Ic. It is seen that even weak quadratic coupling results in very strong depolarization dispersion.

$\tilde{X}^2E'' \leftarrow \tilde{B}^2A_2''$ for sym-$C_6X_3A_3^+$) is observed in the emission spectra. The weak vibronic coupling scheme in the ground state $E-e$ problem has been suggested as an interpretation. Assuming that the dimensionless vibronic constant λ_E is known, one can use the results of Section 3.1, in particular equation (3.1) and the vibronic energy level schemes in Figures 3.1 and 3.2, to estimate the expected line positions and their relative intensities in the relevant electronic transition.

The scheme of transitions in the band $\tilde{X}^2E'' \leftarrow \tilde{B}^2A_2''$ of the $C_6F_3H_3^+$ ion and estimates of the parameters for $\lambda_E = 0.125$ obtained by Cossart-Magos et al.[172] are presented in Figure 4.23, while in Figure 4.24 a part of

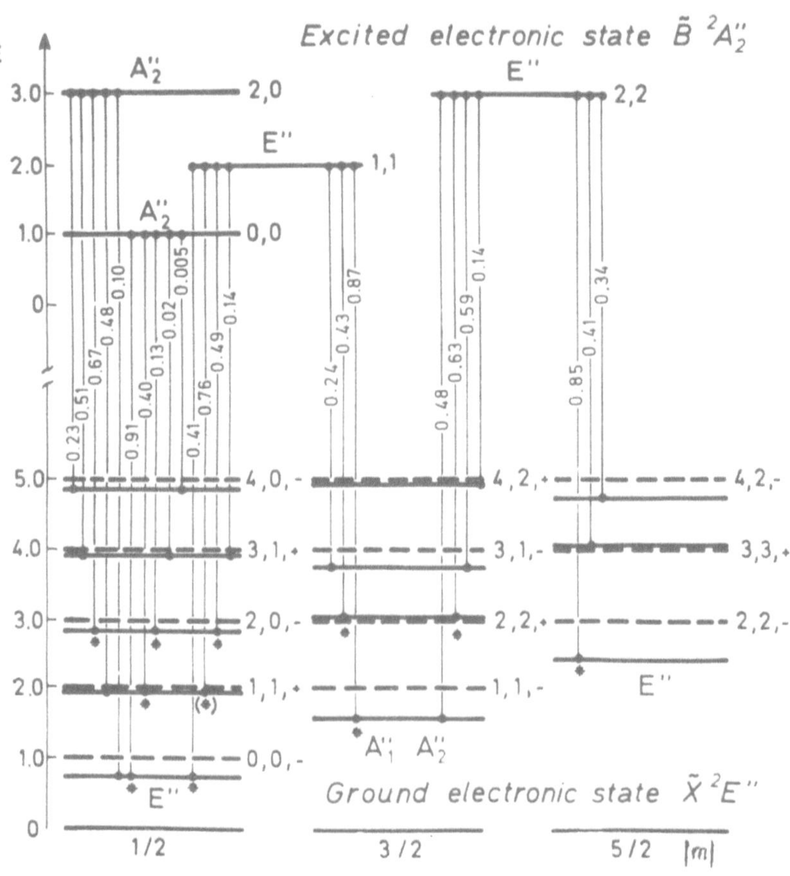

Figure 4.23. Scheme of electronic transitions in the $\tilde{X}^2E'' \leftarrow \tilde{B}^2A_2''$ band of the 1,3,5-$C_6F_3H_3^+$ ion and estimates of parameters for $\lambda_E = 0.125$ (after Cossart-Magos et al.[172]). Energies are in units of the unperturbed vibration frequency ω, their energy levels being shown by dashed lines. Identified transitions are marked by stars. Figures on vertical lines are relative intensities.

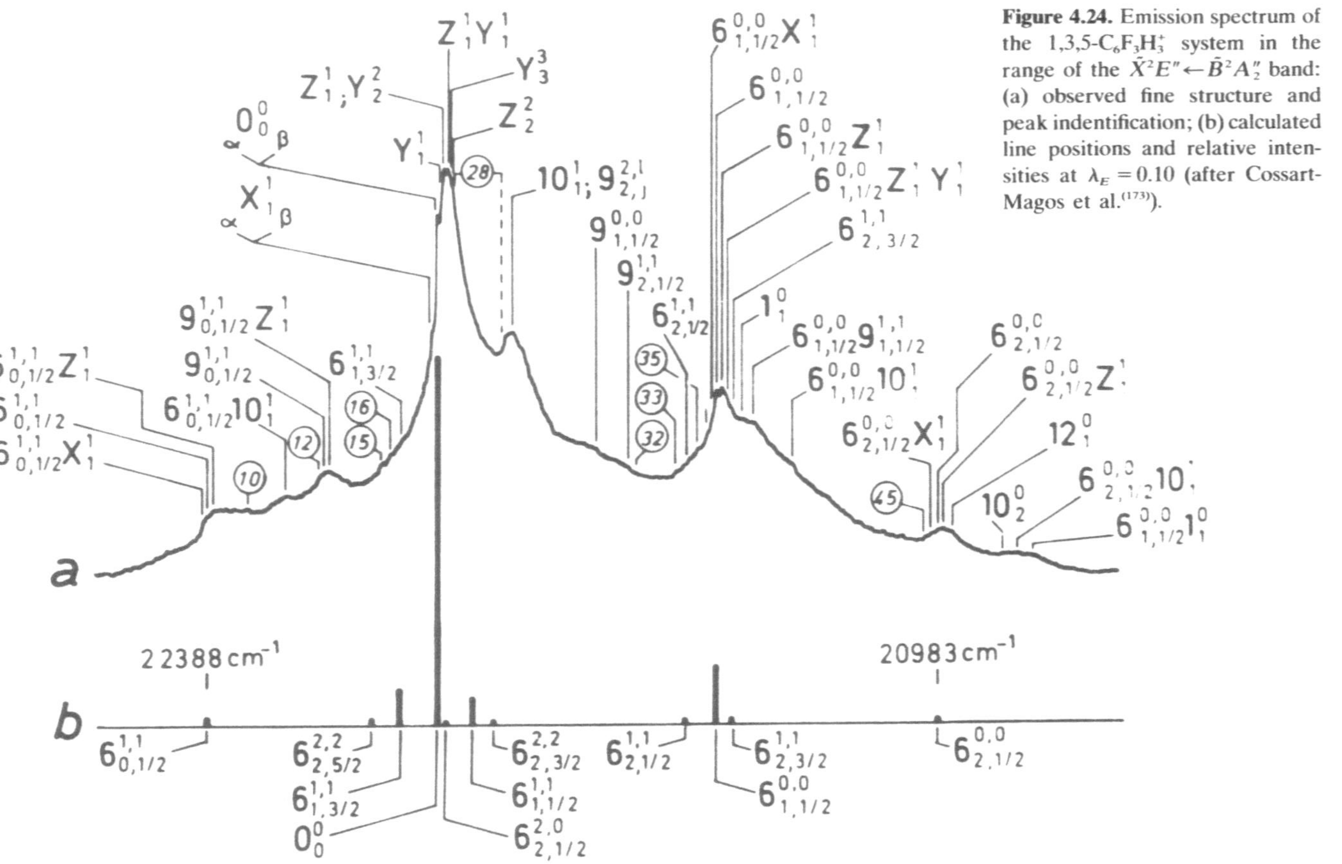

Figure 4.24. Emission spectrum of the 1,3,5-$C_6F_3H_3^+$ system in the range of the $\tilde{X}^2E'' \leftarrow \tilde{B}^2A_2''$ band: (a) observed fine structure and peak indentification; (b) calculated line positions and relative intensities at $\lambda_E = 0.10$ (after Cossart-Magos et al.[173]).

the observed spectra with assigned peaks is shown together with calculated relative intensities for $\lambda_E = 0.10$. In the D_{6h} and D_{3h} symmetry systems under consideration there are four JT-active E type modes, labeled below 6, 7, 8, and 9. The best resolution was attained for mode 6.

In order to estimate the value of λ_E for which the vibronic level scheme (Figures 3.1 and 4.23) conforms to the observed spectrum, the ratio of the first well-resolved intervals in the progression of this mode, $R_{1/2} = (0_0^0 - 6_{2,1/2}^{0,0})/(0_0^0 - 6_{1,1/2}^{0,0})$, is compared with the theoretically expected value. The latter have been calculated numerically for a large range of λ_E from 1 to 9. The results[173] are shown in Figure 4.25. It is seen that, except at the maximum and minimum points (for which $R_{1/2} = 2.062$ when $\lambda_E = 1.78$ and $R_{1/2} = 1.724$ when $\lambda_E = 0.20$, respectively), there are two possible values of λ_E for each value of $R_{1/2}$.

The choice between the two possible λ_E values for a given experimental value of $R_{1/2}$ may be made using additional considerations. For a known value of λ_E, the distribution of the intensities in the $6_{n,1/2}^{0,0}$ progression, as well as the fundamental (undisturbed) frequency $\omega_E^{(6)}$ (equal to the fre-

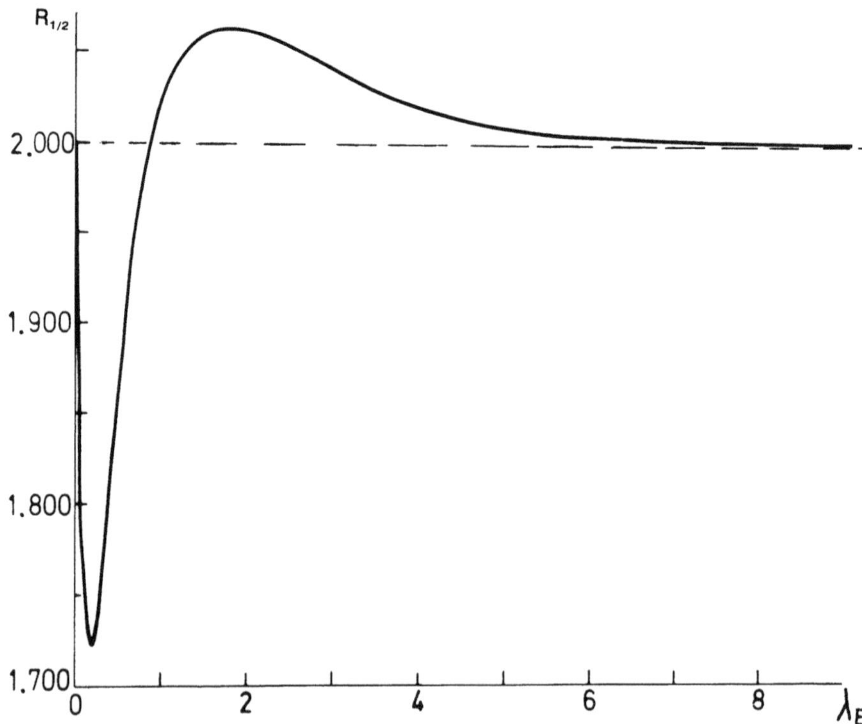

Figure 4.25. Ratio of the fequency intervals between the lines, $R_{1/2} = (0_0^0 - 6_{2,1/2}^{0,0})/(0_0^0 - 6_{1,1/2}^{0,0})$, in the mode 6 progression calculated as a function of the dimensionless vibronic constant λ_E (after Cossart-Magos et al.[173]).

quency of the isodynamic molecule having no such vibronic effects; for $C_6F_3H_3^+$ such an isodynamic molecule is $C_6F_3H_3$, which has no degeneracy in its ground state), can be estimated and compared with experimental values. The interaction between active modes 6, 7, 8, and 9 (their progressions are independent in the limiting case of weak vibronic coupling) and the quadratic terms of the vibronic interactions may also influence the assignment.[175,176] Table 4.4 presents the values of λ_E and $\omega_E^{(i)}$ obtained for three ions both allowing for and disregarding quadratic terms. For the latter case, the ratio G_E/K_E for the relative value of the quadratic vibronic constants is also given. For the $C_6Cl_3H_3^+$ ion, two alternative sets of parameters are shown.[176] One question remains unsolved. In Figure 4.24, the theoretically estimated intensities in the vibrational progression decrease more rapidly than those observed experimentally. This discrepancy has yet to be explained.

One interesting conclusion to be drawn from this group of papers on the JTE in electronic spectra is that the line positions are very sensitive to the magnitude of the vibronic constants, even in the limiting case of weak coupling. This conclusion confirms the theoretical results (see Figures 3.8 and 3.9). It follows that *neglect of vibronic effects in the fine structure of electronic spectra is generally unfounded, even if the vibronic coupling is known to be small.*

The fine structure of the spectrum in the general case of arbitrary vibronic coupling can be calculated using the energy levels and wave functions of a numerical solution to the vibronic problem (Section 3.4, Figures 3.1, 3.5–3.10). The first such computations were carried out by Longuet-Higgins, Öpik, Pryce, and Sack.[82] The vibronic structures of the $A \to E$ and $E \to A$ bands for different λ_E values were evaluated using data of the numerically solved linear E–e problem (Figure 4.26). These results are valid for $T = 0$, since the temperature dependence of the

Table 4.4. Vibronic Parameters for the E Term of Some sym-Halobenzenes Taking into Account both Linear and Quadratic Coupling with Mode $\omega_E^{(6)}$ (after Cossart-Magos and Leach[176])

	Linear coupling		Linear and quadratic coupling			
	$\lambda_E = E_{JT}/\hbar\omega_E^{(6)}$	$\omega_E^{(6)}$, cm^{-1}	G_E/K_E	λ_E	$\omega_E^{(6)}$, cm^{-1}	$\omega_E^{(6)}$(exp), cm^{-1}
1,3,5-$C_6F_3H_3^+$	0.100	484	0.006	0.118	478	481
1,3,5-$C_6F_3D_3^+$	0.100	463	0.006	0.118	458	464
1,3,5-$C_6Cl_3H_3^+$	0.45	395	0.005 0.0486	0.426 0.200	397 432	425

vibronic-level population was disregarded in the calculations. The vibronic spectrum is seen to form a group of equidistant lines having different intensities, their envelope characterizing the transition band shape. The equispacing of the lines is due to the condition $T = 0$, since transitions from the ground state are allowed only to equidistant levels; for $T \neq 0$ the lines are not equispaced.

The fine structure of the electronic spectra may also be affected by vibronic mixing of one of the combining terms with a third. An example of such a PJTE in the band structure is given by Cossart-Magos et al.,[182] where the fine structure of the $1\,^2A_2 \leftarrow 1\,^2B_2$ band in the benzil radical is

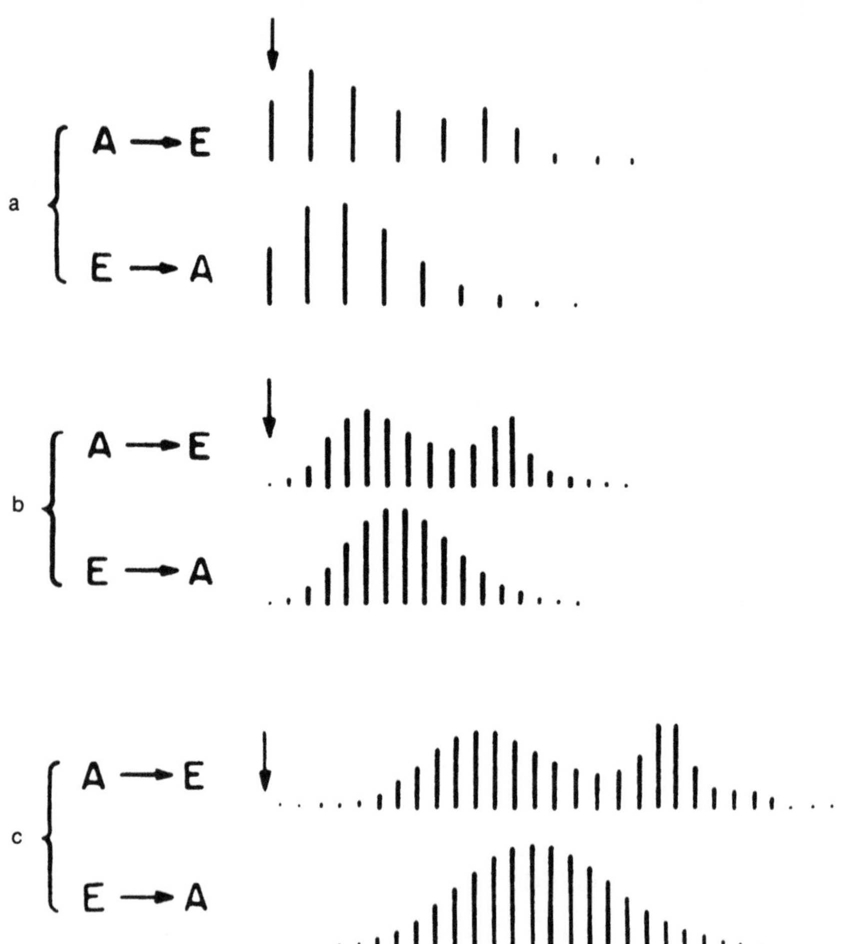

Figure 4.26. Positions and relative intensities of vibronic components and the band shape (envelope) for $A \rightarrow E$ and $E \rightarrow A$ transitions calculated at $T = 0$ and the following values of the dimensionless VC: (a) $\lambda_E = 2.5$; (b) $\lambda_E = 7.5$; (c) $\lambda_E = 15$. The position of the zero-phonon line is shown by an arrow (after Longuet-Higgins et al.[82]).

attributed to vibronic mixing of the $1\,^2A_2$ term with the near-lying $2\,^2B_2$ term by the appropriate $6b$ and $18b$ modes. Analysis of the spectrum allows us to estimate the energy gap between the $1\,^2A_2$ and $2\,^2B_2$ terms lying between 430 and 485 cm^{-1} in $C_6H_5CH_2$, and between 335 and 370 cm^{-1} in $C_6D_5CD_2$. The nondiagonal VC of coupling to the active modes indicated were also estimated. Similar, but weaker vibronic mixing of the $^3n\pi^*$ and $^3\pi\pi^*$ states in dimethylbenzaldehydes is evident in the phosphorescence $\pi\pi^*$ spectra.[183] Vibronic $^2B_{3g}$–$^2B_{3u}$ mixing in butatriene was considered elsewhere.[183a] For other examples see the Bibliographic Review.[14]

Structure of the Zero-Phonon Line. Along with gas phase spectra or spectra of molecules in neutral crystal matrixes, the fine structure of the vibronic spectra can be observed in a large number of classes of crystals at low temperatures, including transition metal and rare-earth coordination compounds, paramagnetic and other impurity centers in crystals. As with gas phase systems, if one or both of the combining terms are degenerate or pseudodegenerate, the vibronic structure of these terms (Chapter 3) determines the fine structure of the spectra. In many cases, the analysis of the vibronic structure of the spectra is similar to that given above for gas phase systems. However, there may be complications due to the multimode nature of the JT crystal problem (see Section 2.4). These difficulties can be avoided only if there are almost isolated (weakly interacting) molecular groups, which make it possible to apply the cluster model.

Experimental information about the fine structure of the spectra of JT crystal systems at low temperatures is in general very poor. JT impurity centers in crystals have been well studied, but for them the cluster model is inadequate. For example, an attempt to use the numerical results obtained for the vibronic structure of the T term energy spectrum of an "ideal" system[90] in order to explain the origin of the fine structure of the $A \rightarrow T$ absorption band in Ni^{2+} : CdS, has not been successful due to the multimode nature of the JTE in this system.

The most studied of the absorption spectra in its fine structure of crystal systems are the zero-phonon lines,[13,59] i.e., lines of electronic transitions having no changes of vibrational states (cf. the 0_0^0 line in the aforementioned spectra of benzene derivatives, Figure 4.24). Usually (even at not very low temperatures) such a line forms an acute peak of considerable intensity against the background of the wide phonon band. This is due to the fact that, in the absence of vibronic interactions, the vibrational levels in the harmonic approximation form an equispaced set (Figure 4.27), and if the difference in the vibrational frequencies of the ground and excited states (the frequency effect $\Delta\omega$) is neglected, all the transitions 0–0, 1–1, 2–2, etc., have the same frequency and form one

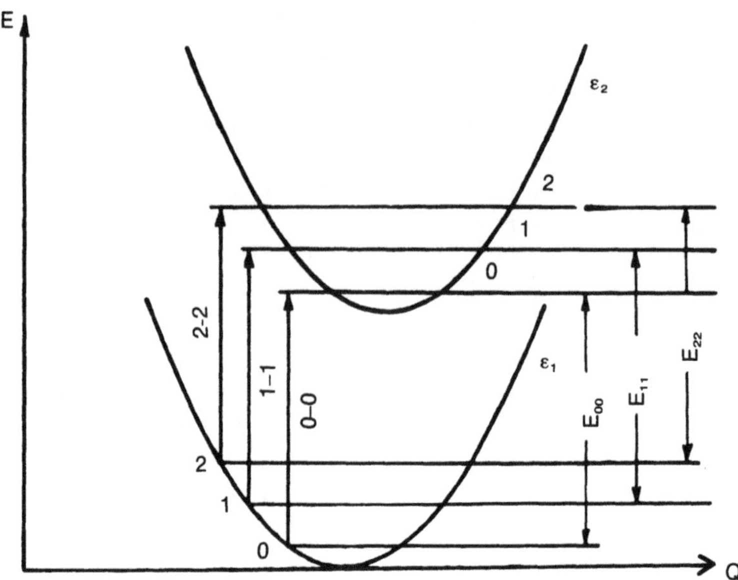

Figure 4.27. Zero-phonon 0–0, 1–1, 2–2,... transitions in a one-mode problem. In the absence of the "frequency effect" the frequencies of all these transitions coincide, giving rise to one intense line.

narrow intense line. The difference in the frequencies of these transitions $\Delta\omega \neq 0$ results in broadening of the line with temperature. One important feature of the zero-phonon line is the coincidence of its position in absorption and emission. Transitions involving vibrational quanta form broad crystal bands.

Taking into account vibronic effects at $T \neq 0$, the zero-phonon line acquires a complicated structure[59] (at $T = 0$ the line does not change, since the nondegenerate ground state is not split by VI). For $A \rightarrow E$ transitions in the case of weak vibronic coupling in the E term, the structure of the zero-phonon line can be presented as a superposition of two equidistant spectra separated by an interval equal to $\hbar\Delta\Omega = 2E_{\mathrm{JT}}$ (cf. equation (3.1)), and with an intensity distribution determined by the Boltzmann population of the energy levels (Figure 4.28a).[184] When the frequency effect $\Delta\omega_E \neq 0$ is taken into account this interval changes: $\hbar\Delta\Omega_{\pm} = |\hbar\Delta\omega_E \pm 2E_{\mathrm{JT}}|$. The resulting spectrum is illustrated in Figure 4.28b for the following numerical values: $\Delta\omega_E/\omega_E = -0.98 \cdot 10^{-3}$, $\lambda_E = 0.25 \cdot 10^{-2}$, $\beta_E = \hbar\omega_E/kT = 0.26$.[184]

In the case of an $A \rightarrow T$ transition, when the coupling to the E vibrations in the T term is predominant (T–e problem), the zero-phonon line is only shifted but has no additional structure. If, however, coupling to the T_2 vibrations is predominant (T–t_2 problem), the zero-phonon line

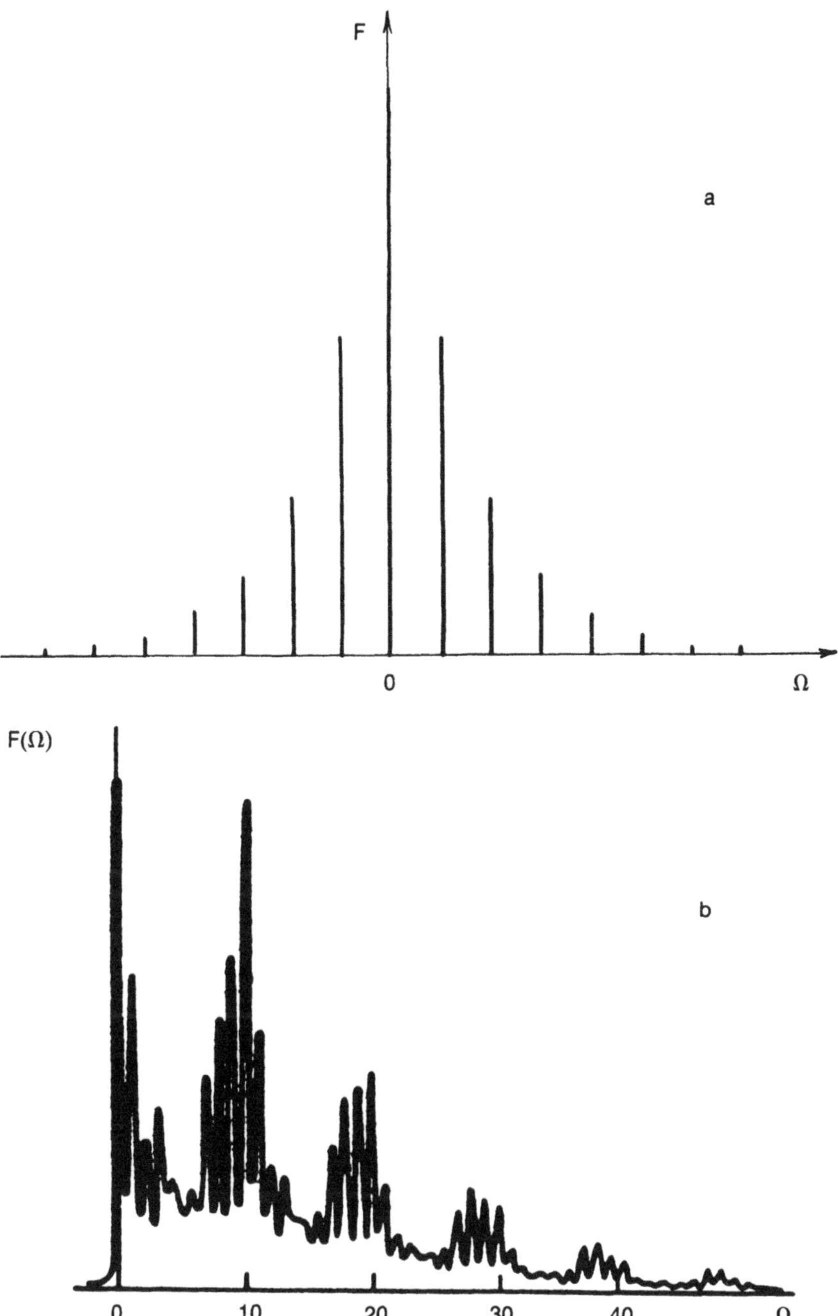

Figure 4.28. Fine structure of the zero-phonon line of the $A \rightarrow E$ transition in the case of weak vibronic coupling: (a) in the absence of the "frequency effect"; (b) with a strong frequency effect: $\Delta\omega_E/\omega_E = 0.98 \cdot 10^{-2}$, $\hbar\omega_E/kT = 0.26$, and $\lambda_E = 0.25 \cdot 10^{-2}$ (calculated numerically[184]).

acquires the structure of a temperature-dependent equispaced (with step equal to $\hbar\Delta\Omega = \frac{3}{4}E_{JT}$) spectrum which may be of three kinds: a low-temperature spectrum, in which the 0–0 line predominates (Figure 4.29a), intermediate-temperature spectrum (Figure 4.29b), and high-temperature spectrum, in which the line shift by $\hbar\Delta\Omega = \frac{3}{4}E_{JT}$ from the $0\rightarrow 0$ line is predominant (Figure 4.29c).[184] These results are concerned with cluster-type JT centers in crystals when multimode interactions are neglected. In the multimode problem, the zero-phonon line becomes rather complicated following the deductions of Section 2.4 and Chapter 3.

The effect of external perturbations on the zero-phonon line properties, in particular on its splitting under stress, may be used as an interesting

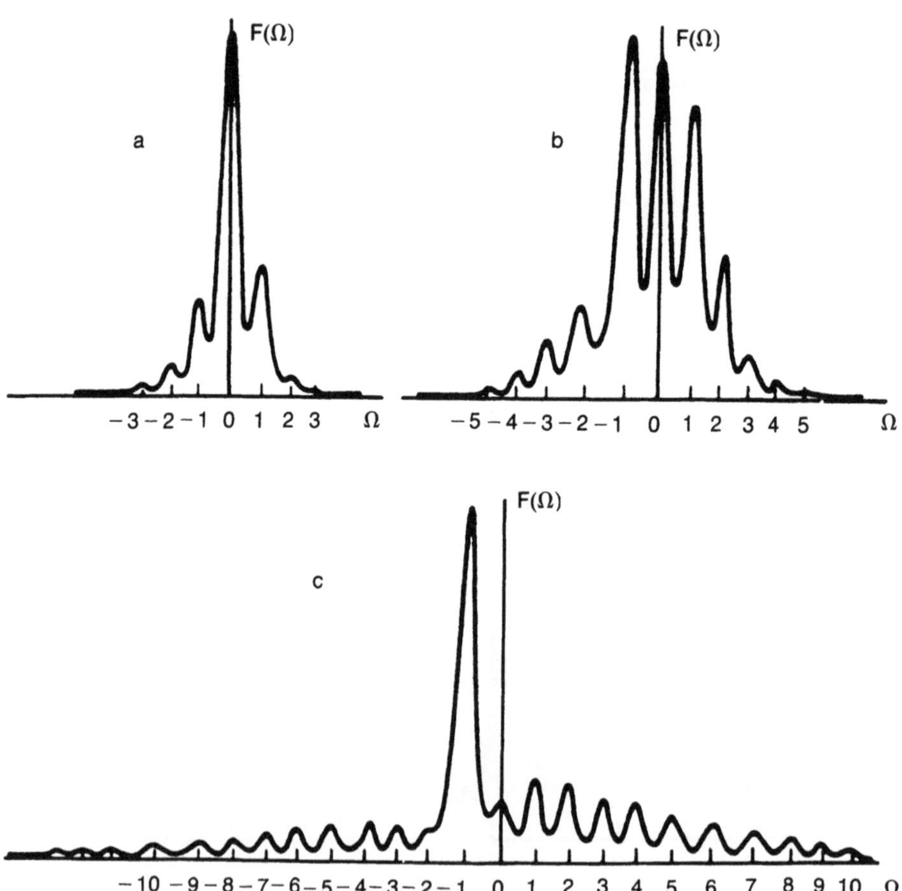

Figure 4.29. Vibronic fine structure of the zero-phonon line of the electronic $A \rightarrow T$ transition at different temperatures: (a) $\hbar\omega_T/kT = 2.74$ (low temperatures); (b) $\hbar\omega_T/kT = 1.45$ (intermediate temperatures); (c) $\hbar\omega_T/kT = 0.37$ (high temperatures).

source of information about vibronic interactions in the system. By way of example, consider the electronic ${}^1A \rightarrow {}^1T$ transition in a cubic symmetry system having an inversion center. Under the influence of a tetragonal uniaxial stress, the ${}^1T$ term splits into ${}^1A$ and ${}^1E$ terms of the D_{4h} group. Without vibronic interactions, the splitting is $\hbar\Omega_E = \frac{3}{2}P_E e_\theta$, where e_θ is the component of the deformation tensor and P_E is the constant of electron–deformation interaction (cf. Section 4.1 and equation (4.14)).

This splitting can be found from the polarized spectra. Transitions from the A term are allowed to the A term in the parallel polarization and to the E term in the perpendicular polarization. In the linear vibronic approximation the center of gravity of the band (the first moment of the band; see below, equation (4.60)), denoted by $\langle\Omega\rangle$, is independent of the value of the vibronic constants (see below). It follows that the difference in the moment shifts for the parallel $\Delta\langle\Omega\rangle_\parallel$ and perpendicular $\Delta\langle\Omega\rangle_\perp$ polarizations equals the total shift under the uniaxial stress:

$$\hbar\Delta\Omega'_E = \hbar\,|\Delta\langle\Omega\rangle_\parallel - \Delta\langle\Omega\rangle_\perp| = \tfrac{3}{2}|e_\theta P_E| \tag{4.48}$$

In the absence of vibronic coupling the transition to each of the A and E terms gives its own zero-phonon line, and under stress the splitting $\Delta\Omega'_E$ is seen in the spectrum. Taking account of vibronic interactions, the electron–deformation interaction is reduced (Section 3.3) and splitting of the zero-phonon line becomes $K_T(E)$ times smaller:

$$\Delta\Omega_E = K_T(E)\Delta\Omega'_E \tag{4.49}$$

where $K_T(E)$ is the vibronic reduction factor (equations (3.38), (3.43)). Thus, if we know the shifts of the first moments of the band (which are not reduced) in different polarizations (the polarization dichroism), and hence $\Delta\Omega'_E$, as well as the splitting of the zero-phonon line, then the vibronic reduction factor $K_T(E)$ can be determined.

Similar relations for uniaxial stress along the trigonal axis

$$\begin{aligned} \Delta\Omega_T &= K_T(T_2)\Delta\Omega'_T \\ \hbar\Delta\Omega'_T &= \hbar\,|\Delta\langle\Omega\rangle_\parallel - \Delta\langle\Omega\rangle_\perp| = \sqrt{3}\,|e_T P_T| \end{aligned} \tag{4.50}$$

allow us to determine the vibronic reduction factor $K_T(T_2)$ from electronic spectral data. Further information about vibronic reduction in optical spectra appear elsewhere.(142,59,14)

In the case of strong vibronic coupling, transition to upper tunneling levels falls near the main 0–0 transition and results in an apparent splitting of the zero-phonon line with a structure repeating the structure of the group of near-lying tunneling levels. Some of these transitions are forbidden by selection rules, but there may be additional perturbations in the crystal (or such perturbations can be produced from outside, e.g., by uniaxial stress), which makes them allowed. The first experiments in which tunneling splitting was observed in the zero-phonon line of $A \to E$ transitions were performed by Caplianskii and Prjevusskii.[185,186] The splitting δ in $Eu^{2+}:CaF_2$, $Eu^{2+}:SrF_2$, $Sm^{2+}:CaF_2$, and $Sm^{2+}:SrF_2$ are (in cm^{-1}) 15.3, 5.5, 27, and 26, respectively. These experiments have been explained in terms of tunneling splitting by Chase.[112,187]

Tunneling splitting of the zero-phonon line of the ${}^5T_{2g} \to {}^5E_g$ magnetic dipole transition in $Fe^{2+}:MgO$, observed at $T = 1.5$ K under uniaxial stress along [001], is illustrated in Figure 4.30.[188] It is seen that by raising the stress, a new line corresponding to the forbidden transition to the tunneling A level (of the excited E term) grows up on the high-frequency side of the 0–0 line. The observed tunneling splitting in the ${}^5E_g$ excited state of the $Fe^{2+}:MgO$ system is $\delta \approx 14\ cm^{-1}$, while the reduced electron–deformation interaction constant equals $qP_E \approx 21 \cdot 10^3\ cm^{-1}$, the ratio of second-order to first-order reduction factors being $|r/q| \simeq 1.2$.

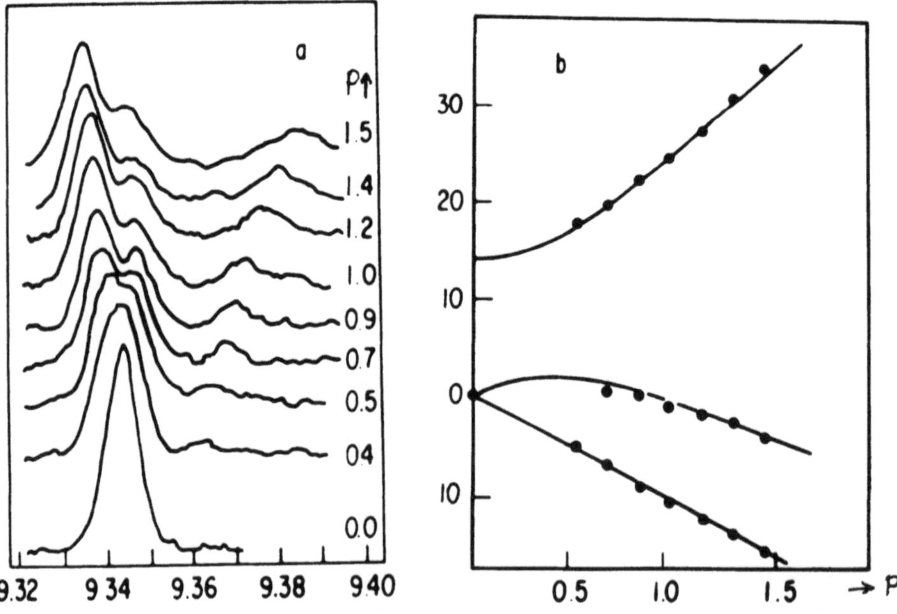

Figure 4.30. Splitting of the zero-phonon line of the ${}^5T_{2g} \to {}^5E_g$ transition in $MgO:Fe^{2+}$ under uniaxial stress P along [001]; (a) line shape at $T = 1.5$ K for different stress values (shown on the right side in 10^8 Pa) — the frequency in $10^3\ cm^{-1}$ is plotted on the abscissa; (b) absorption maximum relative positions (points) and theoretical predictions (solid lines). Splitting at zero stress (tunneling splitting) equals $\delta \approx 14\ cm^{-1}$.

Direct observation of tunneling splitting of the zero-phonon GR1 absorption line for a neutral vacancy in diamond gives $\delta \approx 64$ cm^{-1}.[189,190] Splitting of the zero-phonon line of the $^4A_2 \rightarrow {}^4T_2$ transition in V^{2+} : MgO (~ 40 cm^{-1}) with tetragonal minima in the T_2 state[191,142] proved to be the reduced spin–orbit splitting of the T_2 term.[192] The origin of this splitting has been the subject of lively discussion.[5,6,75,142,191–192a]

Band Shape of Electronic Absorption. Semiclassical Approach. In many cases, especially for condensed matter (crystals, liquids, dense gases) and at not very low temperatures, the absorption and luminescence spectra have the form of bands in which the fine structure, as a rule, is not resolved. Even when the zero-phonon line can be observed, the remaining (phonon) spectrum consists of a continuous band. In these cases the vibronic effects become evident in the band shapes. An example of the absorption band shape formation with vibronic interactions has been given above for the $A \rightarrow E$ transition (Figure 4.26). Some general features of this effect are presented below.

The term *band shape* means an envelope of elementary transitions between vibronic states each of which has a specific width so that the vibronic lines merge into a continuous band. Suppose the light absorption coefficient $K_{12}(\Omega)$ due to the electronic $1 \rightarrow 2$ transition ($K_{12}(\Omega)$ is determined by the relation $I = I_0 \exp[-K_{12}(\Omega) \cdot l]$, where I_0 and I are the intensities of the incident and transmitted light, respectively, and l is the absorption layer thickness) has the form[193,59]

$$K_{12}(\Omega) = (4\pi^2 N\Omega/3C)F_{12}(\Omega) \tag{4.51}$$

where

$$F_{12}(\Omega) = \sum_{\kappa,\kappa'} \rho_{1\kappa} |\langle 1\kappa | \boldsymbol{M} | 2\kappa' \rangle|^2 \delta(E_{2\kappa'} - E_{1\kappa} - \hbar\Omega) \tag{4.52}$$

is the form function of the band, N is the number of absorbing centers per unit volume, $\boldsymbol{M}_{1\kappa 2\kappa'} = \langle 1\kappa | \boldsymbol{M} | 2\kappa' \rangle$ is the matrix element of the transition moment (equation (4.33)), κ and κ' label the vibronic ground and excited states, respectively, $\rho_{1\kappa}$ is the probability that the ground energy state $E_{1\kappa}$ is populated, taking into account the Boltzmann temperature dependence,

$$\rho_{1\kappa} = \exp(-E_{1\kappa}/kT)/Z_1 \tag{4.53}$$

$Z_1 = \Sigma_{\kappa''} \exp(-E_{1\kappa''}/kT)$ is the sum of states of the ground term, and the δ-function takes account of the energy conservation law, due to which the transition $1\kappa \rightarrow 2\kappa'$ is possible only if $\hbar\Omega = E_{2\kappa'} - E_{1\kappa}$.

A simple relationship links the form function for the emission band due to the $2 \to 1$ transition $F_{21}(\Omega)$ with the above function $F_{12}(\Omega)$ for the corresponding absorption[59]:

$$F_{21}(\Omega) = (Z_2/Z_1)\exp(-\hbar\Omega/kT)F_{12}(\Omega) \tag{4.54}$$

where Z_2 is the sum of states of the excited term.

The main features of the form function can be determined by the semiclassical approximation. In this approximation it is assumed that during the electronic transition the nuclei remain fixed at their positions Q of the initial state (in accordance with the Franck–Condon principle), and therefore the energies $E_{1\kappa}$ and $E_{2\kappa'}$ can be taken approximately equal to the respective values of the AP at the points $Q: E_{1\kappa} = \varepsilon_1(Q)$ and $E_{2\kappa'} = \varepsilon_2(Q)$. Then, passing from summation to integration in equation (4.52), we obtain

$$F_{12}(\Omega) = \int \rho_1(Q)|\boldsymbol{M}_{12}(Q)|^2 \delta[\varepsilon_2(Q) - \varepsilon_1(Q) - \hbar\Omega]dQ \tag{4.55}$$

where

$$\boldsymbol{M}_{12}(Q) = \int \Psi_1^*(r, Q)\boldsymbol{M}\Psi_2(r, Q)d\tau_r \tag{4.56}$$

For nondegenerate states,

$$\varepsilon_1(Q) = \tfrac{1}{2}KQ^2, \qquad \varepsilon_2(Q) = \hbar\Omega_0 + \tfrac{1}{2}K(Q - Q_0)^2$$

where Q is the totally symmetric coordinate, Q_0 is the shift of the minimum position in the excited state (the other coordinates remaining unshifted), and Ω_0 is the frequency of the zero-phonon line. Substituting these expressions into equations (4.55) and (4.56) and neglecting the weak dependence of $\boldsymbol{M}_{12}$ on Q, we obtain

$$F_{12}(\Omega) = |\boldsymbol{M}_{12}^{(0)}|^2 \exp[-\hbar^2(\Omega - \Omega_0)^2/2kTQ_0^2K]/Q_0(2\pi kTK)^{1/2} \tag{4.57}$$

It follows that the band has a Gaussian form.

Assume now that the excited state is degenerate. Then, besides the totally symmetric coordinates, the nontotally symmetric JT active coordinates are displaced. For the $A \to E$ transition in the linear vibronic coupling approximation with polar E coordinates ρ and ϕ (see equations (2.4) and (2.7)) we have

$$\begin{aligned} \varepsilon_1(Q, \rho, \phi) &= \tfrac{1}{2}K_A Q^2 + \tfrac{1}{2}K_E\rho^2 \\ \varepsilon_2(Q, \rho, \phi) &= \hbar\Omega_0 + \tfrac{1}{2}K_A(Q - Q_0)^2 + \tfrac{1}{2}K_E\rho^2 \pm |F_E|\rho \end{aligned} \tag{4.58}$$

When these expressions are substituted into equation (4.55) with $Q_0 = 0$, for simplicity,

$$F_{12}(\Omega) = \frac{|M_{12}^{(0)}|^2 \hbar |\Omega - \Omega_0|}{4kTE_{JT}} \exp\left[-\frac{\hbar^2(\Omega - \Omega_0)^2}{4kTE_{JT}} \right] \tag{4.59}$$

The relationship (4.59) is presented graphically in Figure 4.31a. It has a symmetric shape with two humps and a dip at $\Omega = \Omega_0$. *This band shape can be interpreted as being due to* JT *splitting of the nonvibronic band.* The splitting (distance between the two maxima) equals $(8E_{JT}kT)^{1/2}$.

The first calculations of the electronic absorption band shapes of JT systems in a semiclassical approximation were conduced by O'Brien,(194) Moran,(195) and Toyozawa and Inoue.(196) Since then, studies on this subject have advanced considerably, improvements being introduced by several authors(197–215) (see the reviews(59,210,13)). In particular, if one takes into account that $Q_0 \neq 0$ (i.e., the totally symmetric coordinate contribution to the band shape is nonzero), the "acute elements" of the curve in Figure 4.31a are smoothed and the curve itself assumes the form given in Figure

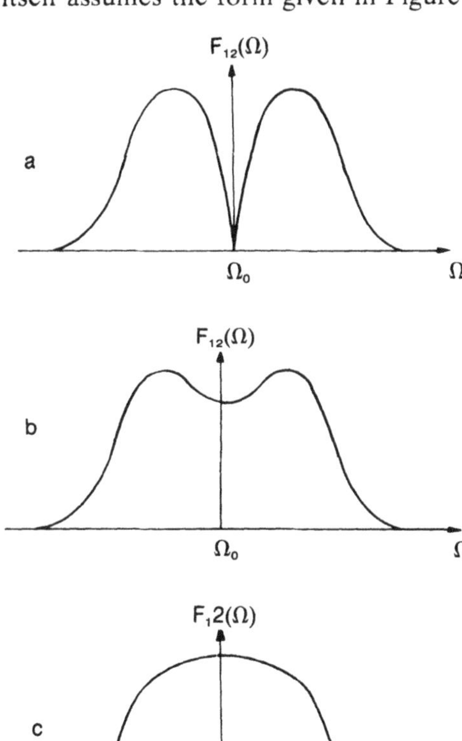

Figure 4.31. Band shape of the electronic $A \to E$ transition calculated by the semiclassical approximation taking into account linear coupling with E and A vibrations, $X_E = F_E^2 \coth(\hbar\omega_E/2kT)$, $X_A = F_A^2 \coth(\hbar\omega_A/2kT)$: (a) $X_A = 0$; (b) $X_A \neq 0$, $X_A < X_E$; (c) $X_A > X_E$.

4.31b. If the totally symmetric vibrations predominate, the dip in the curve is completely filled and so disappears (Figure 4.31c). For comparison, the temperature dependence of the $A \rightarrow E$ transition band as determined by numerical solution[198] is given in Figure 4.32. It is noteworthy that for very large (almost improbable) values of the dimensionless VC more humps occur at the high-frequency edge of the band shape of the $A \rightarrow E$ transition[198–199a] (Figure 4.33).

The corresponding band shapes for $T \neq 0$ are shown in Figure 4.34 for the $E \rightarrow A$ transitions. They differ from the results based on numerical solutions of the E–e problem presented in Figure 4.26, where $T = 0$ is assumed (the semiclassical calculations for $T = 0$ give the same type of band shapes as the numerical data). For $E \rightarrow E$ transitions the band shape becomes more complicated (Figure 4.35).

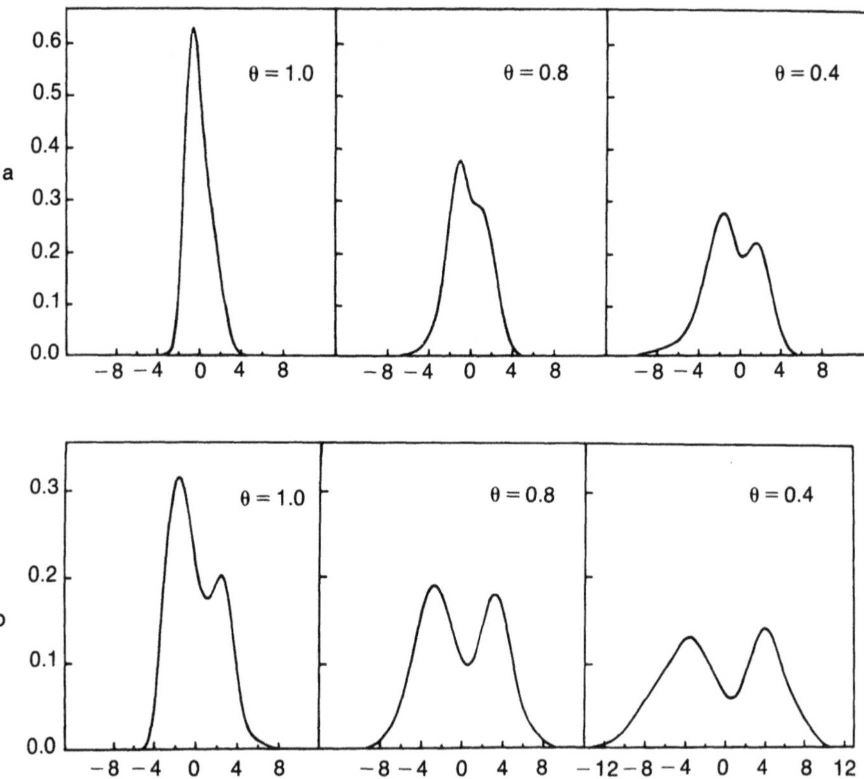

Figure 4.32. Temperature dependence of the absorption band shape of the $A \rightarrow E$ transition obtained by numerical calculations of the linear $E \rightarrow e$ problem taking into account coupling with E and A vibrations at $\lambda_A = 0.5$ and (a) $\lambda_E = 0.5$; (b) $\lambda_E = 2.5$. The frequency $\Omega - \Omega_0$ is given in ω_E units; $\theta = \hbar\omega_E/kT$. In the case of strong vibronic coupling and increasing temperature, the absorption curve approximates the semiclassical one (Figure 4.31) (after Muramatsu and Sakamoto[198]).

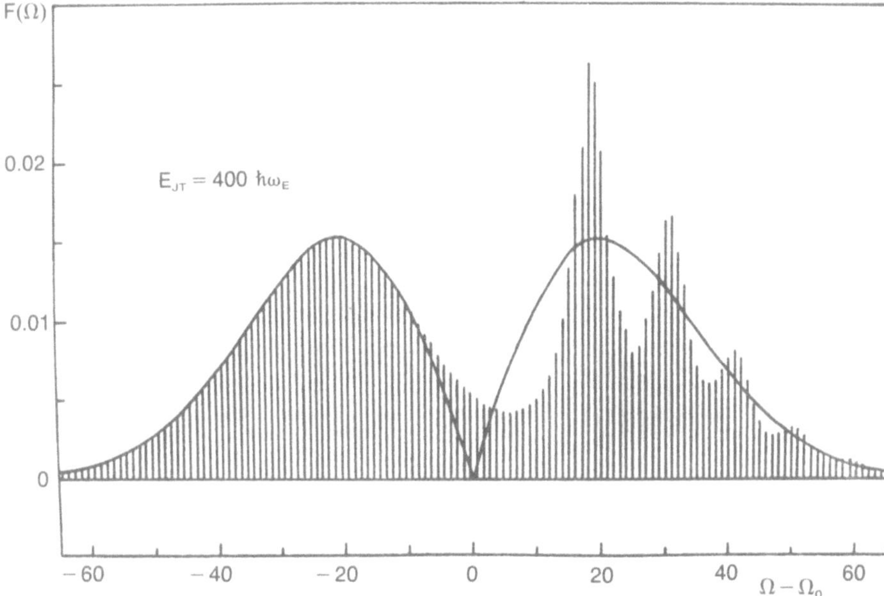

Figure 4.33. Absorption band shape of the $A \rightarrow E$ transition in the case of very strong vibronic coupling $\lambda_E = 400$ (the frequency $\Omega - \Omega_0$ is given in ω_E units). Numerically calculated positions and relative intensities of the elementary transitions are shown by vertical lines, while the solid line correponds to the semiclassical approximation (Figure 4.31) (after Loorits, and Köppel et al.[199a]).

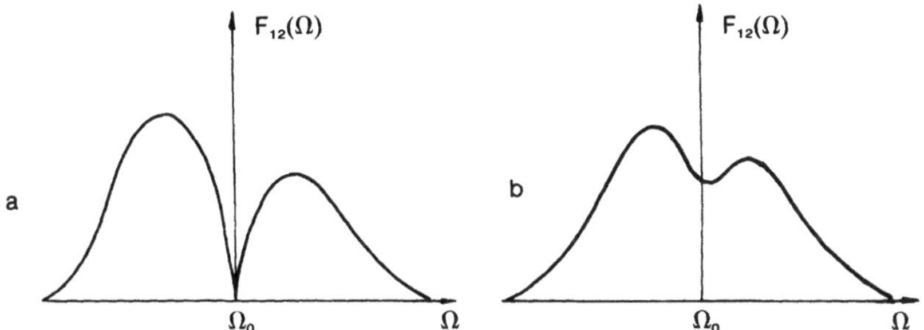

Figure 4.34. Absorption band shape of the $E \rightarrow A$ transition under the same conditions as in Figure 4.31 for: (a) $X_A = 0$; (b) $X_A \neq 0$.

Similar results were obtained for transitions to orbitally triplet electronic terms. The band shape in these cases primarily depends on the relationship between the vibronic constants F_E, F_T, and F_A which characterize vibronic coupling of the T term states to E, T_2, and A vibrations, respectively.

If coupling to E vibrations is predominant (T–e problem), no splitting of the $A \rightarrow T$ band occurs, although the AP of the T term is split (Figure 2.7). This case illustrates how carefully the visual pictures should be used in the analysis of complicated phenomena. In general, it can be said that for absorption transitions from nondegenerate to degenerate terms, the band does not split if the point of degeneracy on the AP is a point of actual crossing of the surfaces, as in the T–e problem (Figure 2.7). This is in contrast to the case when this point is a branching point of the surface, as in the E–e problem, for which the band splits.

If the coupling to T_2 vibration predominates (T–t_2 problem), the $A \to T$ absorption curve has three humps, the band is split into three components (Figure 4.36), but the interaction with the totally symmetric A_1 vibrations smoothes the curve. The latter becomes also asymmetric when temperature factors are taken into account (Figure 4.37).

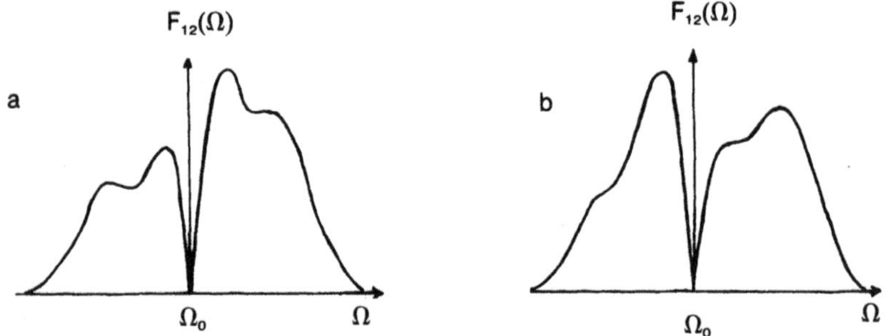

Figure 4.35. Absorption band shape of the $E \to E$ transition calculated in the semiclassical approximation ($F_E^{(1)}$ and $F_E^{(2)}$ are the VC of the ground and excited states, respectively): (a) $F_E^{(1)} > F_E^{(2)}$; (b) $F_E^{(1)} < F_E^{(2)}$.

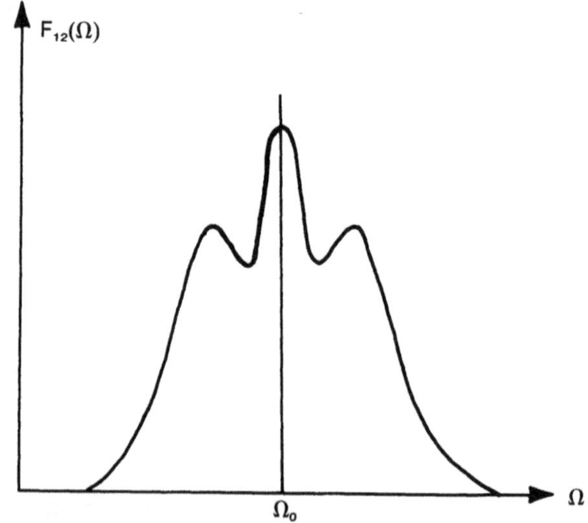

Figure 4.36. Absorption band shape of the electronic $A \to T$ transition calculated in the semiclassical approximation of the T–t_2 problem.

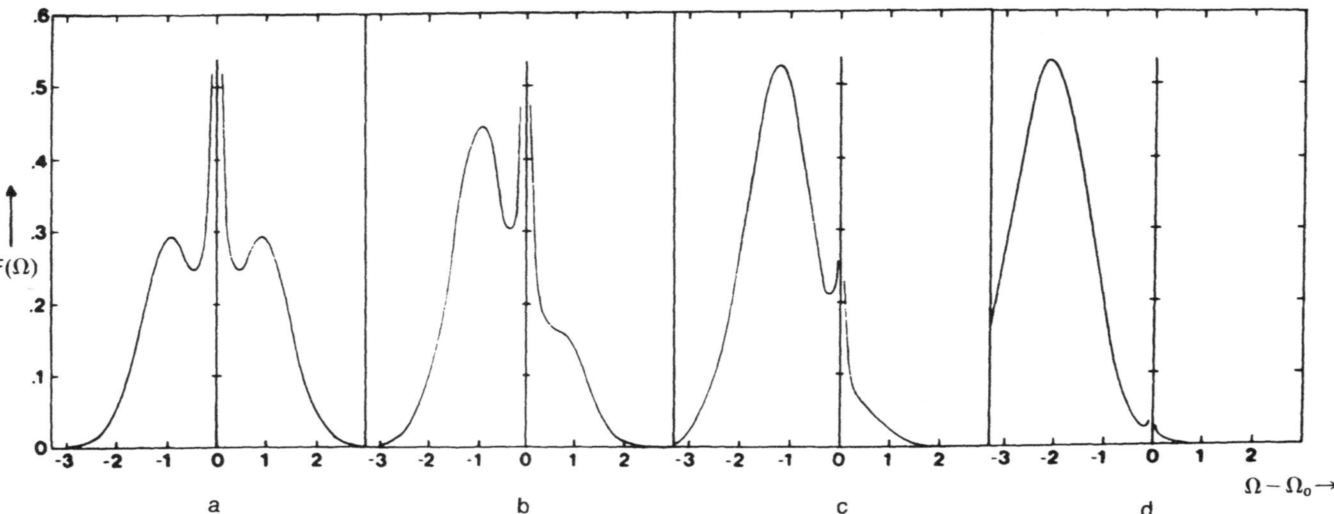

Figure 4.37. Temperature dependence of the band shape of the $T \rightarrow A$ transition calculated in the semiclassical approximation of the T–t_2 problem: (a) $X = Q_0^T/\sqrt{kT} = 0$; (b) $X = 0.2$; (c) $X = 0.5$; (d) $X = 1.0$ (Q_0^T is determined by equation (2.25)) (after Baci et al.[213]). At high temperatures the curve is similar to that of the $A \rightarrow T$ transition (Figure 4.36), whereas at lower temperatures it becomes asymmetrical, as predicted by numerical calculations (Figure 4.38).

Numerical results for the $A \rightarrow T$ band shape were also obtained from the exact solution of the T–t_2 problem[201] discussed in Section 3.4; examples of the resulting absorption and emission curves are given in Figure 4.38. Absorption and emission band shapes for a Renner-type system are given elsewhere [202,203] and in Table 5.1.

In the general case of the T–$(e + t_2)$ problem, the results depend on the aforementioned relationship between the vibronic constants. Illustrations for several cases obtained by means of numerical evaluation of the semiclassical expressions[200] are given in Figure 4.39. Also important are the cases mentioned above when there is mixing of one of the combining terms with a third. In particular, forbidden transitions become allowed, their band shapes being completely determined by the vibronic effects. *The intensities of these normally forbidden but vibronically allowed transitions have a particular temperature dependence: they increase with temperature,* in contrast to the nonvibronic transitions which decrease with temperature. This feature easily distinguishes vibronically allowed transitions from the others.

Some of the other important studies of band shapes, obtained mainly using the semiclassical approximation, are listed below. The ${}^1A \rightarrow {}^2E$ transition was considered for trigonal systems with a spin–orbit interaction.[204] The $A \rightarrow E$ transition was investigated for tetragonal systems, where the E–$(b_1 + b_2)$ problem is realized in the excited state.[205] The absorption band shapes in the two-mode E–$(e + e)$ problem were considered by Haller et al.[205a] The case of strong spin–orbit interaction in the T term, resulting in its splitting into Γ_8 and Γ_6, and the influence of this splitting on the $A \rightarrow T$ band shape are discussed by Moran.[195] The ${}^1A \rightarrow ({}^1T_{1u} + {}^3T_{1u})$ transition, resulting in the A and C bands in the impurity absorption of mercury-like centers in alkaline halides, was also

Figure 4.38. Band shape (envelope) of the electronic $A \rightarrow T$ transitions in absorption and of the $T \rightarrow A$ transitions in emission obtained by numerical solution of the linear T–t_2 problem[201] using the following 12 sets of parameters (α is the elementary transition band width in ω_T units):

	I. Absorption $A \rightarrow T$; $E_{JT}/\hbar\omega_T = \frac{2}{3}$; $\alpha = 0.5$	II. Absorption $A \rightarrow T$; $E_{JT}/\hbar\omega_T = 2$; $\alpha = 1.0$	III. Emission $T \rightarrow A$		
Figure	$kT/\hbar\omega_T$	$kT/\hbar\omega_T$	$E/\hbar\omega_T$	α	$kT/\hbar\omega_T$
a	0	0	2	0.1	0
b	0.5	0.5	2/3	0.5	0.5
c	1.0	1.0	2/3	0.5	4.0
d	4.0	4.0	2	0.5	4.0

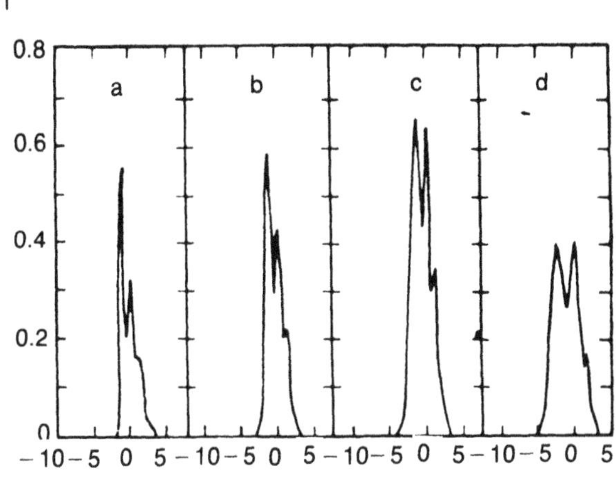
I
0.8
0.6
0.4
0.2
0
a
b
c
d
−10−5 0 5 −10−5 0 5 −10−5 0 5 −10−5 0 5

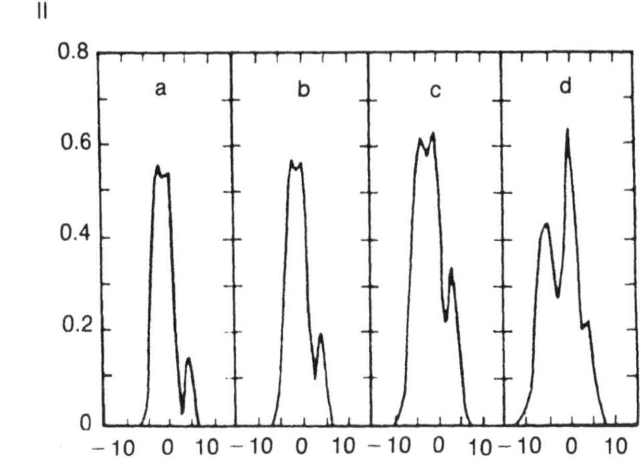
II
Intensity
0.8
0.6
0.4
0.2
0
a
b
c
d
−10 0 10 −10 0 10 −10 0 10 −10 0 10

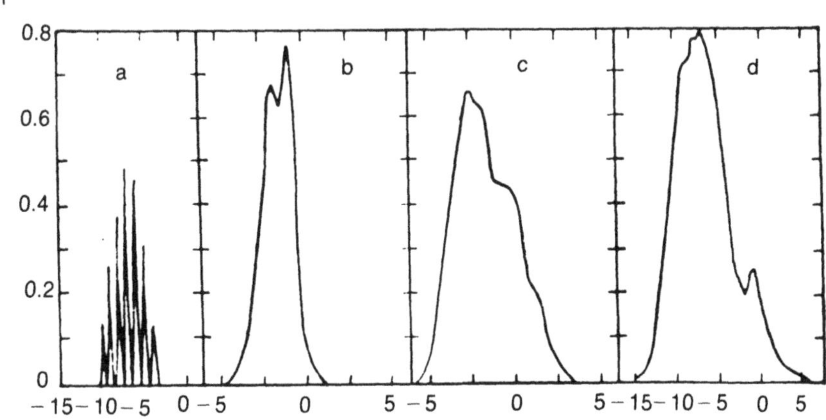
III
0.8
0.6
0.4
0.2
0
a
b
c
d
−15−10−5 0 −5 0 5 −5 0 5 −15−10−5 0 5
Frequency

considered.[206] The B band in these systems, the ${}^1A_{1g} \to ({}^3A_{1u} + {}^3T_{1u} + {}^3E_u + {}^3T_{2u} + {}^1T_{1u})$ transition, has been considered by Matsushima and Fukuda.[207] The cases of $A \to T$, $T \to A$, and $\Gamma_6 \to \Gamma_8$ transitions in the d mode approximation (when $\omega_E = \omega_T$ and $E^E_{JT} = E^T_{JT}$ are assumed, Section 2.2) have also been studied.[87,208] One example approximately obeying these conditions is the F^+ center in CaO. Other examples are Au : KCl, Fe^{2+} : MgO, Ca^+ : KBr, and Tl : KCl (see also Duran et al.[210]). Other $A \to T$ transition investigations appear elsewhere.[211–215,59,14]

It follows from this brief discussion that *vibronic effects strongly influence the band shapes of electronic transitions and that the commonly used interpretation of electronic spectra disregarding vibronic interactions is invalid if the ground or excited state, or both, are degenerate, or if at least one of them is vibronically mixed with a third term.*

Polarized Luminescence. Additional information about the nature of the AP minima of electronic degenerate states can be obtained by analyzing the dependence of luminescence polarization on the polarization of incident light.[216–218a] This possibility is based on the selection rules due to which transition between the minimum of the initial and final terms in the case of strong vibronic coupling is allowed only for certain polarizations. For example, the magnetic dipole transition from the nondegenerate state to the tetragonal minimum [100] of the excited T term is allowed only for polarization in the [100] direction ($\langle d_x \rangle = d$, $\langle d_y \rangle = \langle d_z \rangle = 0$, see Table 4.5), while for incident light polarized in the [111] direction, transitions are allowed to all the three tetragonal minima simultaneously. The transition from the [100] minimum of the ground T term to the tetragonal [100] minimum of the excited T term is forbidden, whereas from the same minimum, transitions to the [010] and [001] minima are allowed only for incident light polarized in the [001] and [010] directions, respectively. The polarization prohibitions are quite different for transitions to trigonal minima. The corresponding rules are given in Table 4.5.

Table 4.5. Matrix Elements of the Components of the Electric Dipole Moment $\langle A_{1g(u)} | d | T_{1u(g)} \rangle$ in the Case of Tetragonal, Trigonal, and Orthorhombic Minima of the T Term Adiabatic Potential

	Tetragonal minima			Trigonal minima				Orthorhombic minima					
	[100]	[010]	[001]	[111]	$[\bar{1}11]$	$[1\bar{1}1]$	$[11\bar{1}]$	[110]	$[1\bar{1}0]$	[101]	$10\bar{1}]$	[011]	$01\bar{1}]$
$\langle d_x \rangle$	d	0	0	d	$-d$	d	d	d	d	d	d	0	0
$\langle d_y \rangle$	0	d	0	d	d	$-d$	d	d	d	0	0	d	d
$\langle d_z \rangle$	0	0	d	d	d	d	$-d$	0	0	d	d	d	d

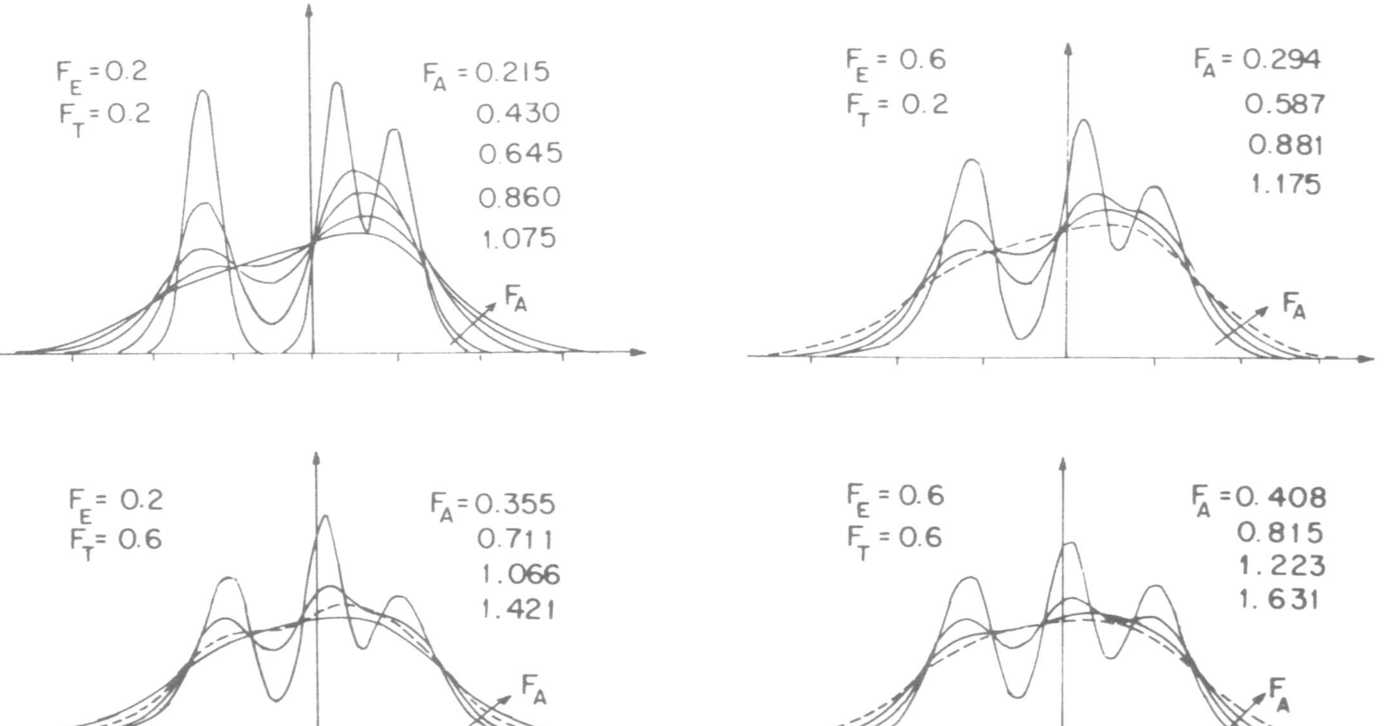

Figure 4.39. Band shape of the electronic $A \rightarrow T$ transition taking into account vibronic coupling with E, T_2, and A vibrations and spin–orbit interaction in the semiclassical approximation of the T–$(e + t_2)$ problem.[200] F_E, F_T, and F_A are the dimensionless VC. The three-humped curve becomes essentially smoothed when coupling with totally symmetric vibrations increases.

Assume that the lifetime τ at the minimum is much greater than the lifetime τ' of the excited electronic state with respect to spontaneous optical transition to the ground state (cf. the relativity rule concerning the means of observation, formulated in Section 5.1). If this is the case, the luminescence should have a definite polarization determined by the polarization of the incident light. Indeed, dependent on this latter polarization the excited state minima will be differently populated, the population relationship being easily obtained from the known AP minima electronic wave functions and Table 4.5. Each of these minima gives rise to a definite polarization of the luminescence light.

Some general conclusions concerning the nature of the AP minima in an excited T term state (when the ground state is nondegenerate), based upon the above analysis, may be formulated as follows[217,218]:

(1) If, for incident light polarized in the [100] direction, the luminescence is polarized in the same direction, and for incident light polarized along [111] the luminescence is completely depolarized, then in the excited state tetragonal absolute minima occur.

(2) If, for incident light polarized along [111], the luminescence is partly polarized in the same direction, and for the [100] polarized light the luminescence is completely depolarized, then trigonal minima occur in the excited T term state.

(3) If, for the [100] polarization of incident light, the luminescence is partly depolarized with an intensity ratio $I_x : I_y : I_z = 2:1:1$, but for the [110] polarization this latter becomes $I_x : I_y : I_z = 3:2:2$, being completely depolárized for incident light polarized along [111], then the six orthorhombic minima occur in the excited state.

The polarization relationships become more complicated for transitions between two degenerate electronic terms.

When the lifetime of the minimum state is much shorter than that of the excited electronic state, $\tau \ll \tau'$, then during time τ' the populations of the equivalent minima have a chance to equalize, and completely depolarized luminescence will be observed in all cases.

Although the most important features of the optical absorption and emission band shapes are determined by the semiclassical approximation, its approximate nature and the possible errors introduced must be taken into account. In particular, this approximation is not valid for very low temperatures and for those regions of the spectra where mixing of different sheets of the AP is strong, i.e., where these sheets are closely spaced. Therefore, the semiclassical approximation is more adequate for strong vibronic coupling in which the energy gap between the AP sheets is large over the major portion of the frequency range (see also Ranfagni and Englman[218a]).

The Method of Moments in the Analysis of Band Shapes. The integral

$$\langle \Omega^n \rangle = \int_{-\infty}^{+\infty} \Omega^n F_{12}(\Omega) d\Omega \tag{4.60}$$

is called the nth moment of the optical band. The first few moments have a specific physical meaning: the zeroth moment $\langle \Omega^0 \rangle$ equals the integral intensity of the band (the area limited by the band curve), the first moment $\langle \Omega^1 \rangle$ gives the center of gravity of the band, the second moment $\langle \Omega^2 \rangle$ is equal to the half-width of the band, the third moment $\langle \Omega^3 \rangle$ characterizes the asymmetry of the band, and so on.

It appears that if the initial state of the transition is nondegenerate, the moments of the band can be calculated exactly.[106,219-222,59] Since these moments can also be determined experimentally from the band shape, some new possibilities arise to evaluate the system parameters. In particular, the vibronic constants of the excited degenerate electronic state can be estimated from the absorption band shape and its behavior under external perturbations.[219]

Important relationships can be obtained by examining the polarization dichroism caused by the uniaxial stress, partly discussed above in considering the zero-phonon line properties. As indicated, the T term of a cubic symmetry system under tetragonal uniaxial deformation splits into the A and E terms of the D_{4h} group. The transition from a nondegenerate A term is allowed to the A term under parallel polarization and to the E term under perpendicular polarization. The change of the first moments $\Delta\langle \Omega \rangle = \langle \Omega \rangle - \langle \Omega \rangle_0$ relative to the initial value $\langle \Omega \rangle_0$ for the unsplit T term is

$$\Delta\langle \Omega \rangle_{\parallel} = -e_\theta P_E/\hbar, \qquad \Delta\langle \Omega \rangle_{\perp} = e_\theta P_E/2\hbar \tag{4.61}$$

The second moments do not change under the influence of stress:

$$\langle \Omega^2 \rangle_{\parallel} = \langle \Omega^2 \rangle_{\perp} = \langle \Omega^2 \rangle_0 \tag{4.62}$$

It is important that the contributions of the totally symmetric and JT active modes to the second moment (half-width) of the band can be separated with respect to their symmetries (A, E, and T_2 for the T term, A and E for the E term). For the T term in question

$$\langle \Omega^2 \rangle_0 = \sigma_2(A) + \sigma_2(E) + \sigma_2(T_2) \tag{4.63}$$

where

$$\sigma_2(\Gamma) = C_\Gamma \omega_\Gamma^2 \coth(\hbar\omega_\Gamma/2kT) \tag{4.64}$$

and $C_\Gamma = F_\Gamma^2/2\hbar\omega_\Gamma K_\Gamma$ if $\Gamma = A, E$, and $C_{T_2} = F_T^2/\hbar\omega_T K_T$. In particular, at zero temperature $\sigma_2(\Gamma) = C_\Gamma \omega_\Gamma^2$.

For the third moments (η is the sign of the polarization, $\parallel$ or $\perp$)

$$\langle \Omega^3 \rangle_\eta = \langle \Omega^3 \rangle_0 - \tfrac{3}{2}\Delta\langle \Omega \rangle_\eta \sigma_2(T_2)$$

or

$$\Delta\langle \Omega^3 \rangle_\eta = -\tfrac{3}{2}\sigma_2(T_2)\Delta\langle \Omega \rangle_\eta \tag{4.65}$$

Under uniaxial stress along the trigonal axis

$$\begin{aligned} \Delta\langle \Omega \rangle_\parallel &= 2e_T P_T/\hbar\sqrt{3} \\ \Delta\langle \Omega \rangle_\perp &= -\Delta\langle \Omega \rangle_\parallel/2 \end{aligned} \tag{4.66}$$

For the second moments, the same relationships (4.63) and (4.64), as for tetragonal stress, are valid, while for the third moments

$$\Delta\langle \Omega^3 \rangle_\eta = -\tfrac{1}{2}\Delta\langle \Omega \rangle_\eta [3\sigma_2(E) + \sigma_2(T_2)] \tag{4.67}$$

The general group theory formula[220,221] for the change of the moments of the $A \to \Gamma$ transition under external influence is

$$\begin{aligned} \Delta\langle \Omega^3 \rangle_\eta &= \Delta\langle \Omega \rangle_\eta \sum_{\bar{\Gamma}} \alpha(\bar{\Gamma})\sigma_2(\bar{\Gamma}) \\ \langle \Omega^2 \rangle &= \sum_{\bar{\Gamma}} \sigma_2(\bar{\Gamma}) \\ \alpha(\bar{\Gamma}) &= [\bar{\Gamma}](-1)^{J(\bar{\Gamma})+J(\tilde{\Gamma})} \begin{Bmatrix} \Gamma\Gamma\tilde{\Gamma} \\ \Gamma\Gamma\bar{\Gamma} \end{Bmatrix} - 1 \end{aligned} \tag{4.68}$$

where $\bar{\Gamma}$ is the representation of the JT active modes, $[\bar{\Gamma}]$ is its dimension, $\tilde{\Gamma}$ is the representation of the external influence, $J(\Gamma)$ is the quasimoment of the Γ representation, and the braces denote the 6Γ symbol.[25,222a]

From (4.67) and (4.65) it can be found that

$$\begin{aligned} -\tfrac{3}{2}\sigma_2(T_2) &= \Delta\langle \Omega^3 \rangle_\eta^{\text{tetr}}/\Delta\langle \Omega \rangle_\eta^{\text{tetr}} \\ -\tfrac{3}{2}\sigma_2(E) - \tfrac{1}{2}\sigma_2(T_2) &= \Delta\langle \Omega^3 \rangle_\eta^{\text{trig}}/\Delta\langle \Omega \rangle_\eta^{\text{trig}} \end{aligned} \tag{4.69}$$

It follows that if the changes of the first and third moments of the band under induced polarization dichroism can be estimated from the experimental absorption curves, the contributions of vibrations of different symmetry $\sigma_2(\Gamma)$ and the corresponding vibronic parameters can be obtained from equation (4.64). Since the above relationships are precise, the accuracy of the VC determined in this way depends only on the precision of the experimental determination of the band shape and its first and third moments.

By way of example, we shall illustrate the evaluation of the numerical values of $\sigma_2(\Gamma)$ from the optical absorption band shape of the F center in $SrCl_2$ ($A_1 \rightarrow T_2$ transition) by means of the method of moments.[221] In this case a complication arises due to the presence of the term A_1, which mixes strongly with the excited T_2 term by T_2 vibrations. Therefore, besides the T_2 term JTE, the PJTE, caused by mixing the pair of terms A_1 and T_2, contributes to the band shape. As a result the second moment does not contain three terms as in equation (4.63), but four terms: $\langle \Omega^2 \rangle = \sigma_2(A_1) + \sigma_2(E) + \sigma_2(T_2) + \sigma_2(T_2')$, the last term taking into account the A_1 and T_2 mixing contribution. By means of trigonal and tetragonal stress dichroism and magnetic circular dichroism, the following equations can be derived:

$$\begin{aligned} 2\Delta\langle \Omega^3 \rangle_\eta^{\text{tetr}} &= \Delta\langle \Omega \rangle_\eta^{\text{tetr}}[3\sigma_2(T_2) + 2\sigma_2(T_2')] \\ 2\Delta\langle \Omega^3 \rangle_\eta^{\text{trig}} &= \Delta\langle \Omega \rangle_\eta^{\text{trig}}[\sigma_2(T_2) + 3\sigma_2(E) + 2\sigma_2(T_2')] \\ 2\Delta\langle \Omega^3 \rangle_\pm &= \Delta\langle \Omega \rangle_\pm[3\sigma_2(T_2) + 3\sigma_2(E) + 2\sigma_2(T_2')] \end{aligned} \tag{4.70}$$

Now, using the empirical data for the band moment changes one obtains (in units of the second moment $\langle \Omega^2 \rangle$) $\sigma_2(A_1) = 0.025$, $\sigma_2(E) = 0.159$, $\sigma_2(T_2) = 0.36$, and $\sigma_2(T_2') = 0.456$. It is seen that in this case the main contributions to the second moment of the band are from the JTE and the PJTE. The contribution of the totally symmetric vibrations A_1 is unimportant.

If spin–orbit splitting is taken into account, additional parameters appear in equations (4.69) and (4.70). However, the number of equations can also be increased by introducing additional perturbations which induce polarization dichroism, as in the above example.[221,222]

The method of moments has been developed mainly for application to the optical properties of impurity centers in crystals and to exciton absorption. *The application of this effective method to molecular spectroscopy, to spectroscopy of transition metal and rare-earth coordination compounds, etc., seems to be very hopeful.*

For other studies of vibronic effects on optical spectra, see the Bibliographic Review.[14]

5

Stereochemistry and Crystal Chemistry

The origin of molecular and crystal atomic structures, including structural phase transitions, is determined mainly by vibronic interaction effects.

5.1. Stereochemistry Rules with JTE and PJTE

Notion of Nuclear Configuration. Semiclassical Approach. The notion of nuclear configuration, widespread in chemistry, is based on the assumption that nuclei are heavy classical particles performing small vibrations near some fixed equilibrium positions. In general, however, a nuclear subsystem, as any other set of microparticles, must be treated as a quantum system and characterized by a wave function, the square of its modulus determining the probability of the nuclei location. In stationary states with given energy, nuclear coordinates (similar to electronic coordinates) are not observable quantities but operators. The mean values of these coordinates, calculated with the wave function of the given state, are experimentally observable.

However, in the absence of electronic degeneracy, and if criterion (1.10) is valid, the AP of the system determined by the solution of electronic equation (1.3) has the sense of the potential energy of the nuclei in the field of the electrons. In these cases, for various purposes, the nuclei may be considered approximately as moving classically along the potential surface (semiclassical approximation). This approximation was used in the previous chapter to determine the optical band shapes; it is successfully used in solutions of many other problems, such as computation of adiabatic chemical reaction rates.

In the case of electronic f-fold degeneracy (or pseudodegeneracy) there are f sheets of the AP intersecting (or branching) at the point of degeneracy. Each of these sheets (and all of them together) loses the meaning of nuclear potential energy, especially in the region near the point of intersection where the energy gap between different sheets is small (see also Section 1.6). In this region, the quantum effects cannot be ignored and the semiclassical approximation is invalid. The behavior of the nuclear framework is determined by the solution of vibronic equations (1.6).

On the other hand, the AP considered in Chapter 2 for different cases of degeneracy and pseudodegeneracy (Figures 2.1–2.7) shows that if vibronic coupling is sufficiently strong, regions exist (far from the degeneracy point and near the ground sheet minima) where the energy gap to the next sheet is large (equal to $4E_{JT}$ in the E–e problem). In these regions, again, it may be possible to use the semiclassical approximation.

The possibility of using the AP in chemical problems without resolving the vibronic equations justifies consideration of the AP problems in Chapter 2 apart from the solution of the vibronic equation (Chapter 3).

In this chapter we consider stereochemistry and crystal chemistry problems with vibronic effects using the above semiclassical approximation, sometimes also taking into account necessary quantum effects which emerge from the results obtained earlier in Chapters 3 and 4.

Relativity to the Means of Observation. Consider a free JT or PJT system. We shall clarify the conditions under which JT distortions are manifest in stereochemistry. As shown in Chapter 2, the AP of these systems have several (or a continuum of) minima, corresponding to equivalent but differently oriented distortions of the nuclear configuration. The equivalent distortions are always oriented so that they complement each other up to the initially assumed high symmetry of the system for which the electronic degeneracy occurs (see, e.g., Figure 2.3). Therefore, in the stationary state the nuclear configuration distortion (averaged over all the minima) is zero.

In fact, however, if there are several deep minima, the system is most commonly obtained first in the distorted configuration of the minima (as a result of a chemical reaction, collisions, etc.). The state at the minimum, due to the presence of other equivalent minima, is quasi-stationary. After a period of time τ the state changes to that of another minima, and so on; in other words, the system performs pulsating motions described in Sections 2.1 and 3.2.

During this time τ, the system vibrates near the minimum position with period $T = 2\pi/\omega$, where ω is the vibration frequency, determined by the AP curvature at the minimum point. It is clear that if $\tau \leqslant T$, the notion of distorted configuration loses its real meaning. The above consideration has a physical sense only when $\tau \gg T$, i.e., when the system performs many (at least several) vibrations in the distorted configuration before making a transition to another configuration (τ is inversely proportional to the tunneling splitting δ).

If τ' is the characteristic time of the elementary act of the measurement process, one can distinguish two limiting cases. If

$$\tau' \ll \tau \tag{5.1}$$

the system remains at one of the minima and the corresponding distorted configuration becomes evident in the measurement under consideration. The system appears distorted (static JTE) from the viewpoint of such a method of measurement. In the other limiting case, when

$$\tau' \gg \tau \tag{5.2}$$

the system performs many (multiple) transitions between the differently oriented distortions during the act of measurement, the latter resulting in an averaged undistorted picture (dynamic JTE). Thus the conclusions about the symmetry and stereochemistry of the JT or PJT system depend on the method of measurement. Several examples confirming this statement are given above in the discussion of ESR, optical Mössbauer spectra, etc. Thus we arrive at the "relativity rule" concerning the means of observation, which can be formulated as follows[8,53,223,224]: *in a free JT and PJT system with strong vibronic coupling, the observable nuclear configurations depend on the means of observation and may vary by changing the method of measurement.*

This rule may have practical significance for qualitative analysis of vibronic effects, and so contribute to a better understanding of their origin. Some difficulties may be encountered in determining the value of τ'. As seen in Section 4.1, in ESR measurements τ' is inversely proportional to the Zeeman interaction with the external field, i.e., inversely proportional to the field intensity $\mathcal{H}$. Therefore τ' can be varied easily by changing the value of $\mathcal{H}$. For large $\mathcal{H}$ the relation $\tau' \ll \tau$ may be satisfied, so that the distorted configuration (an anisotropic spectrum) is observed, while for small $\mathcal{H}$ we may have $\tau' \gg \tau$, in which case an averaged spectrum is to be expected. A similar transition from an anisotropic to an isotropic spectrum can be obtained for the same value of τ' (the same $\mathcal{H}$) by varying τ with temperature, since the tunneling frequency (as stated above) depends strongly on temperature. This temperature aspect of the relativity rule is generally valid.

In polarized luminescence measurements, τ' is determined by the lifetime of the excited electronic state relative to spontaneous irradiation. In polarizability measurements (see below) τ' is determined by the rate of molecular reorientation in the applied electric field. Other examples may be found in different sections of the book.

Stabilization of Static Distortions by Weak Perturbations. Vibronic Amplification. So far we have considered only free JT and PJT systems. For stereochemistry and crystal chemistry, the behavior of these systems subject to small perturbations distorting the nuclear configuration is of great importance. Such perturbations may arise, for instance, either from small differences in the ligands which lower the initially assumed symmetry, or from the influence of the next coordination spheres in the crystal structure.

If these low-symmetry perturbations are large compared with vibronic interactions (i.e., if they produce an energy splitting greater than the Jahn–Teller stabilization), they may completely reduce (quench) the JT effects. In the other limit, when perturbations are smaller than vibronic interactions, they do not remove, but modify the JT effects. We shall now examine this latter case.[8,225]

Consider the simplest E–e problem in the linear approximation, i.e., a system with an AP in the form of a "mexican hat" (Figure 2.1). For the free system, the averaged picture displayed in the experiment is an undistorted nuclear configuration. Under the influence of a small distorting perturbation, say elongating the system in the Q_θ direction, the circular trough will be distorted, namely, an additional potential well in the Q_θ direction (and a hump in the opposite $-Q_\theta$ direction) will appear. If the well depth is greater than the kinetic energy of the circular motion in the trough, then the nuclear motions are localized in this well and a corresponding distorted nuclear configuration will be observed in the experiments.

The most exciting result in this picture is that the *magnitude of the distortion is determined mainly by the vibronic interactions and is almost independent of the perturbation magnitude.*[8,53,225] Indeed, the additional well is formed at a point of the circular trough with coordinates (ρ_0, ϕ_0), where ϕ_0 mainly depends on the perturbation, and therefore the magnitude of the distorted configuration, determined by ρ_0, is independent of the perturbation (in fact, the perturbation slightly distorts the circular trough too, see below). This ρ_0-value distortion is often called the *static limit of the JTE.*

Thus a small perturbation W acting upon a JT system produces a distortion which is determined by the static limit ρ_0 of the JTE (it stabilizes the static distortions). Since the kinetic energy of the motion along the trough $E_k = \hbar^2/8M\rho_0^2$ is of the order of several cm^{-1},[64] the condition $W \gtrsim E_k$ is satisfied even for small perturbations. Meanwhile, the static JTE distortions ρ_0 may be sufficiently great. Hence we obtain the "*amplification rule*" in JT distortions: *a small distorting perturbation can be amplified by vibronic interactions.*

Let us make some qualitative (or semiquantitative) estimations of the coefficient of vibronic amplification P_a. In the absence of vibronic interactions the distortion Q_0 can be found by comparison of the perturbation energy W with the elastic distortion, where $W = \frac{1}{2}K_E Q_0^2$; $Q_0 = (2W/K_E)^{1/2}$.

If vibronic effects are taken into account, $Q_0^{JT} = \rho_0 + Q_0$ and the amplification coefficient, equal to the ratio of the corresponding distortions, is given by

$$P_a = Q_0^{JT}/Q_0 = 1 + (E_{JT}/W)^{1/2} \tag{5.3}$$

where the relationship $\rho_0 = (2E_{JT}/K_E)^{1/2}$ has also been used. The maximum amplification is attained when $W = E_K$, in which case

$$P_a^{max} = 1 + 4(E_{JT}/\hbar\omega_E) = 1 + 4\lambda_E \tag{5.4}$$

It follows that the vibronic amplification may be very large, since λ_E may be substantial in some cases. For example, if we assume (as expected for octahedral compounds of Cu(II), Mn(III), Cr(II)) that λ_E is about 5–10 (although it can be larger), then $P_a^{max} \sim 20–40$.

In the quadratic $E–e$ problem with a more complicated AP having three equivalent tetragonal minima, *a low-symmetry perturbation* makes them nonequivalent, and *the system* becomes "*locked*" *at that minimum, which is deeper.* It follows that the pulsating (and hence dynamically symmetric) system under small perturbations becomes strongly distorted statically, the amplification coefficient of the distorting perturbation being determined by equation (5.3). Such an effect was encountered earlier when we determined from vibronic ESR spectra (Section 4.1) that small perturbations in the form of random strain locked the system in the distorted configuration of the minima.

Temperature effects were omitted in the above consideration. The temperature dependence of the vibronic amplification is illustrated here by considering the $T–t_2$ problem for tetrahedral symmetry with four minima of the AP, the system having a dipole moment p_0 in the minimum configuration. An estimation of the vibronic amplification coefficient will be carried out in a model calculation when external perturbation is approximated by means of an electric field.[225]

Without the electric field, the mean value of the T_2 type coordinate Q equals zero. In an electric field of intensity $\mathscr{E}$, but without vibronic interactions, the induced distortion in the $\mathscr{E}$ direction can be found from the relationship $K_T Q_0 = e^* \mathscr{E}$, where e^* is the effective charge corresponding to the Q coordinate. Using the results of Chapter 3 on vibronic interactions in the $T–t_2$ problem and performing statistical averaging over the ground triplet state,[225] we obtain

$$\begin{gathered} K_T Q_0^{JT} = 2F_T K_T^v(T_2) \sinh x [2\cosh x + 1]^{-1} \\ x = p_0 \mathscr{E}/kT \end{gathered} \tag{5.5}$$

where F_T is the linear vibronic constant, K_T is the force constant, and $K_T^v(T_2)$ is the corresponding vibronic reduction factor (Section 3.3).

In the case of strong vibronic coupling, when the vibronic singlet A approximates the ground vibronic triplet (Figure 3.2) and becomes populated,

$$K_T Q_0^{JT} = F_T \tanh x \tag{5.6}$$

and hence the amplification coefficient is given by

$$P_a = Q_0^{JT}/Q_0 = F_T \tanh x / e^* \mathscr{E} \tag{5.7}$$

At very low temperatures when $kT \ll p_0 \mathscr{E}$, $P_a = F_T / e^* \mathscr{E}$. Since $p_0 \mathscr{E}$ is small, the other limit $kT \gg p_0 \mathscr{E}$ is actually realized in most cases, for which $\tanh x \approx p_0 \mathscr{E} / kT$. Taking into account that $p_0 \sim e^* F_T / K_T$ and $E_{JT} = F_T^2/kT$, we obtain

$$P_a \approx F_{JT}^T / kT \tag{5.8}$$

Consequently, if $E_{JT} \sim 10^3$ cm^{-1}, for instance, then $P_a \sim 10$ even at room temperature. Other models for estimating the effect may be efficient, but the conclusion about *an uncommonly large "susceptibility to distortions"* of vibronic systems *due to vibronic amplification seems to be quite general.*

The notion of vibronic amplification contributes substantially to a better understanding of JTE. In particular, this notion rejects completely the incorrect statements often encountered in the literature, that JTE is not expected in systems where differences in the ligands or other low symmetry perturbations formally remove the electronic orbital degeneracy. On the contrary, we show in this section that it is only such low symmetry perturbations (small, but vibronically amplified) which lead to observable distortions. In the absence of these perturbations the JT distortions are of a dynamic nature and do not become evident in an absolute manner in stereochemistry and crystal chemistry, provided cooperative effects and vibronic structural phase transitions are not essential (see Section 5.2).

Dipolar Instability and Dipole Moments. A special case of molecular stereochemistry with JTE and PJTE is that of dipolar unstable systems, i.e., systems which acquire a dipole moment in the distorted AP minimum configuration[57] (Section 2.3). Dipolar instability is, in principle, possible for any system having two pseudodegenerate electronic terms mixing under dipolar displacements, or for systems with a JT T–t_2 problem but without an inversion center. (It will be recalled that the JTE in systems having an inversion center gives no dipolar distortions, since all active JT modes are of even type whereas dipolar displacements are of odd type.)

Dipolar instability has two, very interesting aspects. First, upon taking into account dipolar instability, molecular systems of high symmetry, which are traditionally considered to have no proper dipole moments, can exhibit properties inherent in molecules with proper dipole moments. Second, molecules with a proper dipole moment, and which are usually regarded as rigid dipole molecules, may have two (or several) equivalent equilibrium configurations with differently oriented dipole moments resulting from the JTE or PJTE. Consideration of the transitions between minima configurations changes the expected dipolar properties.

The ammonia molecule may be considered as a typical example of the second aspect of the problem. Its two AP minima configurations with oppositely oriented dipole moments, together with tunneling splitting resulting from tunneling transitions between the minima configurations, are shown in Figure 5.1. The orientational polarizability of an aggregate of such molecules in an electric field $\mathscr{E}$ differs from that expected for rigid dipole molecules. Computation[226,227] shows that, when the Boltzmann temperature distribution is taken into account as well as the inequalities $p_0\mathscr{E} \ll \delta$ and $p_0\mathscr{E} \ll kT$, where p_0 is the absolute value of the dipole moment in the configuration of the AP minimum, the mean dipole moment in the semiclassical approximation is given by the expression

$$\bar{p} = (p_0^2\mathscr{E}/3\delta)\tanh(\delta/kT) \tag{5.9}$$

A similar expression for a system with four dipolar minima of the AP

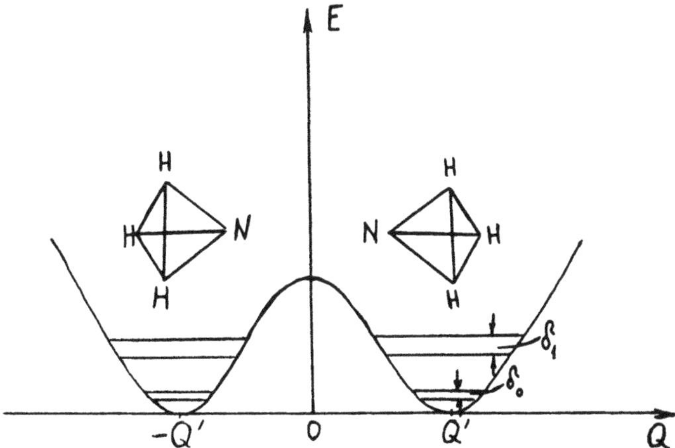

Figure 5.1. Two equivalent minima of the adiabatic potential and two corresponding stable configurations of the ammonia molecule with opposite directions of the dipole moment. Tunneling splitting of the ground (δ_0) and excited (δ_1) vibrational states is also shown.

has the form

$$\bar{p} = \frac{p_0^2\mathscr{E}}{3kT}\,\frac{\exp(-\delta/kT)+(kT/\delta)\sinh(\delta/kT)}{\exp(-\delta/kT)+\cosh(\delta/kT)} \tag{5.10}$$

Expressions (5.9) and (5.10) differ from that obtained for rigid dipole molecules

$$\bar{p}' = p_0^2\mathscr{E}/3kT \tag{5.11}$$

both in absolute value and temperature dependence. They coincide only in the limiting case $\delta \ll kT$ (Figure 5.2). In the opposite case, with $\delta \gg kT$, expressions (5.9) and (5.10) cease to depend on temperature and reduce to the expression for the dipole moment $\bar{p}' = \alpha\mathscr{E}$, characteristic for high-symmetry molecules with no proper dipole moment. These results yield the following principle. *The division of all the molecules into two kinds — rigid dipole molecules and nondipolar molecules — judging by their behavior subject to the electric fields* (the temperature dependence of the polarization), *is rather conventional*: *both kinds of behavior, linear dependence on* $1/T$ *and independence of* T, *are inherent in the same molecules over different temperature ranges* (Figure 5.2), *depending on the strength of the dipolar instability* (cf. with the above "relativity rule"). The two pure limiting cases, when there is no dipolar instability and when the barrier height between the dipolar minima configurations is infinite, may occur as particular cases. For example, heterodiatomics falls into the latter case.

From a practical point of view, since equation (5.9) is more general than equation (5.11), the latter must be used with care when the proper dipole moment is to be evaluated from the temperature dependence of the polarizability. It is seen from Figure 5.2 that the differences between $\bar{p}$ and $\bar{p}'$ are most pronounced in the temperature range for which $kT \leqslant \delta$. In particular, when $kT = \delta$ and $\bar{p}' \cong 1.7\bar{p}$, the determination of p_0 by equation (5.11) leads to a reduction in the result by a factor of 1.3 as compared with the correct value given by equation (5.9).

For weak dipolar instability, when $\delta \gg kT$, the electric properties of the system approach those expected for molecules having no proper dipole moments. If δ is not very large, the temperature dependence of the polarizability may be interpreted as the occurrence of a dipole moment. A systematic experimental study of this effect seems to be of great significance.

Dipolar instability can also arise in other properties of molecular systems, such as IR and Raman spectra, considered above (Section 4.2; see also Ostrovskii et al.,[227a]) dielectric losses in the microwave region,[227b] and spontaneous polarization of crystals and ferroelectric phase transitions (see below).

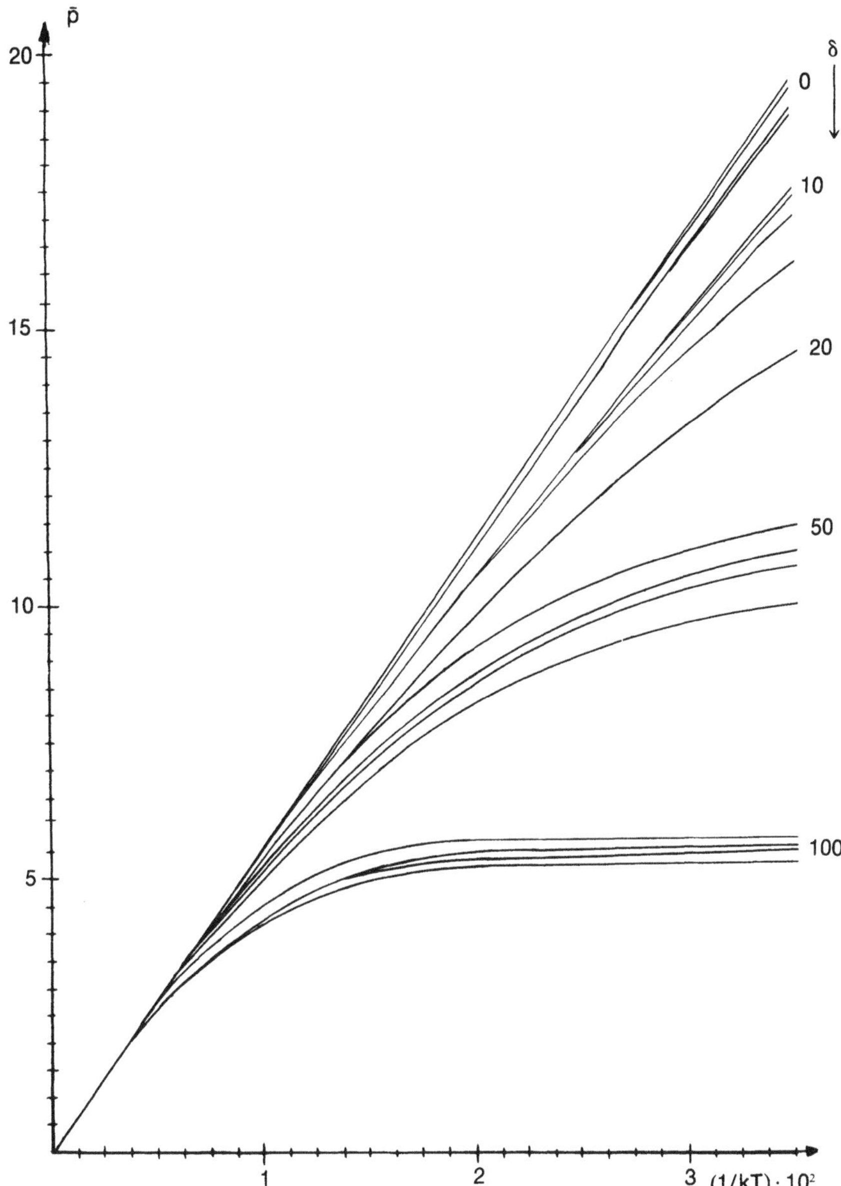

Figure 5.2. Averaged polarizability $\bar{p}$ (in arbitrary units) of dipolar unstable symmetric systems as a function of temperature and tunneling splitting δ (shown on the right in cm^{-1}), the rotational constant (in the quantum consideration) being $0 \leq B \leq 9.3\ cm^{-1}$ (after Bersuker et al.[227]).

Some Qualitative Features of Vibronic Stereochemistry. Suppose the above conditions for vibronic distortion are satisfied (and/or static distortions are stabilized). One can then ask: what kind of molecular stereochemistry is to be expected and what is the role of vibronic effects at its origin? To answer this question, the conditions of the AP minimum formation have to be examined. The major part of this problem has been solved in Chapter 2, especially for degenerate electronic terms. We have seen that a C_{3v} symmetry molecule of type X_3 in an electronic E state is unstable and distorted to the C_{2v} symmetry of the AP minima configurations, while an octahedral O_h symmetry system with an E term is distorted toward the D_{4h} symmetry. Tetragonal, trigonal, and orthorhombic distortions are possible for cubic symmetry systems with T terms. Pseudodegenerate terms may lead to any distortion. We shall consider some examples of stereochemistry with application of these vibronic rules.

First consider linear molecules. The stereochemistry of the NH_2 radical is determined by the Renner effect.[228] In the ground state of the linear H–N–H molecule, the unpaired π electron forms a Π state (E term) which, in accordance with the Renner effect (Section 1.6), is dynamically unstable with respect to the bending, this result being also confirmed by MO LCAO calculations of the AP. The molecule remains linear in the excited nondegenerate state. Analogous thinking applied to AlH_2, BH_2, and HCO leads to similar results: these molecules are bent in the ground state and linear in the lowest excited state.[229,230]

On the other hand, the CO_2 molecule in the linear configuration has a nondegenerate ground state whereas its first excited state is a twofold degenerate E term which leads to a bent configuration.[32,231,231a] The same can be said about the CH_2 radical,[232] which shows a strong Renner effect upon the absorption of light.

The simplest JT molecule of C_{3v} symmetry is H_3.[233] As is known, H_3 is unstable and decomposes into $H_2 + H$. On the other hand, the H_3^+ ion is stable. This fact can be explained easily by means of the rules of vibronic stereochemistry. Indeed, the three $1s$ functions of the hydrogen atoms at the corners of a perfect triangle, $|\mathrm{I}\rangle$, $|\mathrm{II}\rangle$, and $|\mathrm{III}\rangle$, form three MO belonging to the A_1 and E representations of the C_{3v} group (S is the overlap integral):

$$\begin{aligned}
\psi_{A_1} &= [3(1+2S)]^{-1/2}(|\mathrm{I}\rangle + |\mathrm{II}\rangle + |\mathrm{III}\rangle) \\
\psi_{E\theta} &= [6(1-S)]^{-1/2}(2|\mathrm{I}\rangle - |\mathrm{II}\rangle - |\mathrm{III}\rangle) \qquad (5.12) \\
\psi_{E\varepsilon} &= [2(1-S)]^{-1/2}(|\mathrm{II}\rangle - |\mathrm{III}\rangle)
\end{aligned}$$

the A_1 state being lower in energy. In the ground state two electrons occupy the lowest A_1 energy level, and the third one is in the E state.

Therefore the ground state of the H_3 molecule with three electrons is an E term, whereas in the H_3^+ ion with only two electrons the ground A_1 state is nondegenerate. In the E state, according to the JTE, the C_{3v} symmetry configuration is unstable, distortions following the symmetrized E coordinates (Figure 1.1) which, in the given case, correspond to the formation of an isosceles triangle with one acute angle. Since only repulsing protons remain in the core, it may be assumed that $K_E \leqslant 2G_E$ (Section 2.1, equation (2.11)), and the minimum position corresponds to a very large distance from the third hydrogen atom. This is equivalent to a decomposition of the system into $H_2 + H$. However, the H_3^+ system with a nondegenerate ground state remains stable.

The Li_3 and other related systems, in which the main bonding electrons form a degenerate E term (and hence the condition $2G_E \gtrsim K_E$ in equation (2.11) may be satisfied), can be treated similarly. In more complicated triangular molecules, where the chemical bond is formed by other (internal) electrons (and hence the condition $2G_E < K_E$ is more likely to be valid), instability does not lead to complete decomposition but distorts the system. The cyclopropane ion $C_3H_6^+$, having a calculated JT stabilization energy E_{JT} equal to 3260 cm^{-1} [234] (according to other data[235] $E_{JT} = 3630$ cm^{-1}), is such an example. The ground E term of this system is also confirmed by photoelectron spectra.[236] The above JT distortions have been observed in a number of C_{3v} symmetry molecules (see Table 5.1 and the Bibliographic Review[14]).

An important problem discussed below concerns X_4 and MX_4 type molecules. The question is: why are some of these molecules square planar, whereas others are tetrahedral, or distorted tetrahedral? A qualitative answer to this question can be given considering JT and PJT instability. It is known that *ceteris paribus* the steric repulsion (including repulsion of noninteracting atomic cores) is lower in the tetrahedral than in the square planar configuration. Consequently, deviations from tetrahedral T_d symmetry may occur only as a result of more detailed consideration of the electronic distribution. The latter influences the nuclear configuration by vibronic coupling.

Simple systems of this type are hydrides MH_4. For these, the σ type MO formed by the hydrogen $1s$ functions and their counterparts on the M atom in the tetrahedral configuration transform as A_1 and T_2. If M is not a transition or rare earth atom, its four orbitals (three np and one ns) together with the four hydrogen $1s$ orbitals form four bonding MO ($a_1 + t_2$) and four antibonding ones ($a_1^* + t_2^*$). When the symmetry is lowered from T_d to D_{2d}, D_{4h}, or C_{3v}, these MO levels split, as shown in Figure 5.3. For CH_4, eight electrons in T_d symmetry occupy the bonding a_1 and t_2 MO resulting in a nondegenerate ground term A_1, which is ~ 10 eV away from the first excited T_2 term. It follows that the CH_4 molecule in its ground state is stable in the tetrahedral configuration.

Note that in the square planar D_{4h} configuration the CH_4 molecule has the highest occupied MO (HOMO) of a_{2u} type, close to which is the lowest unoccupied MO (LUMO) of b_{1g} type. It follows that the ground state of the molecule as a whole is a nondegenerate A_{1g}, but there is a near excited term B_{2u} which mixes with the ground one under B_{2u} type nuclear displacements. If this mixing is strong, the square planar D_{4h} configuration becomes unstable with regard to B_{2u} type distortions and converts this configuration into tetrahedral (Figure 1.1). It is seen from the MO energy level correlation scheme in Figure 5.3 that the distortion effect, being weaker the greater the A_{1g}–B_{2u} (or a_{2u}–b_{1g} MO) energy gap (Section 2.3), is least for the tetrahedral configuration. Thus the origin of the stable tetrahedral configuration of the methane molecule can be understood qualitatively from vibronic stereochemical considerations.

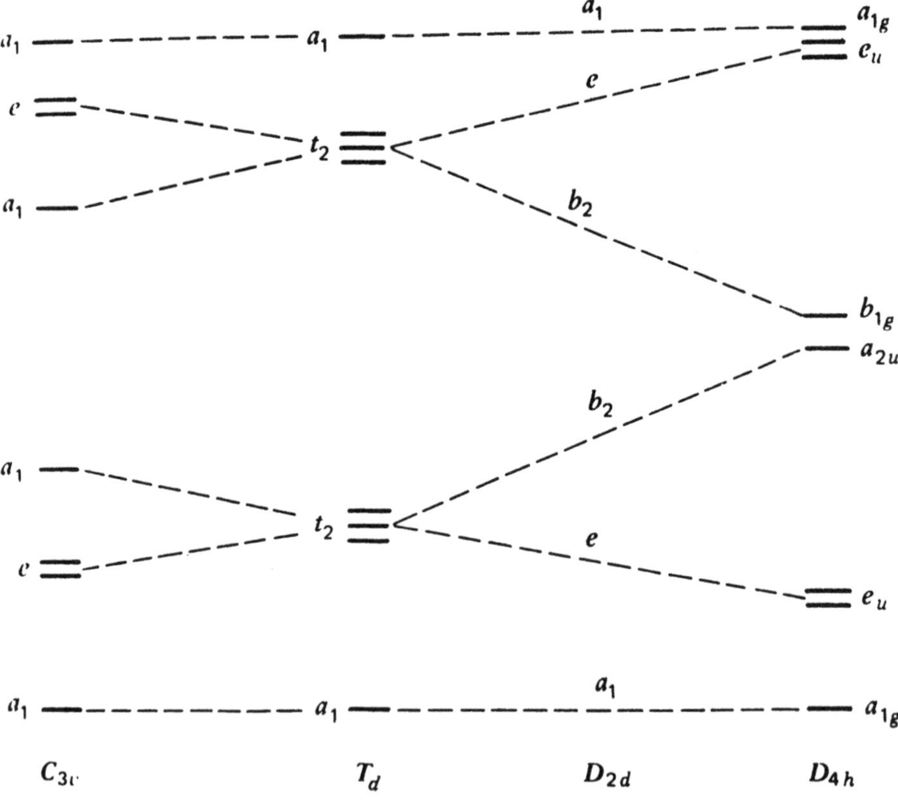

Figure 5.3. MO energy levels of MH_4 type molecules in C_{3v}, T_d, D_{2d}, and D_{4h} symmetry configurations (MO correlation diagram).

The conclusion about the stability of the tetrahedral CH_4 configuration may also be reached without the above vibronic considerations, although the instability of the D_{4h} configuration is unlikely to be understood via other approaches in such a simple and general form. Besides, the vibronic approach allows us to predict also the stereochemistries of excited states and related ions. Indeed, the first excited state T_2 of the CH_4 molecule is unstable in the tetrahedral configuration and results in a series of effects, including optical absorption spectra.[237] The same T_2 term is inherent in the CH_4^- and CH_4^+ systems in the ground state.

The case of the CH_4^+ ion is of special interest due to possible astrophysical applications.[238]* It is easy to be certain that the square planar configuration of CH_4^+ is unstable. Indeed, in the D_{4h} symmetry the ground term is A_{2u} (Figure 5.3) and the near first excited state (corresponding to the one-electron a_{2u}–b_{1g} excitation) is B_{1g}, so their strong mixing leads to a distortion of the square toward the tetrahedral configuration (B_{2u} distortion). Since both the tetrahedral and square planar configurations are unstable, the intermediate tetragonal (D_{2d}) or trigonal (C_{3v}) distorted tetrahedron remains a possible stable configuration. Nonempirical calculations[238,239] show that the stabilization energy of the D_{2d} minima $E_{JT} = 1.41$ eV is somewhat greater than that of the C_{3v} case, $E_{JT} = 1.22$ eV. However, semiempirical calculations[240] and recent experimental data obtained by means of the "Coulomb explosion"[154] show evidence for trigonal distortions. It is possible that orthorhombic minima are actually absolute,[43] for which apparently both theoretical[238–240] and experimental[154] data can be correlated (calculations for orthorhombic minima were not performed by Dixon[239] since at that time the possibility of orthorhombic distortions[43] had not yet been revealed).

Similar consideration may be given to other tetrahedral molecules. In Fig. 5.4 the approximate MO energy level scheme is given for MX_4 molecules, in which both the M and X atoms are *sp* elements, the M atom providing four atomic (one s and three p) orbitals and the X atom three (one σ and two π) orbitals.

It is seen from this scheme that, say, for CF_4 with 24 electrons in these MO, the HOMO t_2 is completely populated and the ground state is A_1. Since the energy gap to the first excited T_2 term (corresponding to the one-electron $t_2 \rightarrow a_1$ excitation) is large (~ 7 eV), the CF_4 molecule is stable in the tetrahedral configuration.

Qualitatively, similar considerations are valid for all the halides CX_4, X = F, Cl, Br, J, as well as for SO_4^{2-}, ClO_4^-, SnF_4, $SiCl_4$, and analogous

* In methane planets the CH_4^+ ion may be present in large amounts because of radiolysis. Due to the JTE and consequent strong reactivity (see Section 6.3) it initiates polymerization reactions or, in the presence of ammonia and water, formation of aminoacids, sugars, etc.

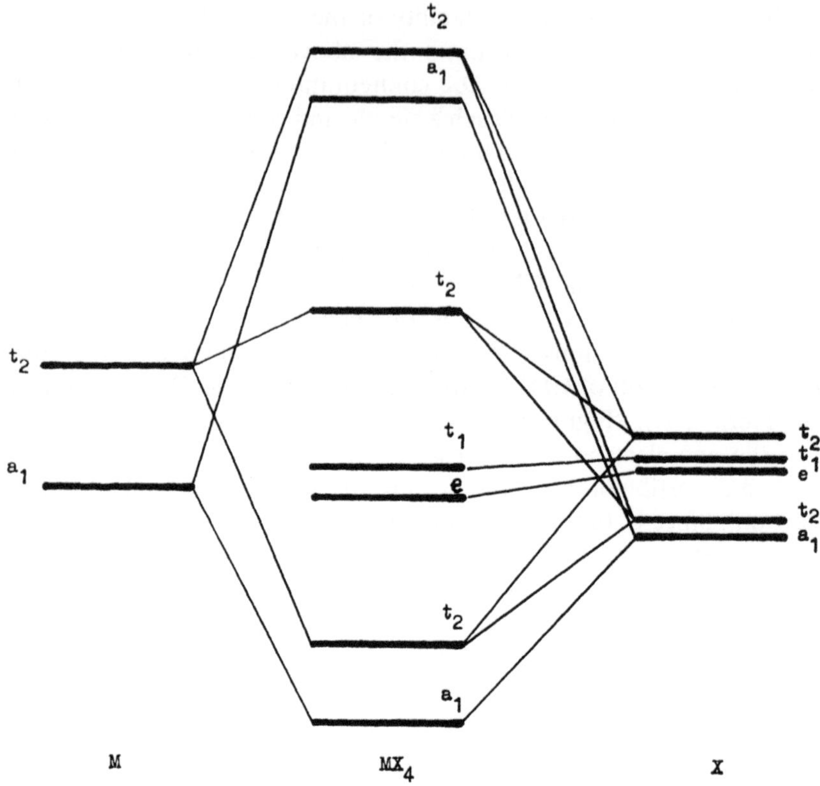

Figure 5.4. MO energy level scheme for MX_4 tetrahedral molecules, where M and X are *sp* elements.

systems. In all these cases the energy gap 2Δ between the ground A_1 and first excited T_2 terms (between the t_2 and a_1 MO) controls the stability of the molecule with respect to T_2 distortions (Figure 1.1): the smaller this energy gap *ceteris paribus*, the softer the molecule (the lower the force constant) in this direction.

However, these same conditions (*ceteris paribus*) may not be satisfied when passing from one molecule to another, even similar ones. Generally, in accord with the requirements of PJTE theory (Section 2.3), there are three parameters which play an important role in the occurrence of instability: the energy gap 2Δ, the force constant K, and the nondiagonal vibronic constant F. Vibronic mixing of the two terms lowers the force constant of the ground state (and increases it for the excited state) by an amount F^2/Δ; when $F^2/\Delta > K_T$, the total force constant K_T becomes negative and the system becomes unstable in the direction of T_2 distortions. In the above series of tetrahedral systems there is no evidence that, besides Δ, the other parameters do not change. On the other hand, it can be shown that, say, CCl_4 (or SO_4^{2-}), similar to CH_4, is unstable in the square planar configuration with regard to B_{2u} nuclear displacements, which convert the square toward a tetrahedron via the intermediate D_{2d} symmetry.

In the SF_4 molecule with 26 electrons in the above-mentioned MO, the HOMO is a_1, the ground term being A_1, with a near T_2 term (originating from the $2a_2 \rightarrow 3t_2$ transition). Therefore this molecule is expected to be unstable in the tetrahedral configuration. It can be shown that it is unstable in the square planar configuration, too. The observed geometry is indeed intermediate between a square and a tetrahedron.[241] In the case of XeF_4 the corresponding number of *sp* electrons is 28, and in the tetrahedral configuration the HOMO t_2 is occupied, but not filled, resulting in a T term ground state. This makes the system vibronically unstable. *Ab initio* calculations for the square planar configuration D_{4h} show that since the energy gap between the ground A_{1g} and first excited E_{1u} terms is large, this configuration is predicted to be stable.

When passing to MX_4 (as in the more general case of MX_n) molecules, where M is a transition metal, the number of active (valent) MO in a small interval of energy increases sharply on account of the large number of atomic *d* orbitals. This results in a degenerate or pseudodegenerate ground state in most cases. The mutual positions of the HOMO and LUMO, and hence the expected stereochemistry, are dependent on both the number of *d* electrons and the nature of the ligand X (note that the JTE and PJTE may both act here simultaneously).

For instance, for a square planar structure the following electronic configuration is typical:

$$\ldots(a_{1g})^2(b_{1g})^2(e_u)^4\{(b_{2g})(e_g)(2a_{1g})(2b_{1g})\}(a_{2u})^0(3a_{1g})^0,$$

where the orbitals in braces have to be occupied by the *d* electrons. If the ground electronic state is nondegenerate, the distortion of the square toward D_{2d} and tetrahedral T_d symmetry requires strong mixing of the ground state with the near excited state by B_{2u} nuclear displacements. This excited state can be obtained by one of the following one-electron transitions: $e_u \rightarrow e_g$, $b_{1g} \rightarrow a_{2u}$, or $b_{2u} \rightarrow 2a_{1g}$ (b_{2u} is an inner MO). Therefore, if the e_g and $2a_{1g}$ MO are fully occupied by *d* electrons, but the $2b_{1g}$ MO is empty, the square planar configuration will be stable. In other words, low spin MX_4 d^8 complexes of Ni(II), Pd(II) and Pt(II) are expected to be square planar. On the other hand, high spin d^5 and d^{10} complexes with an occupied $2b_{1g}$ MO may be unstable in the planar configuration due to strong mixing with the near $B_{2u}(2b_{1g} \rightarrow a_{2u})$ term.

Passing to the important octahedral MX_6 systems, consider the interesting example of XeF_6. Nonempirical calculations of the electronic structure of this molecule[242] show that the outer MO are arranged in the following sequence: $(t_{2g})^6(t_{2u})^6(t_{1u})^6(t_{1g})^6(e_g)^4(a_{1g})^2(2t_{1u})^0$ with an energy gap of about 3.7 eV between the a_{1g} and t_{1u} MO. This results in an instability of the system with respect to T displacements. (Note that the energy gap of 3–4 eV is still not the limiting value for which mixing can be regarded as strong enough to result in instability.[28a]) For comparison, computation[243] performed for a similar system SF_6 results in the following sequence of MO: $(t_{2g})^6(e_g)^4(t_{1u})^6(t_{2u})^6(t_{1g})^6(a_{1g})^0(2t_{1u})^0$. Since the energy gap between the HOMO t_{1g} and empty t_{1u} MO is sufficiently large, as distinct from the XeF_6 case, this determines the stability of the O_h symmetry configuration with respect to odd (dipolar) displacements. A series of investigations of vibronic effects in the XeF_6 system, including electronic and spectral measurements[244,245] (see also recent MO LCAO calculations[246]), confirms the presence of a PJTE in the space of odd T_{1u} displacements (dipolar instability).

Interesting conclusions about the vibronic effect in the stereochemistry of transition metal and rare-earth octahedral systems are examined below in the section on crystal chemistry, since they are stable mostly in the solid phase. Many examples of vibronic effects in stereochemistry are given by Pearson.[17] Some examples are given in Table 5.1 (see also the Bibliographic Review[14]).

Displacement of the Central Atom in Coordination Compounds. The displacement of the central atom from its geometric center in coordination compounds is a special case of vibronic effects in stereochemistry. An example of such an effect was considered in Section 2.3 for the TiO_6^{8-} cluster in titanate crystals, where it was shown that, as a result of the PJTE (vibronic mixing of the ground A_{1g} state with the near T_{1u} term), the Ti atom is displaced from the inversion center leading to dipole moment formation. This displacement (as shown in the next section) results in spontaneous polarization and ferroelectric phase transitions, provided cooperative effects in crystals are taken into account.

Consider now another illustrative example: the out-of-plane displacement of the iron atom, as well as of other metal atoms, in metal porphyrins and hemproteins.[247,248] Besides the special interest in biology, this example has general significance which reflects the situation in a great number of corresponding classes of coordination and organometal compounds. The presence of a number of near energy levels of the metal d electrons and porphyrin (or related ligand) π electrons is an important feature inherent in these systems. For instance, suppose the near states mix under A_{2u} nuclear displacements (shifting the metal atom out of and transverse to the porphyrin ring plane), as in the case of metal porphyrins, the system is softened or even becomes unstable with respect to this displacement (PJTE, Section 2.3).

Visually, the out-of-plane displacement of the metal atom with regard to the porphyrin ring can be explained as follows. This displacement causes an additional π type overlap between the d orbitals of the metal and π orbitals of the ligands (Figure 2.13) and the contribution of the additional π bond formation to the bonding energy is greater than the corresponding deterioration of the σ bonds.

According to equation (2.38), the softening and instability of the high symmetry configuration due to the PJTE for a two-level system is determined by the value of F^2/Δ, where $F = \langle 1|(\partial V/\partial Q)_0|2\rangle$ is the vibronic constant of mixing of the two levels, 1 and 2, by the nuclear displacements Q, and 2Δ is the energy gap between these levels. With the PJTE the new force constant is $K - F^2/\Delta$, and if $F^2/\Delta > K$, the system becomes unstable with respect to the displacement Q. For metal porphyrins (MeP) of D_{4h} symmetry, displacement Q shifting the metal atom out of the porphyrin ring plane is of A_{2u} symmetry (Fig. 5.5), which means that F is nonzero if (and only if) the product of the representations of mixing states 1 and 2 contains the A_{2u} representation.

Figure 5.6 shows several HOMO and LUMO energy levels of Mn, Fe, Co, Ni, Cu, and Zn porphyrins, as well as Mn phthalocyanine (MnPc)

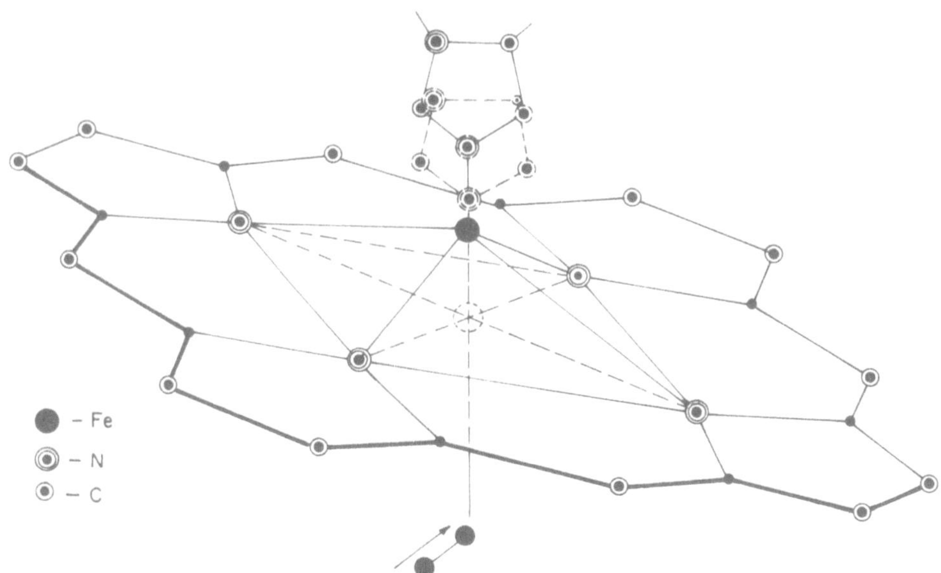

Figure 5.5. Structure of metal porphyrins and corresponding hemprotein fragments. In the D_{4h} point group symmetry the out-of-plane displacement of the metal atom is of A_{2u} symmetry.

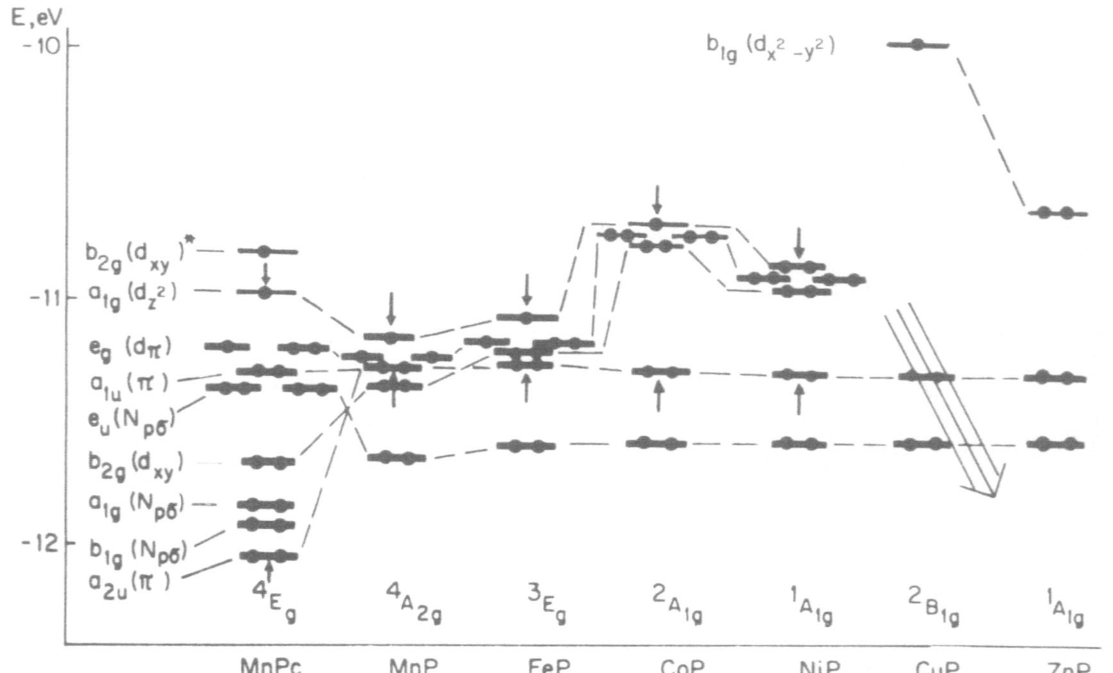

Figure 5.6. Several HOMO and LUMO energy levels of some metal porphyrins (MeP) and manganese phthalocyanine (MnPc) for the metal in-plane configuration. Ground states are indicated below. The energy gap 2Δ between the ground and lowest excited states, which are mixed by the metal out-of-plane displacement, is shown by arrows.

obtained from calculations of the planar configuration.[249,250] It is seen that in FeP, for instance, the calculated ground state is ^3E_g, so the excited state mixing with the ground state by A_{2u} displacements must be ^3E_u (since in the D_{4h} group $E_g \times A_{2u} = E_u$); in MnP, MnPc, and CoP, the corresponding excited states are $^4A_{2u}$, ^4E_u, and $^4A_{2u}$. All these excited states correspond to a one-electron excitation from the $a_{2u}(\pi)$ orbital predominantly of the porphyrin ring to the empty $a_{1g}(d_{z^2})$ predominantly metal MO, the excitation energy $2\Delta \cong \varepsilon(a_{1g}) - \varepsilon(a_{2u})$ being relatively small. In the remaining metal porphyrins, NiP, CuP, and ZnP, the a_{1g} MO is occupied, whereas the next MO of the same symmetry is very high (not shown in Figure 5.6). It follows that in the last three MeP, there is practically no PJTE with regard to the out-of-plane displacement of the metal atom, whereas in the remaining cases the effect may be important.

For a quantitative estimate of the effect the calculated, approximate values of 2Δ can be used; they are 0.15, 0.20, 0.6, and 1.0 eV in MnP, FeP, CoP, and MnPc, respectively. The vibronic constant has been estimated[247] as $F \approx 10^{-4}$ dyn for FeP (for other MeP with approximately the same wave functions F is of the same order of magnitude), and the value of K for the A_{2u} displacements is $K \sim 10^4$ dyn/cm. Hence $F^2/K \sim 0.1$ eV, i.e., F^2/K and Δ are of the same order of magnitude, and this confirms the possibility of a PJTE which softens the ground state in the cases under consideration. Then, in MnP and FeP ($\Delta \sim 0.08$ eV and 0.1 eV, respectively), $\Delta < F^2/K$ and instability may occur, while in CoP and MnPc ($\Delta \sim 0.3$ eV and 0.5 eV, respectively) only softening but not instability of the metal position is expected. These results are in full qualitative agreement with available empirical data.[251,252]

Interest in the metal atom position with respect to the porphyrin ring in biology is due to its significance in determining some biological functions of hemproteins. In particular, according to Perutz,[253,254] in hemoglobin (Hb) the iron atom going out of the porphyrin plane in deoxy form and returning by oxygenation initiates the conformational $T \rightarrow R$ transition (trigger mechanism) which determines the cooperative properties of Hb and its uncommon ("S" type) oxygenation kinetics. Similar heme centers in different proteins perform a wide spectrum of functions, from electron transport to oxidation of saturated hydrocarbons.

The observed out-of-plane displacements of the iron atom in the deoxy form of Hb (and other related systems) may be understood by means of the above PJTE in metal porphyrins.

Under the influence of imidazole moiety of the proximal histidine residue (Figure 5.5) the PJTE is not removed, since the most important a_{1g} and a_{2u} MO energy level positions remain essentially unchanged.[255] It is interesting that the return of the iron atom back to the heme plane as a result of oxygenation may be explained by the same vibronic approach.

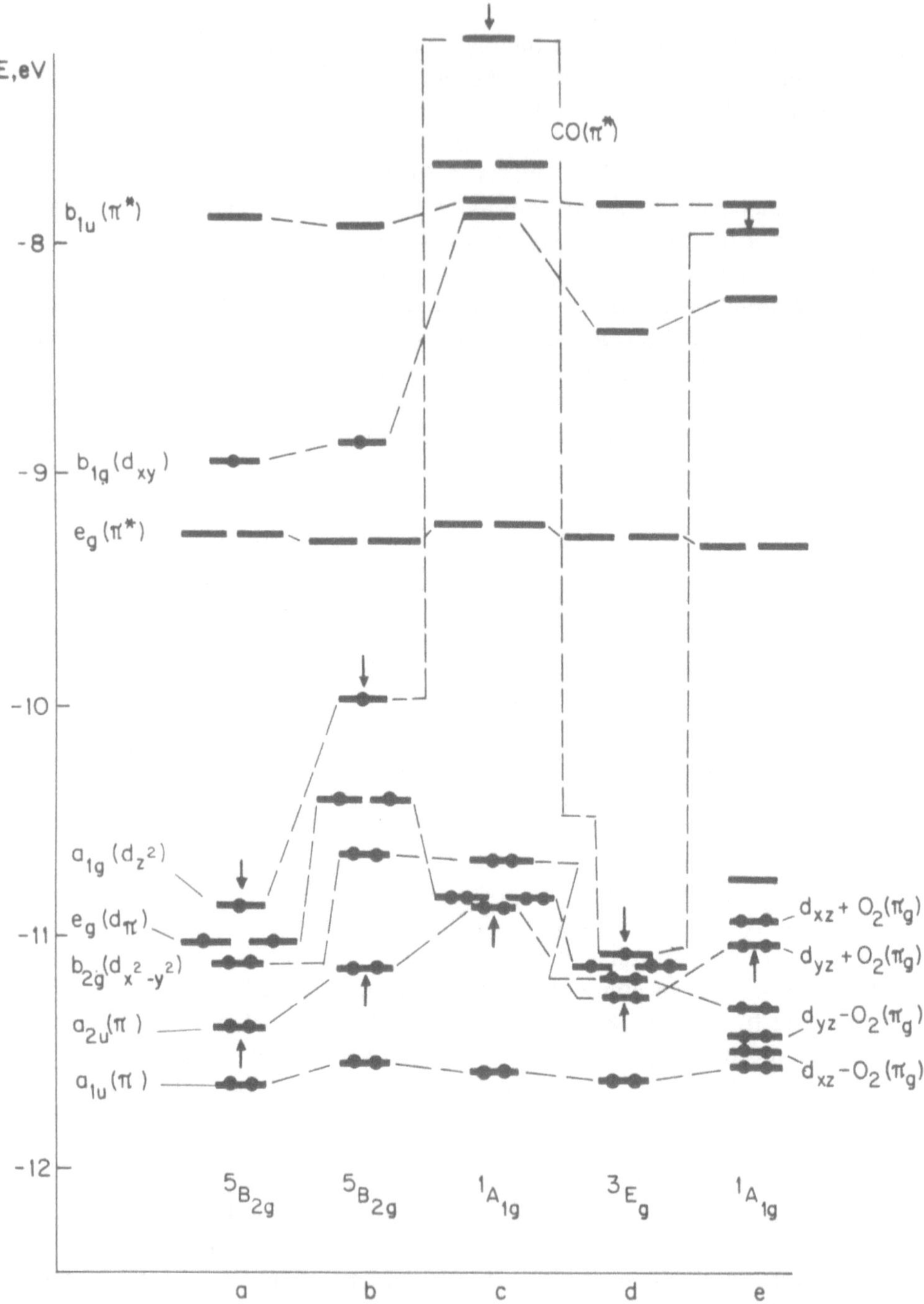

Figure 5.7. Several HOMO and LUMO for: (a) FeP with the metal out-of-plane position at a distance of $R = 0.49$ Å; (b) hemoglobin (Hb) in its deoxy form, $R = 0.62$ Å; (c) Hb–CO; (d) planar FeP, $R = 0$; (e) Hb–O_2. The ground state and the energy gap 2Δ between the states mixing under the Fe atom out-of-plane displacement are also shown.

The MO energy levels are shown in Figure 5.7 for the plane configuration of FeP for the deoxy form of Hb, and for Hb with coordinated O_2 and CO.[249,255] It is clearly seen that the relatively small energy gap $2\Delta = \varepsilon(a_{1g}) - \varepsilon(a_{2u})$, providing for the softness and instability of the first two systems, increases sharply under O_2 or CO coordination ($2\Delta \sim 3$ eV for HbO_2 and $2\Delta \sim 4$ eV for HbCO). Such an increase in Δ completely removes the PJTE. As a result the iron atom position in the heme plane becomes stable. This leads to the mentioned $T \rightarrow R$ conformational transition and triggers the oxygenation mechanism. Simultaneously, the increase in Δ may lead to a change in spin state from $S = 2$ to $S = 0$ (Figure 5.7) typical for iron and some other transition metal compounds and known as the "crossover" phenomena (see Section 5.2; the transition from the high-spin to low-spin state takes place when $\Delta > \Pi$, where Π is the energy of electron pairing on the appropriate MO[53]).

An interesting group of cobalt and rhodium dioxymines with general formula $[Me(DH)_2APh_3Hal]$, where Me = Co, Rh; DH is the dimethylglyoxymat-ion, $A = P$, As, Sb; Ph = $(C_6H_6)_3$, and Hal = Cl, Br, J, has been shown to have the distinctive feature of the central atom going out of the dioxyme plane in one of the two known modifications of these compounds[256–259] (Figure 5.8) (see Section 5.3 about the other modification).

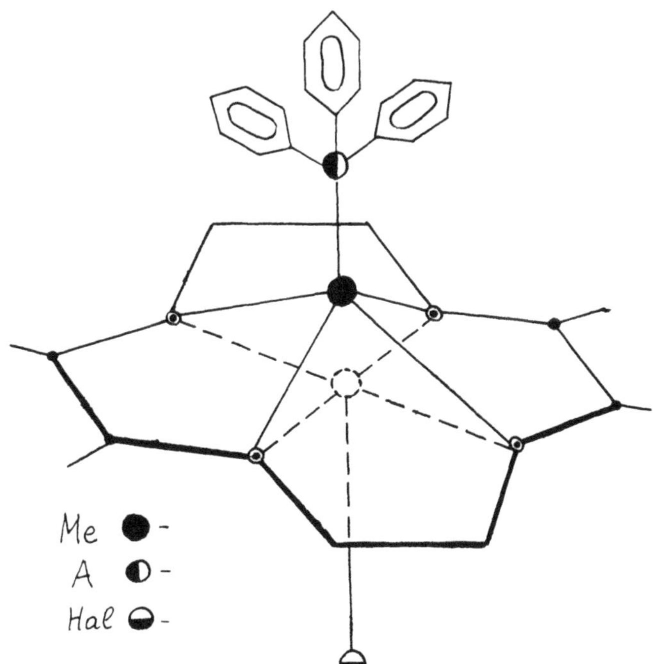

Figure 5.8. Structure of some dioxymines of Co and Rh of general formula $[Me(DH)_2APh_3Hal]$. The out-of-plane displacement of the Me atom is illustrated.

The energy level scheme in Figure 5.9 shows that both in the pure cobalt dioxyme and in its complexes, similar to those under consideration, there is a low-lying ($2\Delta \sim 0.1$ eV)[260] excited electronic state which mixes with the ground state leading to softness and instability of the system with respect to the out-of-plane displacement of the Co atom. The distortion isomers of these compounds arising as a result of this vibronic stereochemistry are discussed below in Section 5.3.

Geometry of Small Ligand Coordination to the Central Atom. The problem of the mode of coordination of small ligands to the central atom in coordination compounds can be successfully examined by means of the vibronic approach as well. Consider, for example, the cases of O_2, CO, and NO coordination to metal porphyrins and hemes in hemproteins (for the coordination of NO to other systems see Enemark and Feltheim[261] and references therein). Four modes of coordination are observed experimentally (Figure 5.10): linear end-on, bent end-on, bent side-on, and symmetrical side-on. The study of the mode of coordination with metal porphyrins in model compounds shows that, depending on the metal, the linear end-on coordination is characteristic for NO and CO, while the bent end-on

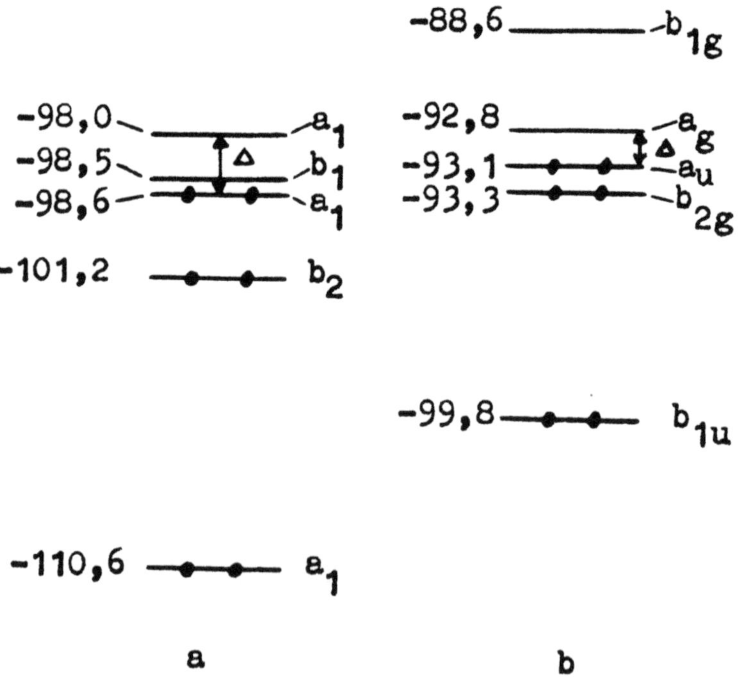

Figure 5.9. Energy levels of several HOMO and LUMO in some metal dioxymines: (a) $Co(DH)_2NH_3Cl$; (b) $Ni(DH)_2$. Values are given in 10^3 cm^{-1}. 2Δ is the energy gap between the states mixing under the metal out-of-plane displacement.

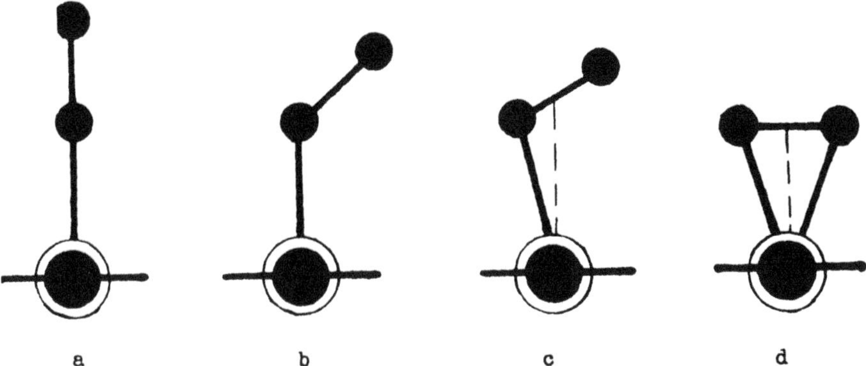

Figure 5.10. Geometry of diatomic ligand coordination with metalloporphyrins: (a) linear end-on; (b) bent end-on; (c) bent side-on; (d) symmetrical side-on.

coordination is observed for NO and O_2, and the symmetrical side-on coordination is seen for O_2.[262–265] The origin of these geometries has been subject to much discussion (see Bersuker and Stavrov[248] and references therein).

Consider this problem using the vibronic approach.[248] The latter implies that we must clarify whether the bent end-on configuration appears as a consequence of the PJT instability of the linear end-on and symmetrical side-on coordinations. In metal porphyrins of D_{4h} symmetry with linear end-on coordination of diatomics at the fifth coordinate, the bending of the ligand is an E type displacement (below, the influence of the imidazol at the sixth coordination position, say, in Hb is disregarded as being unimportant). Therefore the PJTE with respect to bending of the ligand may take place if the product of representations of the ground and near excited states contains the E representation.

Several HOMO and LUMO energy levels for the MeP–CO, MeP–NO, and MeP–O_2 systems calculated under the assumption of linear end-on coordination[255] are given in Figure 5.11. The qualitative changes of some of these levels due to their PJT mixing with the states formed by one-electron transition from the $e(d_\pi - \pi^*)$ or $e(d_\pi + \pi^*)$ to the $a_1(d_{z^2})$ MO are illustrated in Figure 5.12. Here, only the most essential mixing orbitals with sufficiently large vibronic constants are considered. Large vibronic constants occur when the mixing states contribute substantially to the bond formation and are strongly influenced by the E displacements (by ligand bending). For example, the E displacements formally mix the MO $e(P)$ with $a_1(d_{z^2})$, but since $e(P)$ is the state of the porphyrin ring not linked to the metal and diatomic ligand states, the vibronic constant $F = \langle e(P)|(\partial V/\partial Q_E)_0|a_1(d_{z^2})\rangle$ will be near-zero.

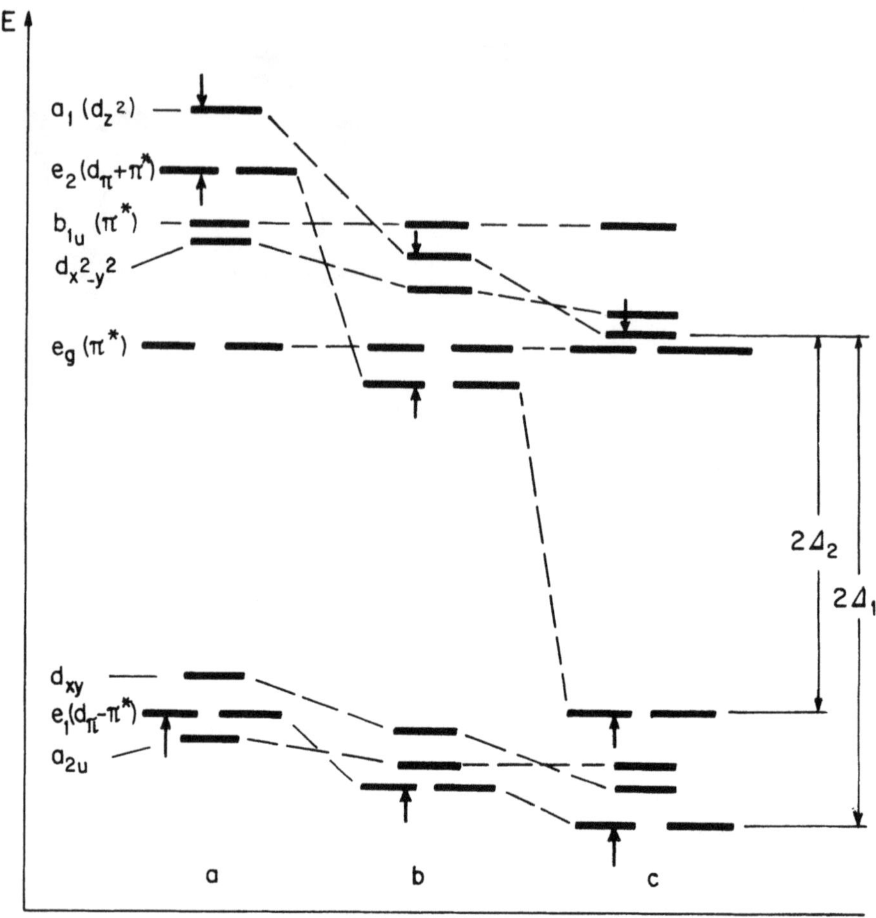

Figure 5.11. MO energy diagram for linear end-on coordination of diatomics CO (a), NO (b), and O_2 (c) with metalloporphyrins. Energy gaps $2\Delta_1$ and $2\Delta_2$ between the corresponding *e* and a_1 MO, mixing under ligand bending (*e* displacements), are shown by arrows.

If the corresponding energy intervals are $2\Delta_1$ and $2\Delta_2$ (Figure 5.11), the vibronic constants F_1 and F_2, and considering vibronic mixing of each pair of MO, $e(d_\pi - \pi^*)$ with $a_1(d_{z^2})$ and $e(d_\pi + \pi^*)$ with $a_1(d_{z^2})$, we conclude (using the expressions of Section 2.3) that each of these mixings lowers the force constant K_E by an amount dependent on the respective MO population numbers. Denoting the latter by q_1, q_2, and q_3, respectively, for levels 1, 2, and 3 (Figure 5.11), we obtain the following condition for instability of the linear end-on configuration:

$$[(q_1 - q_3)F_1^2/\Delta_1] + [(q_2 - q_3)F_2^2/\Delta_2] > K_E \tag{5.13}$$

If this condition is satisfied, the AP of the system with respect to the E displacement in question has a maximum for the linear end-on coordination and a continuum of minima, forming a circular trough. Each point on the latter corresponds to a bent end-on coordination at some angle to the linear end-on line with arbitrary orientation around this line. If the quadratic terms of the VI are included in the calculation, four additional minima are formed along the bottom of the trough, alternating regularly with four saddle points (Section 2.3). Hence the bent end-on coordination of diatomics has four preferable orientations with respect to the pyrrole ring: either toward the nitrogen atoms, or between them, depending on the sign of the quadratic vibronic constants (see the E–A mixing problem considered in Section 2.3).

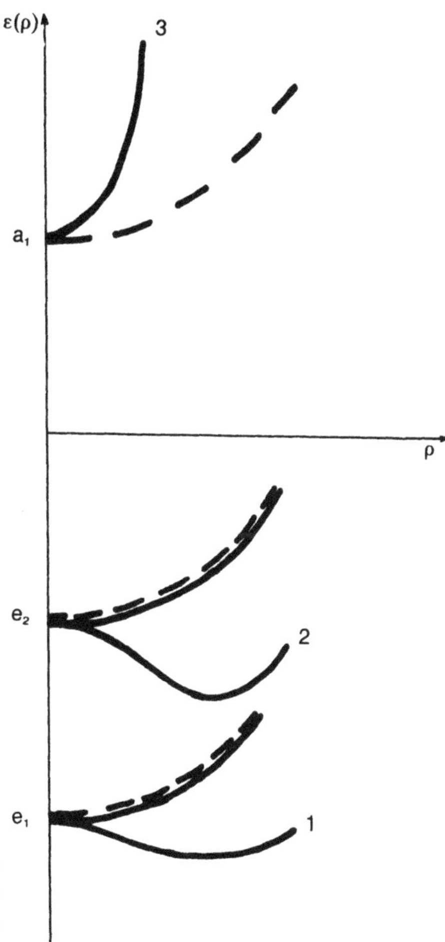

Figure 5.12. Energies of some MO in Figure 5.11 as functions of the absolute value of the angle of ligand bending ρ with (solid line) and without (dashed line) PJT e–a_1 mixing.

Consider now the MeP–CO system, for which $2\Delta_1 \cong 4.0$ eV and $2\Delta_2 \cong 0.3$ eV (Figure 5.11). Using the known order of magnitudes of F and K_E we conclude that if only the lower $e(d_\pi + \pi^*)$ state is populated by electrons and contributes to softening of the system (i.e., $q_2 - q_3 = 0$), inequality (5.13) is not satisfied (Δ_1 is large) and the linear coordination is stable. If the higher MO $e(d_\pi - \pi^*)$ is populated, inequality (5.13) is satisfied due to the small value of Δ_2 and the linear end-on coordination becomes unstable. However, in the linear end-on coordination of MeP–CO the state $e(d_\pi - \pi^*)$ remains unoccupied for all $3d^n$ metals, including $3d^{10}$. Consequently, according to vibronic theory predictions, in $3d^n$ transition metal porphyrins the CO molecule will always be linearly end-on coordinated. This prediction is in agreement with experimental data.[262]

For the MeP–NO system, quantities $2\Delta_1$ and $2\Delta_2$ have the same orders of magnitude, as in the previous case of MeP–CO, but the energy level ordering is different (Figure 5.11). In this case the $e(d_\pi - \pi^*)$ MO becomes populated when the number of d electrons plus the antibonding π^* electrons of the NO molecule exceed six. Thus in the $3d^n$ metal porphyrins, the NO molecule in the linear end-on configuration will be stable for $n \leqslant 5$ and unstable for $n > 5$, in agreement with the above experimental data.[263]

Finally, in the case of MeP–O, both the e MO levels are situated at about the same distance from the a_1 level (Figure 5.11) ($2\Delta_1 \cong 3$ eV and $2\Delta_2 \cong 2.5$ eV) and may make a comparable contribution to softening of the linear coordination Me–O–O with respect to the E bending. Therefore, it can be assumed that the cumulative effect suffices to provide essential softness or even instability of the linear coordination. Besides, softening is expected approximately double when the number of d electrons plus two antibonding π^* electrons of the O_2 molecule exceed six. This result also agrees qualitatively with experimental data.[264,265]

Stability of the symmetric side-on coordination of small ligands can be treated similarly. The displacements toward bent coordination are of B_1 (or B_2) symmetry, for which the nonzero vibronic constants correspond to mixing of the $b_1(d_{xz})$ (or $b_2(d_{yz})$) with the a_1 MO. Both these MO are nondegenerate and hence the PJTE can be obtained directly using the two-level expressions of Section 2.3. Estimates for the Me–O_2 system show that the criterion $F^2/\Delta > K_E$ is not satisfied, and the symmetric side-on coordination of O_2 is expected to be stable.

It is clear that the above vibronic approach can be used to explain the origin and expected geometries of coordination in any other molecular or solid state system and for any kind of ligands. In Chapter 6, some other aspects of ligand coordination are considered in connection with their activation.

Examples of Molecular Systems in Which Vibronic Effects Were Studied. Some examples of molecules and molecular groups in which the appearance of vibronic effects were studied by different methods are listed in Table 5.1. The list is by no means exhaustive, neither are the references complete for all the systems considered. Not every class of JT system studied in the crystal state is listed in this table; some are discussed in the next two sections (see also Tables 5.2–5.5). Also, no examples of the numerous JT and PJT impurity centers in crystals are listed. Since the vibronic effects are in principle inherent in all molecular and crystal systems, this discussion is only about those systems which have already been studied from the standpoint of the vibronic effect, the others have not yet been studied. For a more comprehensive list of vibronic systems studied by different methods, the reader is referred to the Bibliographic Review.(14)

5.2. Cooperative Vibronic Effects in Crystals. Structural Phase Transitions

Cooperative JTE. A new trend in solid state chemistry and the theory of structural phase transitions, based on the cooperative JTE and PJTE (CJTE and CPJTE, respectively), *has developed in the last twenty years.* Consider a crystal in which some of the regular (translational) centers are in degenerate (or pseudodegenerate) states (JT or PJT centers) as described by means of local AP given in Chapter 2. Without interaction between these centers, the motion of each is determined by its own AP and does not depend on the motions of other centers. The dynamic distortions around different centers are not mutually correlated. When the interaction between the centers is taken into account, the mean "molecular field" produced by the environment and influencing each center is not isotropic at any given instant. Therefore, equivalent distortions in free centers become nonequivalent when the interaction is nonzero. When this happens the minimum of the free energy corresponds to the crystal state in which each JT or PJT center is statically distorted and the distortions at different centers are correlated. The structure of such a crystal is considerably different from that expected without vibronic interaction.

Consider a simple case when the JT centers in the crystal possess a twofold orbital degenerate E term vibronically coupled to the B_1 vibrations (for the E–b_1 problem see Section 2.1; the E–b_2 problem is similar). Such a situation may occur, for example, when the crystal E term centers in question possess local D_{4h} symmetry. For simplicity assume that the JT centers are square planar. The AP with two equivalent minima and the rhombic distortions (B_{1g} displacements) of the square at each of the

Table 5.1. Examples of Molecular Systems for Which Vibronic Effects Were Studied

Molecular system	Symmetry, vibronic term (problem)	Method of investigation[a]	Reference
H_3	D_{3h}, $E-e$	*Ab initio* and semiempirical	233, 266
Li_3	D_{3h}, $E-e$	*Ab initio* calculation	267
BH_2	$D_{\infty h}$, RE	Electronic spectra + theory	229, 268
BH_3	D_{3h}, C_{3v}, $E-e$	Calculation of the AP	268–270
CH_2	$D_{\infty h}$, RE	Absorption spectra	232, 268
CH_4	T_d, $^2T_2$	PES	237, 271, 272
		mass spectrometry	273
		Ab initio calculation	338
CH_4^+	T_d, $^2T_2$	PES, mass spectrometry	273
	T_d, C_{3v}, D_{2d}	Calculation of AP and $^2T_2 \rightarrow A_1$ transition	238–240
	C_{3v}	Ionic experiment	154
CF_4^+	T_d, C_{3v}, D_{2d}	Theory	240
CH_3F, CH_3J CF_3J, CH_3Cl^+, CH_3Br^+	C_{3v}, $E-e$	PES	271–275
C_2H_4, C_2D_4	PJTE	Calculation	276
C_2H_6, $C_2H_6^+$	D_{3d}, ^2E_g, $E-e$	PES, *Ab initio* calculation	277, 278
Allene	JTE, $E-e$	PES	279
	PJTE	PES	280
Cyclopropane	D_{3h}, $E-e$	PES, CNDO calculation	281
Cyclopropyllenes	$E-e$	PES	282
Cyclopropenyl		PES	339
Cyclobutadiene	D_{4h}, $E-(b_1+b_2)$, PJTE	Calculation	283–285
C_5H_5	D_{5h}, $E-e$	Calculation, ESR	284, 286, 287
C_6H_6	D_{6h}, $E-e$	PES, Calculation	284, 288

Table 5.1 — *continued*

Molecular system	Symmetry, vibronic term (problem)	Method of investigation[a]	Reference
$C_6H_6^{\pm}$	D_{6h}, $E-e$	Calculation	284
$C_6X_3H_3^+$	D_{3h}, $E-e$	Emission spectra, calculation	170–181
Dedacyclene	C_{3v}, $E-e$	Calculation	289
Adamantan	JTE	PES	290
Cumulenes		PES, calculation	291
Triphenylene ion		MCD, absorption spectra, theory	292 293
Acetophenon	PJTE	Phosphorescence spectra	294
Aromatic aldehydes	$^3A''(n\pi)$, PJTE	IR spectra	295
Phenyl alkyl ketones		Phosphorescence spectra	296
NH_2	RE	Theory	228
NH_3, NH_3^+	D_{3h}, C_{3v}	Theory	240, 269, 270, 297
NO_3^-	D_{3d}, $E-e$	Raman spectra	298
P_4		PES, theory	317
S-triazine, tricyano-S-triazine	$E-e$	ESR, electronic spectra, calculation	170, 299
p-chloraniline	PJTE	Phosphorescence spectra	300
Cinnoline		Phosphorescence spectra	301
Pentachlorides $NbCl_5$, . . . , WCl_5	$E-e$	IR and Raman spectra	302
Acetylacetonato metals, Tropolonato metals	$E-e$	IR spectra, X-ray analysis, theory	303, 304, 304a
α-Thenoyl trifluoroacetonato metals		IR spectra	336

continued

Table 5.1 — *continued*

Molecular system	Symmetry, vibronic term (problem)	Method of investigation[a]	Reference
MX_6, MX_{12}, M = Al, Ga, X = inert gas element	E_{2g}, T_{1u}	Absorption spectra, ESR	305
Metalloporphyrins	C_{4v}, D_{4h}, $E-(b_1+b_2)$, PJTE	Theory	247–249, 306, 337
$Cr(C_6H_6)_2$	D_{6h}, D_{3d}, PJTE	Theory	307
$Co(C_5H_5)_2$	D_{5h}, D_{5d}, $E-e$	Absorption spectra, ESR magnetic susceptibility, electron diffraction	128, 308
Sandwich compounds Me(Cp)$_2$	D_{5h}, D_{5d}, $E-e$	CNDO calculation	309
		ESR	129
		PJTE theory	321
Mn(III) and Cr(II) complexes	O_h, E, T	Absorption spectra	311
$CuCl_6^{4-}$, CuF_6^{4-}	O_h, $E-e$	Calculation	310
$Cu(H_2O)_6^{2+}$	O_h, $E-e$	Calculation	311a
Hexa-fluoride complexes	O_h, T	Electron diffraction, Raman spectra	312, 318
ReF_6	O_h, T_2	All kinds of spectra, electron diffraction, calculation, general theory	242, 244–246, 340
TcO_4^{2-}, ReO_4^{2-}	T_d, $^2T_2$	Electronic and IR spectra, magnetic susceptibility	313
K_2PtCl_4	D_{4h}, E	Absorption spectra	314
$SeCl_6^{2-}$, $SeBr_6^{2-}$, $TeCl_6^{2-}$, $TeBr_6^{2-}$	O_h, $T-(e+t_2)$	Absorption and Raman spectra, theory	315
$Ti(urea)_6J_3$ $Cr(urea)_6(NO_3)_3$	O_h, $E-e$	Polarized luminescence spectra	316

Table 5.1 — *continued*

Molecular system	Symmetry, vibronic term (problem)	Method of investigation[a]	Reference
$Cr(NH_3)_6^{3+}$	O_h, JTE, PJTE	Luminescence spectra	326
$Mn(DMSO)_6(ClO_4)_3$ } $Mn(DMF)_6(ClO_4)_3$ }	O_h, ${}^5E_g$	IR spectra, magnetic susceptibility	319
Fe(II)phen$_2$X$_2$ } Fe(II)dipyX$_2$ }	${}^5E$–${}^5T_2$	Calculation	320
HCN	RE, PJTE	Calculation	322
NCO	RE	Absorption spectra, theory	323
VCl_4	T_d, E–e, T–e	Absorption spectra	324, 325
SO_3^+	D_{3h}, E–e	PES	327
$V(CO)_6$	O_h	Electronic diffraction	328
$CuZn(F_6acac_2)(py)_2$	O_h, E–e	Optics and ESR	329
$Ti(H_2O)_6^{3+}$	O_h, ${}^2T_2$–e	ESR	330
$Co_3(h^5$–$C_5H_5)S_2$ } $Co_3(h^5$–$C_5H_5)S_2^+$ }	T–e	Theory	331
$FeBO_3$	C_{3v}, E–e	Mössbauer spectroscopy	332
O=C(H)H		MCD, calculation	333
K_2OsCl_6 } K_2OsBr_6 }	O_h	Raman spectra	334
Cr_3COCl_5 } Cr_3COBr_5 }	${}^4A_2$, ${}^2E$, PJTE	Zeeman splitting under stress	335

[a] PES — photoelectron spectra, MCD — magnetic circular dichroism.

minima (+) and (−) are illustrated in Figure 5.13 (cf. Figure 2.6b). In the absence of interaction between the centers, the two configurations (+) and (−) are equally probable and the JTE has the above-discussed dynamic nature (the mean distortion equals zero and the initial symmetry is preserved).

The resulting picture changes when interactions between centers are taken into account. Consider two interacting centers at each of which the E–b_1 problem is realized. Under the assumption of parallel orientation of their squares, there may be four configurations in which the one-center distortions are correlated: $(+\,+)$, $(-\,-)$, $(+\,-)$, and $(-\,+)$ (Figure 5.14). It is clear that if the interaction is better with parallel orientation of the distortions (ferrodistortive interactions), the energy of the configurations $(+\,+)$ and $(-\,-)$ is lower than that of the $(+\,-)$ and $(-\,+)$ cases. On the other hand, if the distortions directed antiparallel to each other are

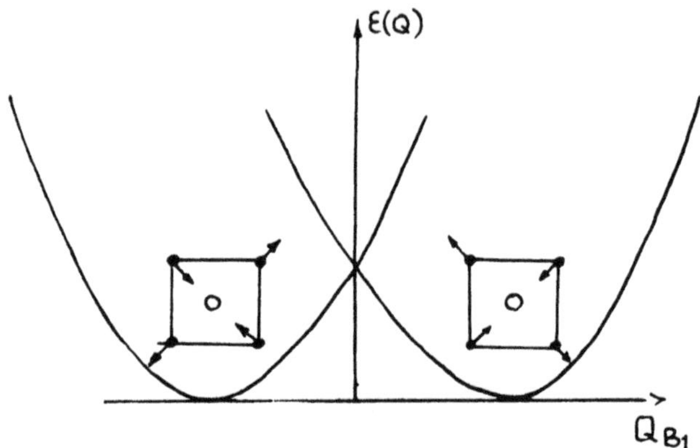

Figure 5.13. Two equivalent minima of the AP for the E–b_1 (or E–b_2) problem of a D_{4h} (C_{4v}) symmetry system with illustration of B_1 type distortion of a square system in minima configurations (cf. Figure 2.6b).

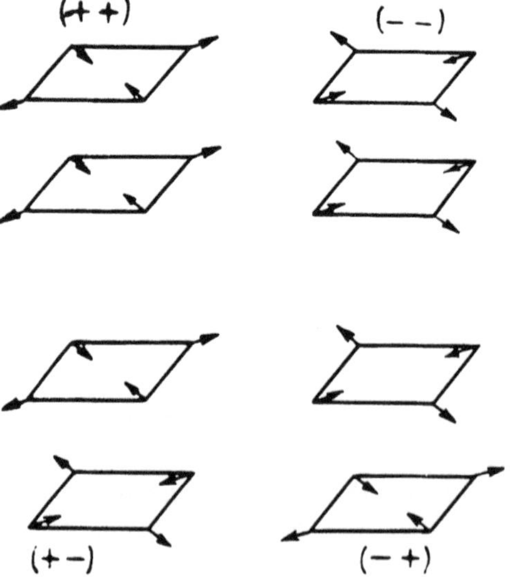

Figure 5.14. Four types of possible packing of two interacting JT centers, for each of which the E–b_1 JTE is realized.

preferred (antiferrodistortive interactions), configurations $(+\,-)$ and $(-\,+)$ possess less energy. In both cases the two configurations of each center are no longer equivalent, since the interaction energy depends on the kind of distortion at the other center. This conclusion can be generalized to many interaction centers and in the limit, to the whole crystal.

It is evident that at suitably low temperatures, the minimum free energy configuration is realized in which the crystal has statically distorted and distortion-correlated centers. In the case of ferrodistortive interactions, such an ordering of local distortions (ferrodistortive ordering) leads to a macrodeformation of the crystal as a whole. New properties of the crystal arising from the correlation (ordering) of the JT (PJT) center distortions, including the formation of new crystal structures and structural phase transitions, are called cooperative JTE (CJTE) or cooperative PJTE (CPJTE).

It may appear that in the above description the CJTE (or CPJTE) leads to a break of the initial symmetry of the system. In fact this is not true since, e.g., the two correlated configurations $(+\,+)$ and $(-\,-)$ are equivalent and, in principle, the pair of correlated centers may resonate between the $(+\,+)$ and $(-\,-)$ states, dynamically preserving the initial symmetry. However, in the case of a large number of centers of the macroscopic crystal the barriers between the equivalently distorted configurations of the entire crystal $(+\,+\,+\cdots)$ and $(-\,-\,-\cdots)$ become so high that transitions between them are practically impossible, and the crystal remains in one of them. The situation here is quite similar to that found in ferromagnetics, in which there are also different equivalent directions of magnetization but no spontaneous inversion of the magnetic moment.

In these cases, as distinct from the one-center problem, the symmetry of the ground state configuration of the crystal is lower than that of the Hamiltonian (the effect of "broken symmetry"). Strictly speaking, such a crystal state is not stationary, but due to the infinitely large barrier heights it may remain there for an infinitely long time.

It is clear that lattice vibrations and temperature fluctuations tend to destroy the correlation between JT distortions. Therefore, in principle, for any given energy of distortion interactions, there is a certain temperature above which the distortion ordering becomes destroyed. The lattice acquires another, more symmetric structure with an independent dynamic JTE at each center (provided the crystal does not melt). This temperature-dependent breakdown of the JT center distortion correlations (disordering) is nothing else than a structural phase transition. The stronger the JTE at each center and the energy of distortion interactions, the higher the temperature of the phase transition to the disordered state. *The structural phase transition to the crystal state with disordered JT or PJT local distortions is one of the most important features of the CJTE or CPJTE.*

The first studies of the CJTE appeared in the late fifties and early sixties.[341,342] Cooperative PJT ordering leading to spontaneous polarization and ferroelectric phase transitions has been established independently using another approach.[52,12] Rapid development of this trend began in the seventies, when the JT origin of structural phase transitions had been confirmed experimentally in a series of rare earth orthovanadates by Elliott et al. The results obtained are summarized in a series of reviews[5,9,10,12,13,14,343–347]; only illustrative examples are given below.

Rare Earth Zircons. The CJTE has been analyzed in more detail in the tetragonal rare-earth zircons of general formula RXO_4, where R is a rare earth element and X = V, As, P, which include the orthovanadates. From the viewpoint of possible experimental research on vibronic effects, these crystals have certain comparative advantages. First, they are transparent and so allow a series of convenient optical and spectroscopic methods of investigation to be used in order to determine the positions of the energy levels and their shifts. Second, the expected CJT structural phase transition occurs at low temperatures (~ 10 K), for which the measurements are not masked by thermal effects. Third, phase transitions at low temperatures mean that the corresponding JT energy level splitting is small (see below) and can be easily changed by external perturbations. Finally, in these crystals the JT ion is in a tetragonal symmetry environment, in which a twofold degenerate E term is coupled to nondegenerate vibrations. This results in the simplest vibronic E–b_1 problem.

The JT ion in rare earth zircons is the rare earth element. Figure 5.15 shows the energy levels of the ground and first excited states in three orthovanadates, $TmVO_4$, $DyVO_4$, and $TbVO_4$. The Tm^{3+} ion has a ground state $^3H_6$ (electron configuration $4f^{12}$), which in the tetragonal D_{4h} field of the VO_4^{3-} ions in TmVO reduces to an orbital doublet E. The latter is coupled predominantly to the B_{1g} vibrations which distort the square of ligands into a rhombus and split the E term into two nondegenerate terms, as shown in Figure 5.15. The Dy^{3+} ion in the same environment has a ground, orbitally nondegenerate Kramers doublet E' and another near E'' excited state at an energy interval of ~ 9 cm^{-1}. The two Kramers doublets E' and E'' mix strongly under the B_{2g} displacement resulting in PJT instability and B_{2g} type distortion. This increases the E'–E'' splitting. For the Tb^{3+} ion in $TbVO_4$ there are four such near energy levels which mix under B_{1g} displacements.

Consider first the simplest case of $TmVO_4$. The AP of the E–b_1 problem in question consists of two intersecting parabolas with two minima corresponding to the possible directions of distortion of a square into a rhombus (two signs of the minima in Figure 5.13). The fact that the B_{1g} vibration is nondegenerate and the states $(+)$ and $(-)$ are described by

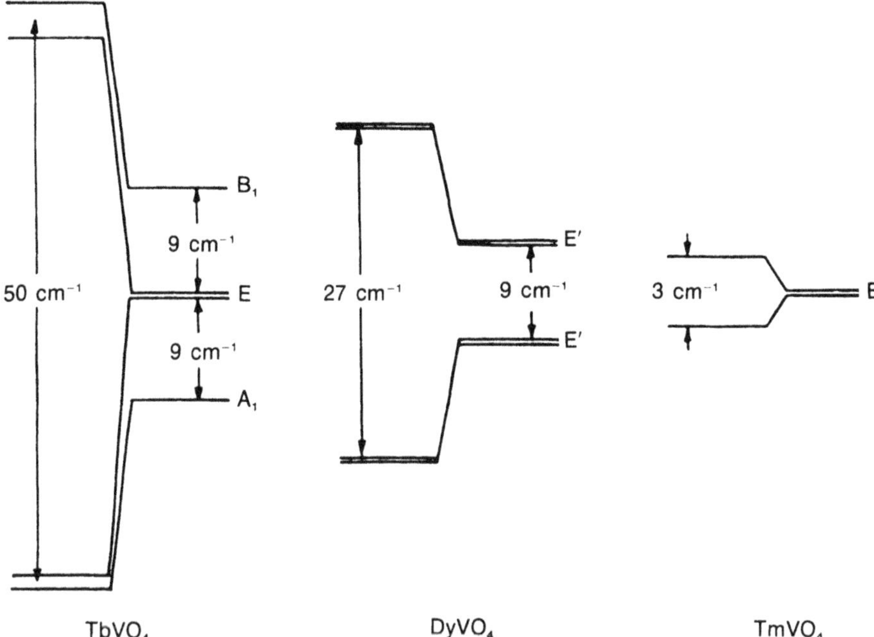

Figure 5.15. Energy levels of the lowest vibronic states of rare earth ions in crystals $TbVO_4$, $DyVO_4$, and $TmVO_4$ with (left) and without (right) CJTE ordering.

orthogonal electronic wave functions (there is no continuous transition from one minimum to another along the AP, see Figure 5.13) essentially simplifies the problem of distortion correlations of the JT centers in crystals. The two AP minimum states are formally similar to two possible spin states for systems with $S = \frac{1}{2}$, which can be described by means of the Pauli matrix $\hat{\sigma}_z$ having two eigenvalues $\sigma_z = \pm 1$ (the pseudospin method).

It can be shown that a relatively simple shift transformation of the Hamiltonian results in a single parameter λ describing ferrodistortive interactions of JT centers through nuclear displacements (through the phonon field). Along with λ, splitting ΔE of electronic energy levels due to ferrodistortive ordering also depends on homogeneous deformation of the lattice as a whole which accompanies ferrodistortive phase transition. Denoting the electron-deformation coupling parameter by μ, one can introduce a total parameter of correlation of different JT center distortions $\gamma = \lambda + \mu$. Introducing also an ordering parameter $\sigma = \langle \sigma_z \rangle$, one can obtain the following expressions[110]:

$$\Delta E = 2\gamma\sigma \tag{5.14}$$

$$\sigma = \tanh(\Delta E/2kT) \tag{5.15}$$

Hence the phase transition temperature T_0 for which the ordering parameter σ falls to zero is given by

$$kT_0 = \gamma \tag{5.16}$$

The temperature dependence of E term splitting has been determined experimentally from the position of the line of optical transition to the lowest excited singlet level $^1G_4$ ($\hbar\Omega = 20940$ cm^{-1}),[348] as well as by magnetic measurements[349] (see below). The results are shown in Figure 5.16. It is seen that at $T = 0$, $\Delta E = 2\gamma = 3$ cm^{-1} and hence $\gamma = \lambda + \mu = 1.5$ cm^{-1}. At $T_0 = 2.1$ K the splitting ΔE becomes zero. At this temperature a structural phase transition from the rhombic to the tetragonal phase is observed. Figure 5.17 illustrates the phase transition by the temperature dependence of heat capacity.[350] This value of T_0 coincides with that predicted by theory, equation (5.16).

The values of λ and μ can be obtained separately by piezospectroscopic measurements. Under external pressure P, the E term splitting is[351] $\Delta E = 2(\lambda + \mu)\sigma + \alpha P$, where α is proportional to μ. Therefore, by measuring the splitting ΔE as a function of P at $T = T_0$,[352] one obtains $\lambda = -0.75$ cm^{-1} and $\mu = 2.25$ cm^{-1}.

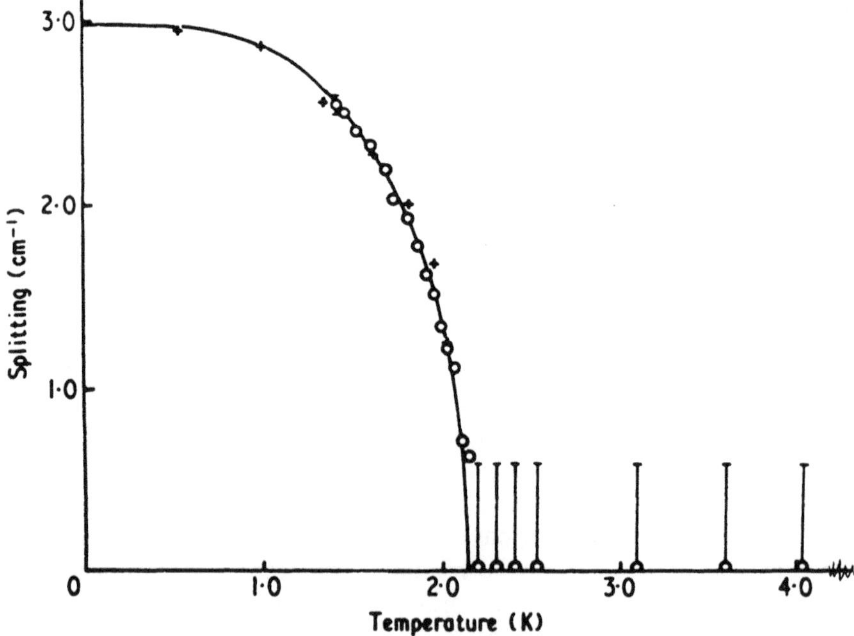

Figure 5.16. Splitting of the ground vibronic E level of the Tm^{3+} ion in $TmVO_4$ as a function of temperature obtained from the position of the line of optical transition to the excited $^1G_4$ level (shown by circles) and magnetic measurements (crosses). Vertical lines at $T > T_0 = 2.14$ K show the experimental error due to crystal imperfections (after Gehring and Gehring[10]).

The orbital doublet E also splits under the influence of magnetic field $\mathcal{H}_z$ in the presence of which expressions (5.15) take the form

$$\begin{aligned} \Delta E &= 2\gamma \tanh(\Delta E/2kT) \\ \Delta E &= [4\gamma^2\sigma_{\mathcal{H}}^2 + g^2K_E^2(T)\beta^2\mathcal{H}^2]^{1/2} \end{aligned} \tag{5.17}$$

where $\sigma_{\mathcal{H}} = \langle \sigma_z^{\mathcal{H}} \rangle$ is the ordering parameter in the magnetic field and, besides the g-factor, the magnetic field influence is determined by the vibronic reduction factor $K_E(T)$ (here, T indicates the symmetry of the magnetic field). It follows that the phase transition temperature T_0 at which $\sigma_{\mathcal{H}}$ decreases to zero depends on the magnetic field:

$$kT_0 = \tfrac{1}{2}gK_E(T)\beta\mathcal{H}_z/\operatorname{arc\,tanh}(gK_E(T)\beta\mathcal{H}_z/2\gamma) \tag{5.18}$$

Hence T_0 decreases with increasing magnetic field intensity $\mathcal{H}_z$. At some (critical) value $\mathcal{H}_z = \mathcal{H}_{cr} = 2\gamma/gK_E(T)\beta$, $T_0 = 0$, and for $\mathcal{H}_z > \mathcal{H}_{cr}$ no phase transitions can occur at real temperatures. The magnetic field also reduces the ordering parameter $\sigma_{\mathcal{H}} = \sigma - (g^2K_E^2(T)\beta^2\mathcal{H}_z^2/4\gamma^2)$. Substitution of this expression into equation (5.17) yields that for $\mathcal{H}_z < \mathcal{H}_{cr}$, E term

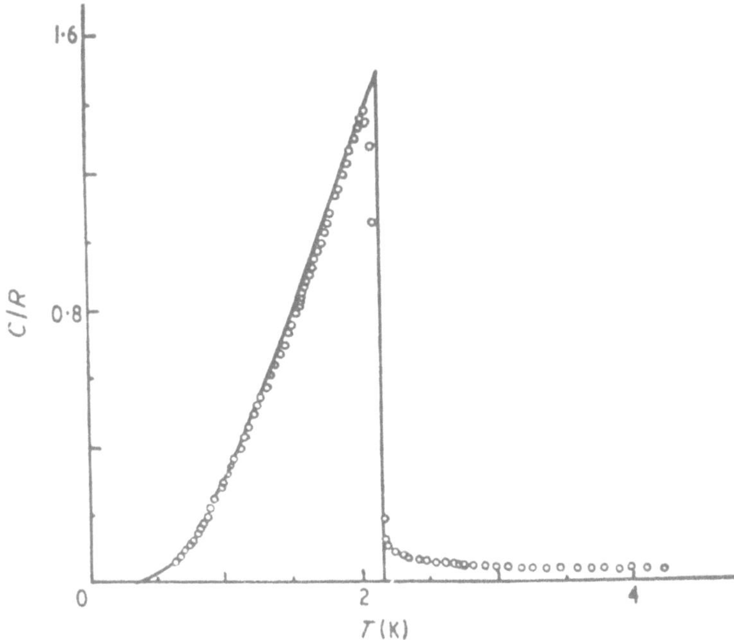

Figure 5.17. Temperature dependence of the heat capacity of the $TmVO_4$ crystal near the phase transition due to CJTE.

splitting ΔE does not depend on $\mathscr{H}_z$. If $\mathscr{H}_z > \mathscr{H}_{cr}$, $\sigma_{\mathscr{H}} = 0$ and ΔE increases linearly with $\mathscr{H}_z$. This result is in complete agreement with experimental data on optical absorption in magnetic fields[(349)] presented in Figure 5.18.

Thus for the $TmVO_4$ crystal under consideration, *comprehensive qualitative and quantitative experimental* (optical, magnetic, piezospectroscopic) *confirmation of the JT nature of low-temperature structural phase transition has been obtained.* These results are also confirmed by Raman spectra, neutron scattering, ultrasound absorption, X-ray analysis, heat conductivity, and so on.

An analogous explanation of the JT origin of structural phase transitions in other rare earth zircons can be given. In $DyVO_4$, as distinct from $TmVO_4$, the PJTE is realized in the Dy^{3+} ion centers (Figure 5.15), and as a result of their ferrodistortive interaction and cooperative structural phase transition, splitting of the energy levels is given by

$$\Delta E = 2[\gamma^2\sigma^2 + \Delta^2]^{1/2} \tag{5.19}$$

where 2Δ is the splitting in the high symmetry phase, for which $\sigma = 0$. On the other hand, as in the case of $TmVO_4$, the condition of ordering in a

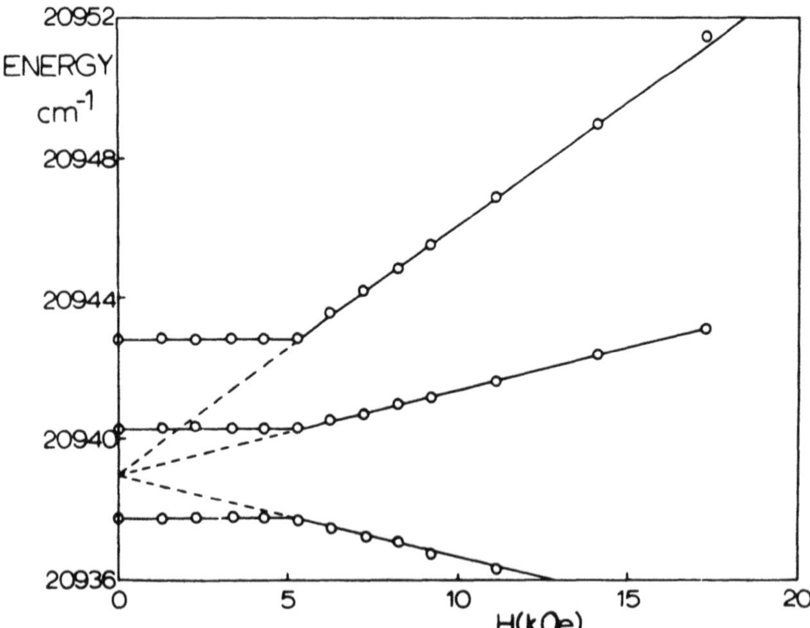

Figure 5.18. Splitting ΔE of the Tm^{3+} ion ground state in $TmVO_4$ in magnetic fields (obtained from the position of the line of optical transition to the $^1G_4$ level, Figure 5.16). When $\mathscr{H} \leqslant \mathscr{H}_{cr} = 5.4$ kOe and $T = 1.4$ K, ΔE is independent of the magnetic field intensity $\mathscr{H}$.

two-level system gives

$$\Delta E = 2\gamma \tanh(\Delta E/2kT) \tag{5.20}$$

and the phase transition temperature T_0 is determined by

$$\Delta = \gamma \tanh(\Delta/kT_0) \tag{5.21}$$

It follows from this equation that if $\Delta > \gamma$, there is no phase transition.

An important consequence of the structural phase transition of ferrodistortive type is the aforementioned homogeneous deformation of the crystal as a whole, due to which the elasticity constants C become temperature dependent near the phase transition. The effect was first observed in $DyVO_4$, for which the cooperative JT structural phase transition takes place at $T_0 = 14$ K. Figure 5.19 presents the observed temperature dependence $C(T)/C_0$ for this crystal as determined by sound velocity measurements[353] and calculated by the formula

$$\frac{C(T)}{C_0} = \frac{\Delta - (\lambda + \mu)\tanh(\Delta/kT)}{\Delta - \lambda \tanh(\Delta/kT)} \tag{5.22}$$

with $\lambda = -3.6$ cm^{-1} and $\mu = 14.7$ cm^{-1} ($C_0 = 0.63\ 10^{11}$ dyn $\cdot$ cm^{-2} is the elasticity constant in the region far from the phase transition). These results are also confirmed by other experiments, in particular by piezospectroscopic measurements.[354] Relationships between elastic and dielectric parameters of this crystal can be found elsewhere.[354a]

In $TbVO_4$ the local vibronic system of the Tb^{3+} center is a four-level one (Figure 5.15). The A_1–B_1 splitting due to CJTE with B_{1g} distortions is similar to the E'–E' one, equation (5.19):

$$\Delta E_1 = 2[\gamma^2\sigma^2 + \Delta^2]^{1/2} \tag{5.23}$$

where 2Δ is the initial splitting. Splitting of the doublet E is

$$\Delta E_2 = 2\gamma\sigma \tag{5.24}$$

It follows that the phase transition temperature T_0 obeys the equation[10]:

$$\frac{\gamma}{\Delta} + \frac{\gamma}{\Delta} \sinh \frac{\Delta}{kT_0} = 1 + \cosh \frac{\Delta}{kT_0} \tag{5.25}$$

Phase transition in $TbVO_4$ has been observed at $T_0 = 33$ K. The temperature dependence of splitting of the phonon E mode in this crystal, determined experimentally by Raman spectra and calculated in the

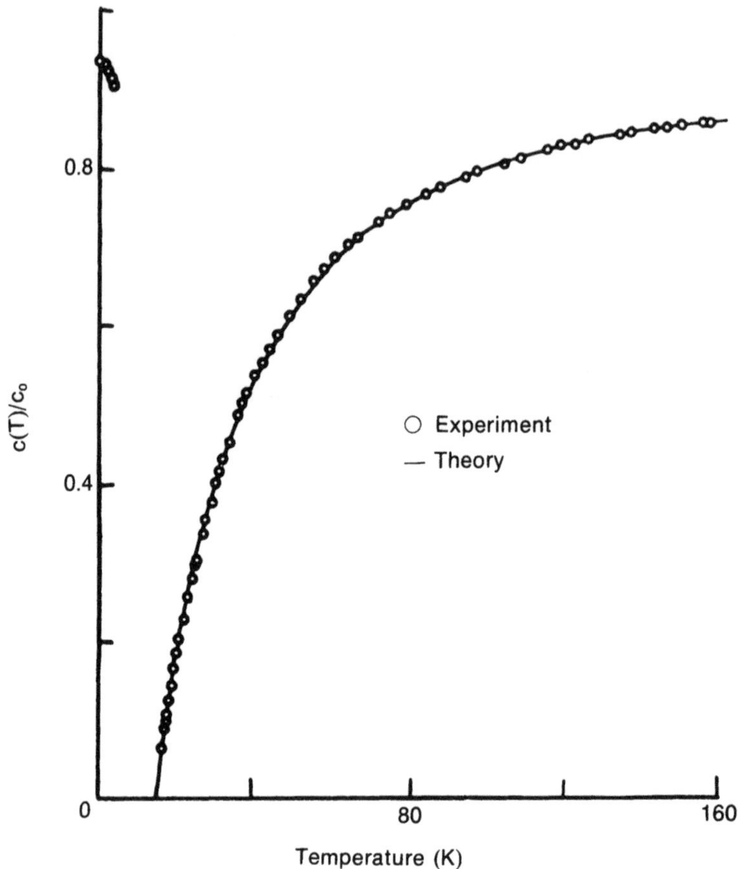

Figure 5.19. Temperature dependence of the elasticity modulus $C(T)$ (in C_0 units) of the $DyVO_4$ crystal near CJTE phase transition taking place at $T = 14$ K (after Melcher and Scott[353]).

"molecular field" approximation, is illustrated in Figure 5.20. Good agreement between theory and experiment confirms the JT origin of phase transition.

Similar conclusions have been drawn for other rare earth zircons mentioned above, including $TbAsO_4$ (27.7 K), $DyAsO_4$ (11.2 K), $TmAsO_4$ (6.1 K), and $TbPO_4$ (3.5 K), where the CJTE or CPJTE phase transition temperatures are given in parentheses. Some other classes of rare earth compounds, such as pictides with general formula RX, X = N, P, As, Sb, sulfides, and others of type TmCd and $PrAlO_3$, have a similar CJTE (or CPJTE). For example, DySb has a structural phase transition at $T_0 = 9.5$ K, its JT nature being confirmed by experiments on magnetic susceptibility, neutron scattering, electric conductivity, etc.[10]

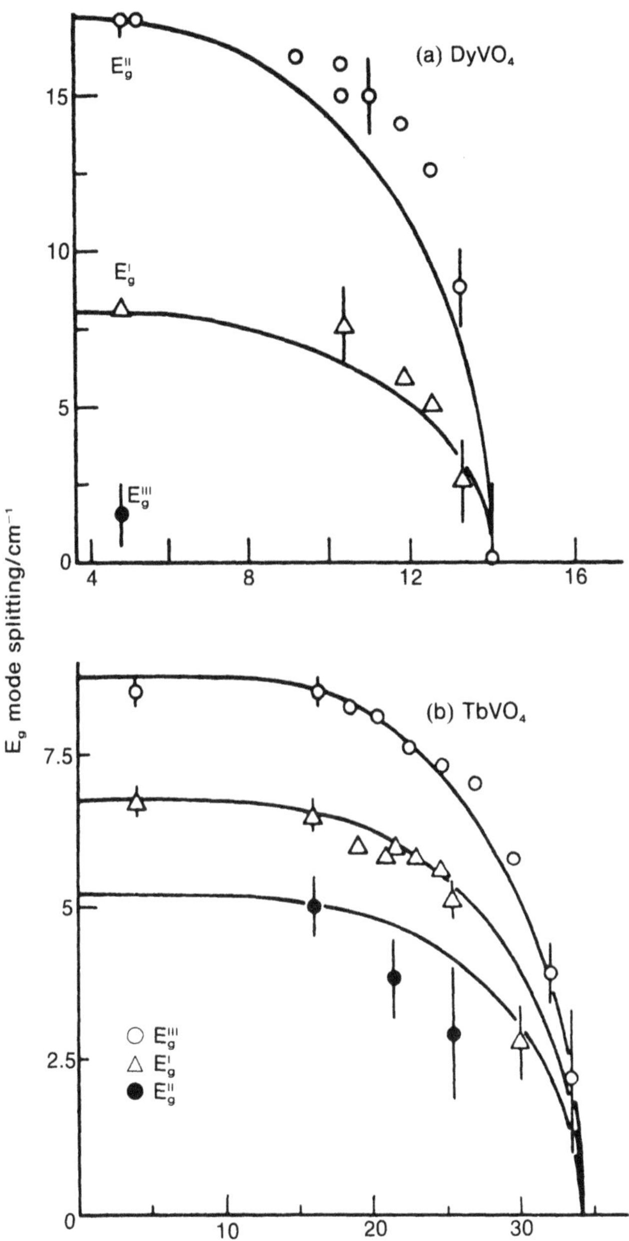

Figure 5.20. Temperature dependence of phonon E mode splitting near CJTE transition in $DyVO_4$(a) and $TbVO_4$(b). The points indicate experimental data obtained from Raman scattering, while the solid line follows theoretical results obtained by the mean field approximation (from Elliot et al.[343]).

Spinels, Perovskites, and Other Crystals. Spinels and perovskites containing JT transition metal ions in octahedral and tetrahedral environments were suggested as subject to the CJTE prior to the investigation of rare earth compounds.[355] However, research on transition metal compounds was delayed by experimental difficulties due to their nontransparency in the optical region, high phase transition temperatures, etc. The successful study of rare earth compounds stimulated more intensive work on spinels, perovskites, and other transition metal crystals.

Two types of spinel may be distinguished: T spinels, in which the ground state of the JT ion is a T term, and E spinels with similar E ground terms. At present, the CJTE structural phase transition has been found in a large number of crystals. The following are listed as examples (transition temperatures are shown in parentheses): T spinels — $NiCr_2O_4$ (274 K), $FeCr_2O_4$ (135 K), $CuCr_2O_4$ (860 K), FeV_2O_4, $FeCr_2S_4$; E spinels — Mn_3O_4 (1170 K), $CuFe_2O_4$ (360 K), $FeTiO_4$ (142 K); perovskites — $KCuF_3$, $KMnF_3$, $PbFeF_3$; other crystal structures — $CsCuCl_3$ (423 K), K_2CuF_4, Ba_2CuF_4, $(NH_4)_6CoCl_2$, $Copy_2Cl_2$, UO_2, $A_2BCu(NO_2)_6$ (A = K, Rb, Tl, Cs; B = Ca, Sr, Ba, Pb), $Cu(ONC_5H_6)_6X_2$ ($X = ClO_4$, BF_4).

The theory of the CJTE and CPJTE for d elements is much more complicated than for rare earth zircons, since in the former degenerate terms are usually coupled to degenerate vibrations and there are several (or an infinite number of) equivalent directions of local distortions (Chapter 2). It can be shown that in these cases the type of ordering and structural phase transition depends on the relative values of linear and quadratic vibronic coupling at each center and the kind of interaction between them. Qualitatively different results may be expected for strong, intermediate and weak vibronic coupling (characterized by dimensionless vibronic constants $\lambda_\Gamma = E_{JT}^\Gamma / n_\Gamma \hbar\omega_\Gamma$, Chapters 2 and 3) in all possible combinations with strong, moderate, and weak correlations between the centers (characterized by parameter γ).

By way of illustration, consider the case of an E–e problem in cubic systems with strong vibronic coupling $E_{JT}^E \gg \hbar\omega_E$ at each JT center. When the quadratic terms of vibronic interaction are taken into account, the local AP has three minima which, in the space of polar coordinates ρ and ϕ of the (Q_θ, Q_ϵ) plane, form a regular triangle with coordinates $(\rho_0, 0)$, $(\rho_0, 2\pi/3)$, and $(\rho_0, 4\pi/3)$ (Section 2.1). In the absence of external low-symmetry perturbations and for high quadratic barriers, the system performs local pulsating motions, i.e., tunneling transitions between the minima. The lowest tunneling levels are E (ground) and A_1 (or A_2), the tunneling splitting $\delta = 3\Gamma$ being much smaller than the vibrational quantum $\hbar\omega_E$. Therefore the problem may be reduced to a three-level one, or to a pseudospin problem with $S = 1$, $2S + 1 = 3$.

Under the influence of the anisotropic "molecular field" of the adjacent center distortions, the three minima become nonequivalent, the vibronic E and A_1 levels being displaced. The calculated dependence of the tunneling levels on the external tetragonal field f_θ (in the Q_θ direction) together with the relative probabilities that the system is at each of the three nonequivalent AP minima, is shown schematically in Figure 5.21.[344] When $f_\theta > 0$, the system in its ground state is in the main situated along the Q_θ axis minimum $(\rho_0, 0)$ (which is deepest). When $f_\theta < 0$, the other two minima, $(\rho_0, 2\pi/3)$ and $(\rho_0, 4\pi/3)$, are deeper in the ground state, and the system occupies them both with equal probability. At $T = 0$, the lowest energy corresponds to the stabilization of the system in distorted configurations for both cases. When $T > 0$, all three vibronic levels become populated after Boltzmann and the mean value $\langle Q_\theta \rangle$, which characterizes the distortion of a given center in the Q_θ direction, becomes temperature dependent.

A simple result can be obtained in the case of ferrodistortive tetragonal interactions between centers in the limiting case of very strong

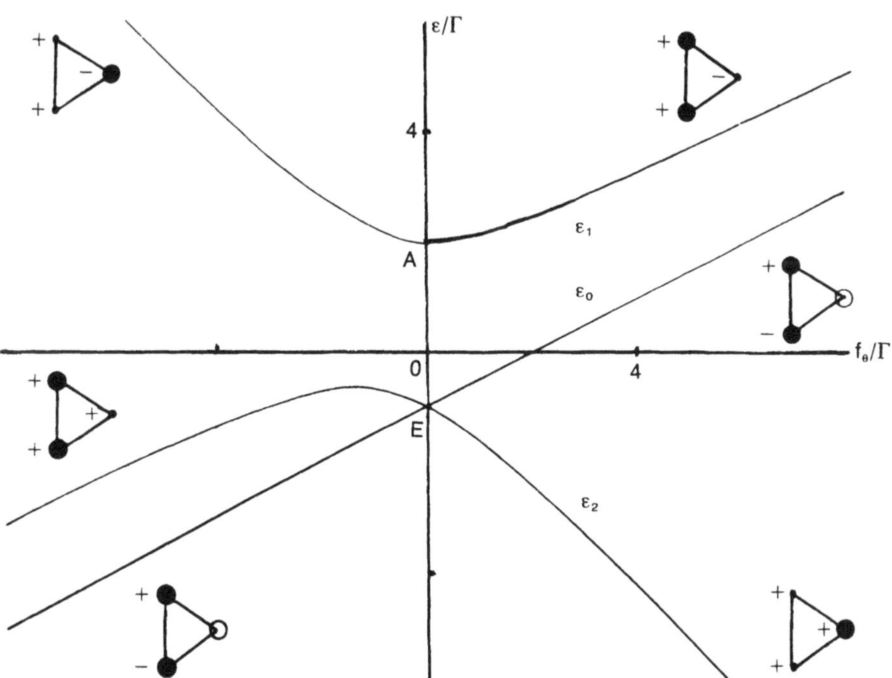

Figure 5.21. Tunneling energy levels of the JT E–e problem in tetragonal fields f_θ. Dimensions of the black circles at the corners of the triangle indicate the relative probability of the corresponding distortion (zero probability is shown by open circles), while + and − indicate the respective phase of the wave function.

vibronic coupling, when $\delta \approx 0$. In this case, for the ordering parameter $\bar{\sigma} = \langle Q_\theta \rangle / \rho_0$, we have[356]:

$$\bar{\sigma} = (e^u - e^{-2u})/(2e^u + e^{-2u}) \tag{5.26}$$

where $u = \gamma\bar{\sigma}/2kT$. Assuming the free energy is identical in both ordered and disordered phases, $F(\bar{\sigma}) = F(0)$, and taking into account the equivalency of all the centers in the crystal, the following expression can be obtained:

$$u\bar{\sigma} = 2\ln[(e^u + 2e^{-u/2})/3] \tag{5.27}$$

Equations (5.26) and (5.27) are solved together to obtain the structural phase transition temperature:

$$kT_0 = 3\gamma/16\ln 2, \qquad \bar{\sigma} = \tfrac{1}{2} \tag{5.28}$$

Antiferrodistortive ordering in the CJTE of the E–e problem in the three-level approximation under consideration results in two low-symmetry phases dependent on the mode of correlation between the JT centers.[357] In one of them the tetragonal distortions of the sublattices Q_A and Q_B are the same, but they are oriented along different axes resulting in a structure similar to that of multiaxial spin ordering (Figure 5.22a). In the other phase, distortions in the two sublattices differ according to the different signs of the tetragonal molecular field. For one of the sublattices we have $f_\theta > 0$, and in accord with the foregoing (Figure 5.21) the $(\rho_0, 0)$ minimum along the Q_θ axis is stabilized. For the other sublattice the inequality $f_\theta < 0$ is valid, and the states of the other two minima $(\rho_0, 2\pi/3)$ and $(\rho_0, 4\pi/3)$ are stabilized, the dynamics between them being preserved since they remain equivalent ($\langle Q_\theta \rangle = \rho_0/2$, $\langle Q_\varepsilon \rangle = 0$) (Figure 5.22b). The resulting structure is a ferridistortive one, due to incomplete compensation of oppositely directed distortions.

All these results are concerned with cubic symmetry crystals.[357] Similar situations in other crystal structures lead to quite different effects. For example, in a trigonal crystal the case illustrated in Figure 5.22a corresponds to a noncollinear weak ferrodistortive ordering.

In the more complicated variants of the E–e problem, as well as in the more complicated T–$(e + T_2)$, Γ_8–$(e + t_2)$, etc., problems, the solution cannot be obtained using the simple model with only several low-lying tunneling levels. Hence the behavior of more vibronic levels of the JT center in the molecular field of the crystal, obtained by numerical solution, must be considered.[356,357] However, even in simple cases more complicated types of ordering may be expected, one of which is discussed below.

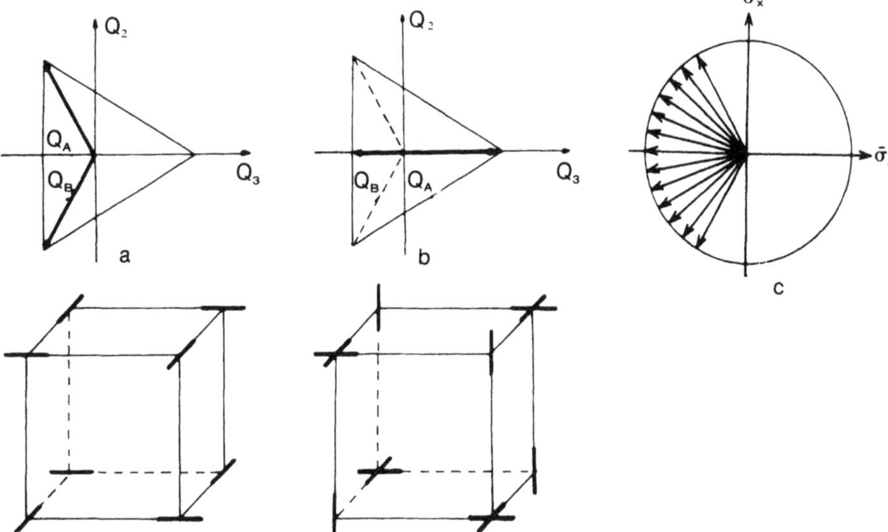

Figure 5.22. Three types of possible ordering of JT distortions due to CJTE in the case of an E–e problem for each center: (a) antiferrodistortive; (b) ferridistortive; (c) helicoidal. Below, possible antiferrodistortive and ferridistortive orderings in a perovskite type structure are illustrated schematically.

$K_2PbCu(NO_2)_6$, *Incommensurate Phases.* Besides the ferro- and antiferrodistortive orderings considered above, the CJTE and CPJTE may lead to phase transitions to more complicated crystal structures. Suppose in the above three-level E–e problem, owing to special distortion correlation interactions, the resulting molecular field is negative and directed at an angle $\alpha < \pi/2$ to the Q_θ axis ($f_\theta < 0$, $f_\varepsilon \neq 0$), and that this angle, and hence the f_ε component, vary smoothly along the crystal, assuming periodically all values between given positive and negative numbers and passing through zero. In this case, the two minima $(\rho_0, 2\pi/3)$ and $(\rho_0, 4\pi/3)$ at each center are no longer equivalent ($\langle Q_\varepsilon \rangle \neq 0$), and the degree of nonequivalency (the $\langle Q_\varepsilon \rangle$ value) changes from one center to another, repeating the periodicity of the f_ε component of the molecular field.

In terms of pseudospin terminology, the three equivalent minima of the E–e problem can be presented by three directions of the pseudospin $S = 1$ at angles 0, $2\pi/3$, and $4\pi/3$. Then, the case $f_\theta < 0$, $f_\varepsilon = 0$ corresponds to the equivalent minima with pseudospin directed at angles $2\pi/3$ and $4\pi/3$, respectively (Figure 5.22a), its averaged value being oriented along the Q_θ axis (in the negative direction). When $f_\varepsilon \neq 0$, the averaged value of the pseudospin acquires a nonzero ε component. In this case the pseudospin behavior can be presented by changes in its direction from one center to another, depending on the absolute value of f_ε. As a result, under the above conditions the pseudospin rotates from one center to another forming a fan-shaped (helical) structure, shown in Figure 5.22c (similar to known helical structures of real spin in the theory of magnetism).

Note that in such a crystal, in which the structure is changed due to the CJTE, the lattice period increases. If the period of f_ε is a multiple of the initial lattice period of the high symmetry phase, a superstructure is formed. In particular, antiferrodistortive ordering results in a superstructure with a double lattice period. However, the periodicity of the f_ε component of the molecular field may not be a multiple of the lattice period. The "superstructure" is then called incommensurate; distortions of the crystal centers are frozen with a wave vector which is incommensurate with its limiting values. By taking into account details of the JT center interactions in some crystals, in particular the interaction of each center with its second neighbors,[359] the helical structure becomes one of the stable phases of the CJTE, and a structural phase transition to the incommensurate phase may take place. The crystal $K_2PbCu(NO_2)_6$, studied comprehensively in recent years, appears to be an interesting example of this kind.

$K_2PbCu(NO_2)_6$ is a representative of the family of $A_2BCu(NO_2)_6$ crystals, in which CJTE takes place. In these crystals the $Cu(NO_2)_6^{4-}$ octahedra have a cubic symmetry environment, which at high temperatures does not remove the pulsating transitions between the three equivalent minima of the $E-e$ problem with strong vibronic coupling (Section 2.1). Due to the CJTE, a structural phase transition to the ordered state is expected at lower temperatures.

Two phase transitions are observed in $K_2PbCu(NO_2)_6$. At room temperature the crystal is cubic (α phase). At 280 K, phase transition to the pseudotetragonal structure (β phase) is observed, and at 273 K transition to a structure with symmetry lower than orthorhombic (γ phase) takes place.[360,361] JT distortions around the Cu^{2+} centers are also confirmed by ESR measurements.[361,362]

The nature of low-temperature β and γ phases has been subject to discussion in the literature. (The discussion is, however, not concerned with the general JT origin of these phases, which is beyond doubt.) X-ray investigation shows that in low-symmetry phases, the elementary cell is pseudotetragonal, $c/a < 1$. Therefore, Harrowfield et al.[360,363] assumed that at low temperatures the crystal consists of compressed octahedra, which are ferrodistortively ordered due to the CJTE. Reinen et al.[364,365] (see also the review article[347]), disputing this point of view, show that the observed crystal structure of the γ phase fits the description of antiferrodistortive ordering of elongated octahedra. As a result of such ordering the mean value of the observed Cu–N distances will appear as if the octahedra are compressed but packed in "parallel" (cf. Figure 5.22a).

This point of view is confirmed by direct and indirect data, including a detailed analysis of the ESR spectra of this and related crystals (as well as other copper compounds with a CJTE) and comparison of the thermal ellipsoids in X-ray analysis of nitrogen and oxygen atoms in NO_2 groups. In the β phase the same JT distortions (elongated octahedra) are assumed but, as opposed to the γ phase, the dynamics of the distortions in the plane are supposed to be preserved, cooperative ordering being only along one direction. In the high-temperature phase (above 280 K) complete dynamics of these elongated octahedra (with no ordering) is attained. A similar viewpoint about antiferrodistortive ordering in the plane has been proved by Joesten et al.[366] with almost the same reasoning as in the studies by Reinen et al.[364,365]

Yamada et al.[367,368] studied in detail the structural phase transitions in question by means of X-ray and neutron scattering techniques. These authors used diffuse X-ray scattering and succeeded in observing and studying in detail the superstructure of the two low-temperature phases. It was shown that the $Cu(NO_2)_6^{4-}$ octahedra of the crystal under consideration are elongated and interaction between them is of an antiferrodistortive nature. In the high-temperature phase the three elongated configurations are equivalent and pulsations between them take place at each center independently. In the low-temperature γ phase (below 273 K), a typical antiferrodistortive structure with alternating parallel and perpendicular packing of the elongated octahedra is realized (cf. Figure 5.22a).

In the intermediate β phase (at temperatures from 280 K to 273 K), an incommensurate helical structure discussed above (Figure 5.22c) is formed.

The structure of the crystal in the cubic phase is given in Figure 5.23 (JT dynamics being disregarded), while observed displacements of NO_2 groups (and Pb atoms) in β (incommensurate) and γ (antiferrodistortive) phases are shown (in projection) in Figure 5.24. In the γ phase the displacements, corresponding to the elongation of the octahedra and their antiferrodistortive packing in the crystal, produce a crystal mode with wave vector $k_0 = (\frac{1}{2}, \frac{1}{2}, \frac{1}{2})$ conforming with the border of the vibrational band. These displacements, as mentioned above, correspond to the formation of a superstructure with a doubled period of the lattice. However, in the β phase the corresponding displacements change from one cell to another not quite "regularly," but forming a wave along the crystal with a period which is not a multiple of (incommensurate with) the period of the initial lattice. The wave vector of these displacements is shown to be $(0.425, 0.425, 0)$. This result confirms the theoretical prediction[357] of possible stabilization of incommensurate phases due to the CJTE in these crystals.

Within the described series of hexanitrocomplexes of general formula $A_2BCu(NO_2)_6$, only in the case $B = Pb$ is the above antiferrodistortive

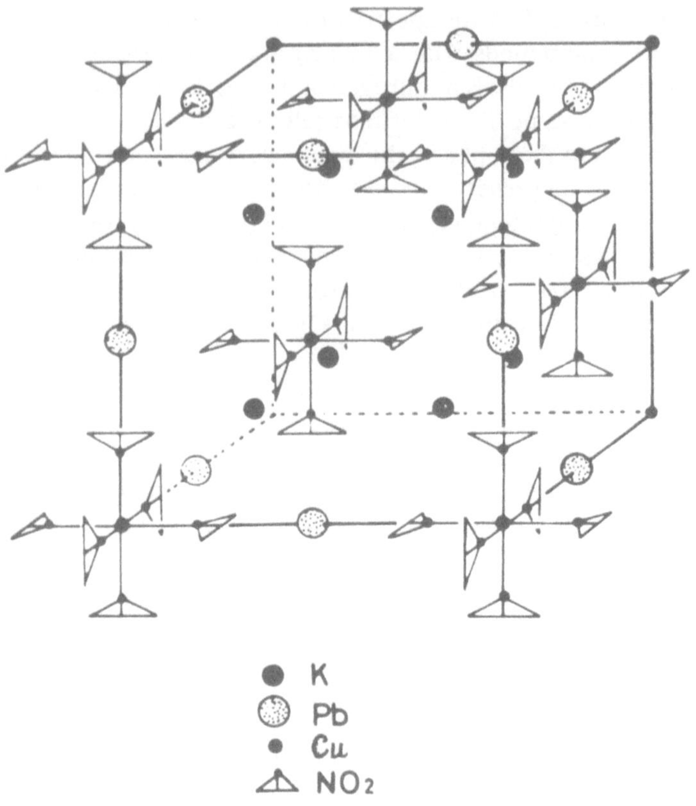

Figure 5.23. Structure of the crystal $K_2PbCu(NO_2)_6$ in the high symmetry (nondistorted by the CJTE) configuration.

ordering observed. The reasons are discussed elsewhere[365,366,347] and explained mainly by the presence of an additional unshared pair of electrons on the Pb^{2+} ion. MO LCAO calculations of the AP of the octahedra of the hexanitrocomplexes of Co(II) and Cu(II) have been carried out recently.[369]

Structural-Magnetic Transitions. Spin and Orbital Orderings. If JT centers in the crystal have unpaired electrons, the CJTE structural phase transition leading to a certain ordering of the JT distortions strongly affects the possible electron orbital momentum and spin orderings. Consider a cubic symmetry crystal with octahedral JT centers in à ground E_g state corresponding to transition metals with electronic configuration d^9 (e.g., Cu^{2+}, Ni^+, Ag^{2+}), d^4 (Mn^{3+}, Cr^{2+}), and low spin d^7 (Co^{2+}, Ni^{3+}). In these

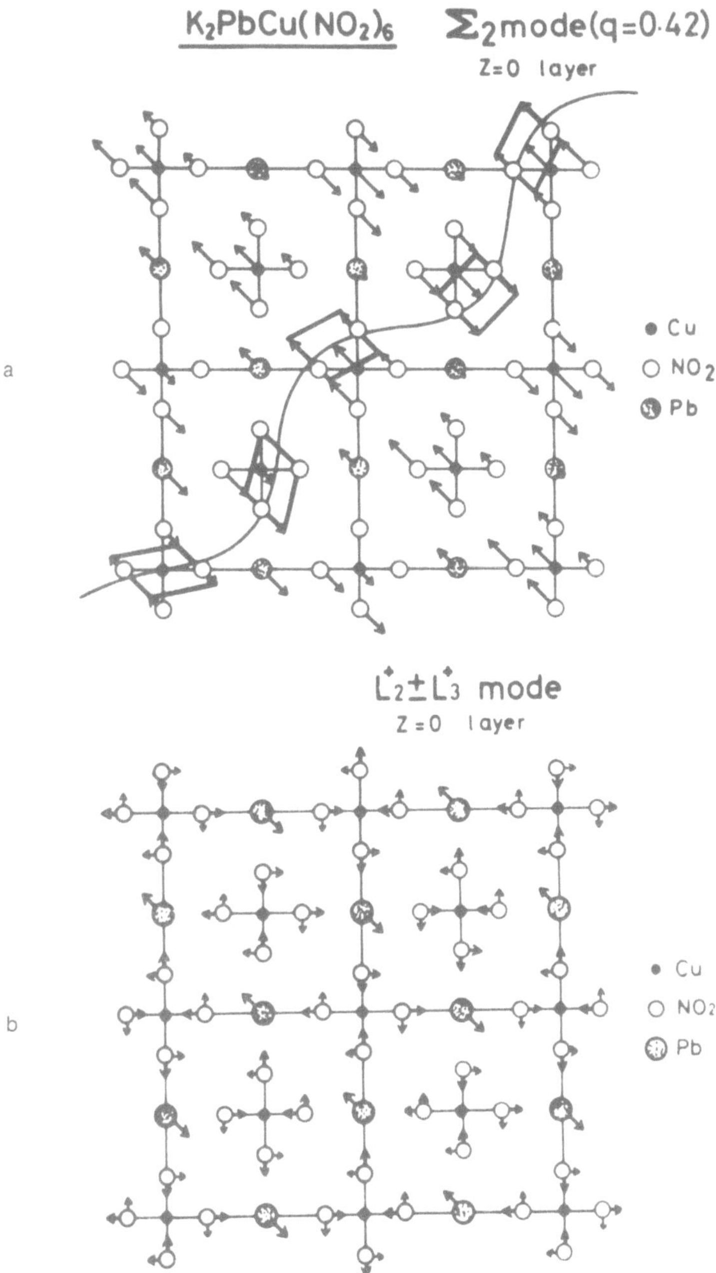

Figure 5.24. Atomic displacements in the $Z = 0$ layer in the incommensurate phase β (a) and in the antiferrodistortive phase γ (b) of the $K_2PbCu(NO_2)_6$ crystal (after Yamada[368]).

cases the E_g term is formed by the contribution of two d functions, d_{z^2} and $d_{x^2-y^2}$. Depending on the nature of the distortion, the unpaired electron of the E_g term falls either in the d_{z^2} state (compressed octahedron for d^9 and elongated for d^4 and d^7), or in the $d_{x^2-y^2}$ state (elongated octahedron for d^9 and compressed for d^4 and d^7).

Consider the K_2CuF_4 crystal with a perovskite structure. The lattice has a layered structure formed by the CuF_6 octahedra, which are interlinked in the (001) plane by common fluorine atoms. If the octahedra are elongated in the ordered phase, the unpaired electron at each center is in the $d_{x^2-y^2}$ orbital, whereas the d_{z^2} orbital is occupied by two electrons. If the ordering of the JT distortions is ferrodistortive, the respective orbital and spin ordering corresponds to that shown schematically in Figure 5.25a. It follows that planar antiferromagnetic ordering of the spins should be expected as a result of superexchange interaction through the common fluorine atoms.

In fact, planar ferromagnetism is observed.[370] Therefore, Khomskii and Kugel[371,372] assumed that in K_2CuF_4, antiferrodistortive ordering of the Cu^{2+} orbitals (and hence of the elongated CuF_6 octahedra) takes place in the (001) plane. The orientations of the orbitals and spins in line with this ordering lead to ferromagnetic interaction of the spins of the occupied d_{z^2} orbital and half-filled $d_{x^2-y^2}$ orbital (Figure 5.25b). In this case planar

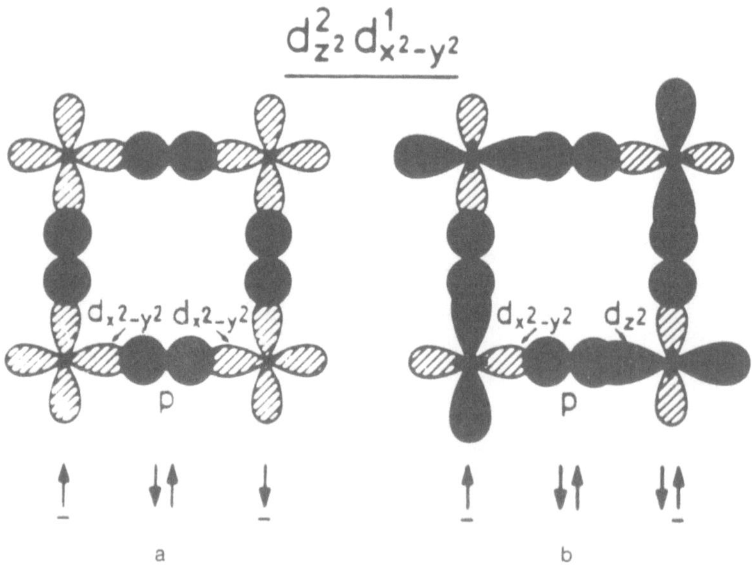

Figure 5.25. Orbital and spin ordering scheme for crystals containing corner connected octahedral CuX_6 clusters of divalent copper in the (001) plane in the case of: (a) ferrodistortive and (b) antiferrodistortive orderings (from Reinen and Friebel[347]).

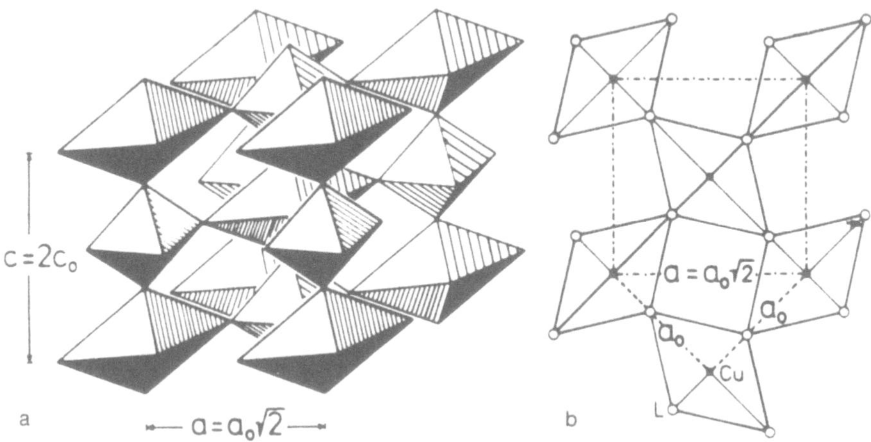

Figure 5.26. Antiferrodistortive order of elongated octahedra in perovskites ABX_3: (a) three-dimensional sketch; (b) projection into the (001) plane (after Reinen and Friebel[347]).

ferromagnetism of Heisenberg type is realized. This assumption has been confirmed by ESR[373] and direct X-ray[374] measurements.

Passing to the $KCuF_3$ crystal, one can see that, as distinct from K_2CuF_4, the CuF_6 octahedra are linked by fluorine atoms in common not only within the (001) plane, but also along the [001] direction (Figures 5.26 and 5.27). Here, the antiferromagnetic exchange interaction along the [001] direction between the unpaired spins of the $d_{x^2-y^2}$ orbitals is much stronger ($\sim 10^3$ times) than the ferromagnetic interaction between the occupied d_{z^2} and half-filled $d_{x^2-y^2}$ orbitals within the (001) plane. Therefore the resulting linear antiferromagnetism is expected in accordance with the experimental data.[370]

These ideas also explain the origin of the magnetic structure of other types of Cu(II) compounds in which CJTE takes place.[347,375–377] In particular, unlike the crystals $KCuF_3$ and K_2CuF_4 in which JT octahedra are strongly coupled by ligands in common, Reinen and Krause[377] considered the compounds $Cu(ONC_5H_6)_6X_2$, $X = BF_4^-$ (I) and $X = ClO_4^-$ (II). In the crystal state the $Cu(ONC_5H_6)_6^{2+}$ octahedra are rather isolated (having no ligands in common) and occupy the apexes of a slightly trigonally distorted cube.[378] The magnetic properties of the two crystals, I and II, are different: at low temperatures ($\leqslant 1$ K) crystal I is planar antiferromagnetic, whereas crystal II is one-dimensionally ferromagnetic.

The superexchange schemes in Figure 5.25 can be used to assume that for elongated octahedra, as a result of the CJTE, ferrodistortive ordering takes place in the fluoborate, I, whereas antiferrodistortive ordering is realized in the perchlorate, II. A study of ESR and electronic absorption

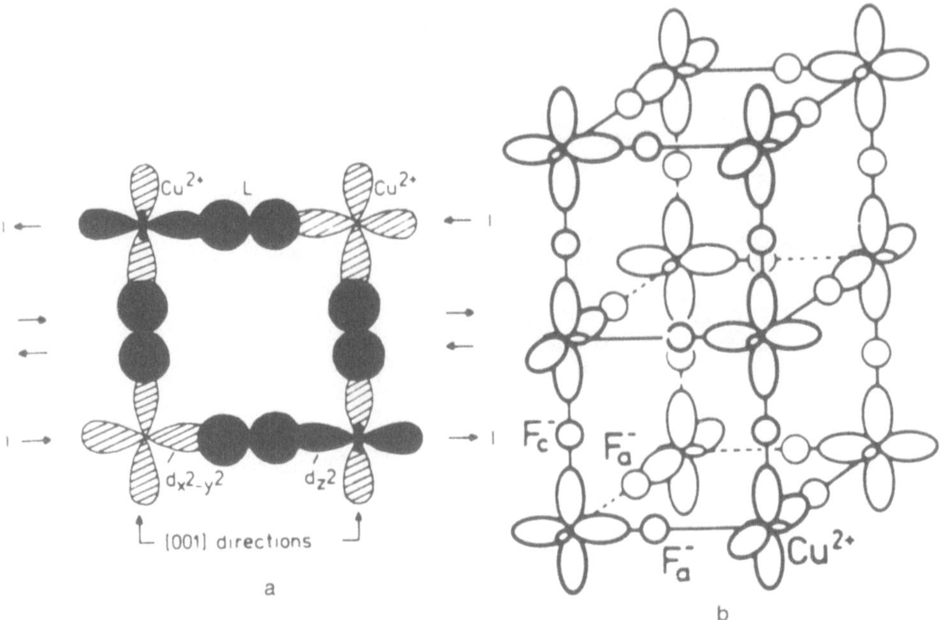

Figure 5.27. Antiferromagnetic ordering of orbitals and spin in $KCuF_3$: (a) ordering scheme along the [001] direction; (b) three-dimensional orbital locations.

spectra at different temperatures confirms these assumptions. At room temperature the $Cu(ONC_5H_6)_6^{2+}$ octahedral distortions are of a dynamic nature, as confirmed by the magnitude and shape of the temperature ellipsoids in X-ray analysis.(378) The octahedral distortions can be evaluated from the temperature ellipsoid dimensions using the technique of Ammeter et al.(130) In the cases under consideration the distortions are $\rho_0 \cong 0.37$ Å for I, and $\rho_0 \cong 0.35$ Å for II; these values agree with spectroscopic absorption band splitting.

Phase transition takes place at lower temperatures (at 90 K in I and 77 K in II). ESR data imply that low-temperature phases correspond to ferrodistortively ordered elongated octahedra in I and antiferrodistortively ordered elongated octahedra in II (compressed octahedra are unacceptable). The difference in the nature of JT center distortion interactions in the two crystals is apparently due to the different orientations of the BF_4^- and ClO_4^- groups in the lattice. The $Cu(ONC_5H_6)(NO_3)_2$ crystal remains ordered up to a temperature of 298 K.

Analogous results for magnetic structure investigations have also been obtained for Mn(III) and Cr(II) compounds having the d^4 electronic configuration.(347,375,376,379) For example, a series of compounds A_2CrCl_4, where A = Cs, Pb, NH_4, K, crystallizes in the same structure as K_2CuF_4 and, similar to the latter, exhibits ferromagnetic properties. The scheme of superexchange interactions, analogous to the octahedral compounds of Cu(II), is shown in Fig. 5.28 (cf. Figure 5.25). It is seen that for the plane layers, ferromagnetism occurs only if an antiferrodistortive ordering of elongated octahedra takes place whereas ferrodistortive ordering leads to an antiferromagnetic structure. The assumption[380] of ferrodistortive ordering of compressed $CrCl_6^{4-}$ octahedra in $(NH_4)_2CrCl_4$, based on neutron powder diffraction data, is unacceptable since, as seen from Figure 5.28a, under this assumption the crystal must be antiferromagnetic. Similar properties were observed in $AMnF_4$ with A = Cs, Pb, NH_4, K.

In the above examples, ordering of the JT distortions in the crystal due to CJTE leads to some orientation of the orbitals of the unpaired electrons which, in turn, determines the nature of the exchange interaction between the spins, and hence the spin ordering. The structural and magnetic phase transition temperatures are not in general necessarily the same, since they are determined by different correlation constants (though there may be conditions under which the two temperatures coincide).

By stabilizing the distorted configuration of each center, cases arise when structural ordering changes its orbital and spin ground state. Con-

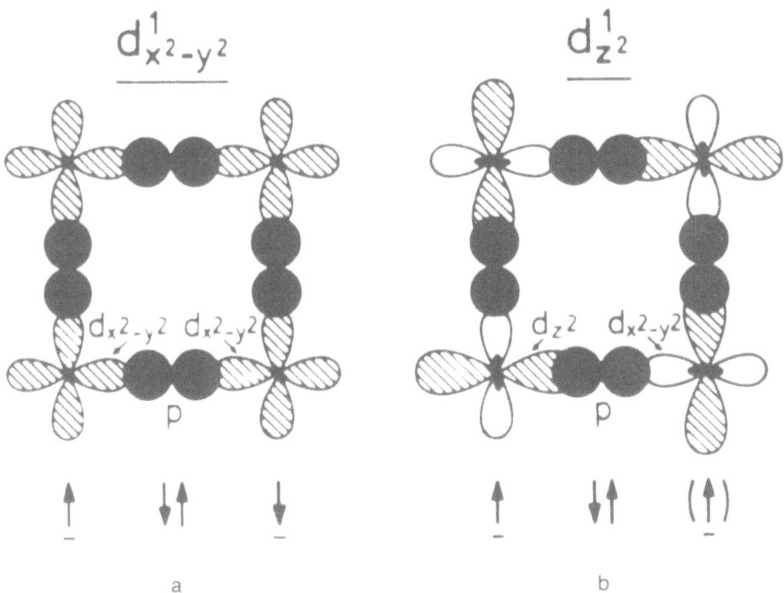

Figure 5.28. Orbital and spin ordering of corner connected MX_6 clusters, where M is a d^4 transition ion, in the (001) plane: (a) ferrodistortive ordering of compressed octahedra; (b) antiferrodistortive ordering of elongated octahedra.

sider crystal centers with an orbitally nondegenerate ground state of type ${}^6A_2$ and a near orbital degenerate excited state of another spin multiplicity, say ${}^2T_2$. Such a situation with two near terms of different multiplicity is in fact often realized as, for example, in a series of iron (as well as other transition metal) compounds. In particular, it is found in H[Fe(5-Clthsa)$_2$], where thsa = thiosemicarbazon. Another situation of this kind, with ${}^1A_{1g}$ and ${}^5T_{2g}$ terms instead of ${}^6A_2$ and ${}^2T_2$, is observed in Fe(III) compounds such as $[\mathrm{Fe(phen)_2(NCX)_2}]^+$, X = S, Se. In these compounds the "crossover" phenomenon takes place for which, at certain temperatures, the magnetic moment (magnetic susceptibility) changes such that the ${}^2T_2$ term becomes the ground one instead of ${}^6A_2$ (or ${}^5T_{2g}$ instead of ${}^1A_{1g}$), or vice versa (see Goodwin's review[381]). In many cases, this transition from one magnetic state to another is of a magnetic phase transition nature accompanied by a structural phase transition.

There have been several attempts to explain the origin of the "crossover." Arguments based on the CJTE will be presented here briefly.[382] Figure 5.29 presents schematically the behavior of the energy levels of the ${}^6A_2$ and ${}^2T_2$ terms as a result of JT dynamic distortions of each center, and the freezing and amplification of this distortion by the molecular field and structural phase transition due to the CJTE. Computation[382] shows that if the energy gap between the two terms is not very large, and vibronic interactions as well as parameter λ of the correlations between the JT centers are substantial, the structural phase transition causes the ground term multiplicity to change (Figure 5.29), in agreement with experimental data.

If the T–e problem is realized for the T term (Section 2.2), the temperature dependence of the ordering parameter $\bar{\sigma}$ is given by[382]

$$\bar{\sigma} = \frac{\exp(-\Delta/kT)[\exp(2\lambda\bar{\sigma}/kT) - \exp(-2\lambda\bar{\sigma}/kT)]}{\exp[-(\Delta - 2\lambda\bar{\sigma})/kT] + 2\exp[-(\Delta + 2\lambda\bar{\sigma})/kT] + 3} \tag{5.29}$$

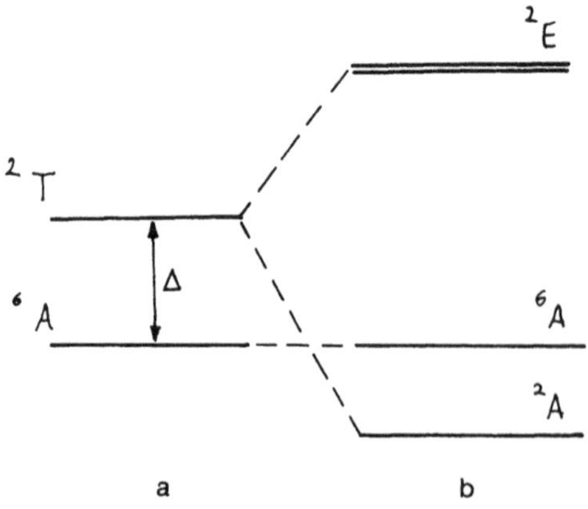

Figure 5.29. Energy terms of a JT octahedral $Fe^{2+}(d^6)$ center in crystals without (a) and with (b) the CJTE.

Together with the condition $\bar{\sigma}=0$, this expression determines the temperature of a structural phase transition to the (each center) tetragonally distorted configuration, for which the magnetic moment changes simultaneously. For a somewhat different version of this idea see Kambara's paper.[383]

Ferroelectric Phase Transitions. Another quite large and very important class of crystals, in which vibronic effects are essential, are ferroelectrics. As is well known,[384] in these crystals at certain temperatures structural phase transitions take place to a spontaneous polarized (ferroelectric) state. If there are molecular groups with proper dipole moments in the initial crystal structure, the transition to the ferroelectric phase may be regarded as an ordering of rigid dipoles (order–disorder transitions). However, in most cases there are no dipole groups in the high-temperature nonferroelectric phase (paraphase). For instance, in ferroelectrics with perovskite type structure (such as barium titanate) the lattice has a high symmetry in the paraphase (the Ti side local symmetry is O_h). A dipole formation during the ferroelectric phase transition is assumed in this case (displacive phase transition).

What is the origin of spontaneous polarization and ferroelectric phase transitions in crystals? It is clear that if there are no rigid dipole groups in the elementary cells of the crystal, but such groups might occur as a result of the JTE or PJTE, *ferroelectric phase transition can take place as a result of the CJTE or CPJTE arising from ordering of dipolar distortions of the interacting centers.* As noted in Section 2.3, the JTE cannot result in dipolar distortions in systems having an inversion center. Dipolar instability in these cases can occur only as a result of the PJTE. Again, the ferroelectrics, being dielectrics, in most cases do not have degenerate ground states. For these cases concern is mainly about the possibility of a PJTE.

Such a PJT mechanism of spontaneous polarization and ferroelectric phase transitions was suggested earlier[52] and led to the vibronic theory of ferroelectricity.[52,385–388] (For a review see Bersuker and Vekhter.[12]) The rigorous version of the problem was solved by modern methods of theoretical physics. Below we give the more simple aspect of the problem, mainly its qualitative features which lead to a physically visual picture of the microscopic origin of ferroelectricity.

By way of example consider the ferroelectric $BaTiO_3$ crystal which has a perovskite structure. The PJTE in the octahedral cluster TiO_6^{8-} is considered in Section 2.3. It has been shown that vibronic mixing of the ground A_{1g} term with the near excited term T_{1u} by odd nuclear displacements of T_{1u} type (dipolar displacements) under certain conditions makes the ground state unstable with respect to these displacements.

The calculated AP has a complicated shape: as well as a maximum at the point of highest symmetry with the Ti atom at the center of a regular octahedron (dynamic instability), there are eight equivalent trigonal minima (at which the Ti atom is displaced along one of the trigonal axes), twelve orthorhombic saddle points (at which the Ti atom is displaced along one of the second-order axes), and six tetragonal saddle points (at which the Ti atom is displaced along one of the tetragonal axes). The condition of instability of the high symmetry configuration is given by equation (2.43):

$$\Delta < 4F^2/K \tag{5.30}$$

where 2Δ is the energy gap between the ground and excited states under consideration which are mixed by T_{1u} displacements, F is the vibronic constant (equation (2.40)), and K is the force constant of the T_{1u} vibrations.

Note that, at the above minima and saddle points, the cluster has a nonzero dipole moment (the cluster is dipolarly unstable). Then, allowing for the strong correlation between these clusters in $BaTiO_3$, through common ligands, the CPJTE can be assumed to result in a structural phase transition to the spontaneously polarized ferroelectric state of the crystal.

In various studies[12,52,385–387] the problem was solved somewhat differently, taking into account the substantially nonlocalized nature of the PJT centers in the crystal (large valent bands). First the PJTE is considered for the crystal as a whole with respect to the bound optical displacements which correspond to the wave vector $k = 0$. These diplacements shift one sublattice relative to the other (the Ti atom relative to the O atom sublattice). Such a "crystal" PJTE results in the same features of the AP as listed above for the cluster. In particular, the condition of dipolar instability has the same form as equation (5.30) with somewhat different effective parameters, modified by electron band formation.

Similarly, the behavior of the crystal has been evaluated with respect to antiferroelectric displacements for which $k = \pi/2a$ (a is the lattice constant).

Once the features of the AP are known with regard to the dipole displacements of the sublattices, the semiclassical approach can be used to consider the motion of the crystal along this AP, assuming that all the remaining vibrations may be regarded as a thermal reservoir. By solving the appropriate kinetic equations it was shown that, in the crystal under consideration, three structural phase transitions should be observed. At low temperatures the crystal is at one of its AP trigonal minima, which corresponds to the ordered displacement of the Ti atoms along one of the trigonal axes, the lattice having rhombohedral structure and being spontaneously polarized along the corresponding trigonal axis.

When the temperature increases, the kinetic energy (amplitude) of motions at the minimum increases, and overcoming of the lowest barrier between nearest-neighbor minima through the orthorhombic saddle points begins. At a specific temperature, there occurs phase transition to the (macroscopically) orthorhombic structure, resulting from averaging over the two neighboring minima. Hence, the JT dipolar distortions corresponding to the displacements of the Ti atom in the trigonal axis directions are not completely ordered in this orthorhombic phase: the lowest barrier between the two nearest-neighbor minima is overcome (directly or by tunneling, fluctuations, etc.), so the crystal becomes disordered in the plane of these two minima in the direction perpendicular to the C_2 axis of spontaneous polarization. Thus in this phase the JT distortions are ordered along two of three independent directions, being disordered along the third.

At higher temperatures, transitions between the minima overcome the higher barrier saddle points of tetragonal type, resulting in phase transition to the (macroscopically) tetragonal phase. In the latter, dipolar distortions are ordered only along one tetragonal axis and the crystal is polarized in this direction. In the plane perpendicular to this direction the distortions are completely disordered, and may be macroscopically described as averaged over four equivalent trigonal minima. Finally, at higher temperatures the AP maximum point (the highest in energy) is overcome by thermal transitions and the crystal becomes completely disordered along all three directions. This corresponds to a phase transition to the paraphase, which is thus a macroscopically averaged state of disordered dynamic dipolar PJTE distortions.

This picture obtained by calculation[52,12] qualitatively reproduces all the observed ferroelectric phases in barium titanate, including the consequence of their occurrence with temperature, the symmetry of the crystal, direction and relative polarization in each of them. Moreover, it follows that only one of the three ferroelectric phases in $BaTiO_3$ — the low-temperature rhombohedral phase — is completely ordered. The other two, orthorhombic and tetragonal, are partly disordered, the former along one direction and the latter along two directions in the plane, the paraphase being completely disordered in all directions. This conclusion is fundamentally novel and does not follow from any other micro- or phenomenological theories of ferroelectricity. It has been confirmed by refined experiments on diffuse scattering of X-rays, as well as light and neutron scattering.[12,389] The picture is somewhat similar to the above-discussed stepwise ordering in $K_2PbCu(NO_2)_6$, though the $BaTiO_3$ crystal was studied long before.

Quantitatively, the phase transition temperatures are determined by the parameters entering criterion (5.30), the calculation of which generally involves difficulties. In some cases they can be determined from empirical data. In particular, if one substitutes the known values of the first two transitions ($T_3 = 393$ K and $T_2 = 278$ K) into the expressions for the three ferroelectric phase transition temperatures in $BaTiO_3$ obtained from the above scheme, then the system parameters can be estimated and the third transition temperature T_1 calculated. Computation yields $T_1 = 201$ K, the experimental value being $T_1 = 183$ K. This reasonable agreement confirms the vibronic nature of the phase transitions in question.

Besides $BaTiO_3$ the cubic crystals of GeTe, as well as other crystals, were studied.[385–387] In particular, a qualitative investigation of the variation of the vibronic parameters in inequality (5.30) for a series of perovskites allows an examination of the origin of observed changes in their ferroelectric poperties. The stronger inequality (5.30), i.e., the greater the difference $4F^2/K\Delta - 1$, the deeper the AP minima and the higher the temperature of phase transition to the paraphase.

For instance, in the series of titanates of Ca, Sr, Ba the Ti–O distance increases (due to the increase in the respective atomic dimensions), and hence the $d_\pi - p_\pi$ overlap and the covalency of the bond decreases from Ca to Ba. Consequently, the force constant K for the corresponding displacements of the Ti atom as well as the effective energy distance between the bands decrease, the vibronic constant F being less influenced. Thus when passing from calcium titanate to strontium titanate and then to barium titanate, inequality (5.30) is strengthened and the conditions of ferroelectricity occurrence are improved, i.e., the temperature of phase transition to the disordered phase increases.

This conclusion is confirmed by empirical data: $CaTiO_3$ is nonferroelectric down to $T = 0$, $SrTiO_3$ is a virtual ferroelectric and with appropriate impurities or under external perturbation becomes a true ferroelectric at low temperatures, whereas $BaTiO_3$ is ferroelectric up to room temperature.

In another series of perovskites of Ti, Zr, and Hf, the diffusiveness of the outer d orbitals ($3d$, $4d$, and $5d$, respectively) increases. Consequently, the $d_\pi - p_\pi$ overlap, the covalency of the Me–O bond, and hence the K and Δ parameters also increase. Therefore, the ferroelectric properties are expected to deteriorate (the phase transition temperature is expected to decrease) in this series from Ti to Hf in accord with the empirical data. Similarly, when passing from GeTe to SnTe the $p_\sigma - p_\sigma$ overlap ($4p - 5p$ and $5p - 5p$, respectively) increases, and therefore the temperature of phase transition in GeTe is higher than in SnTe.

The vibronic theory of ferroelectricity also leads to predictions of other ferroelectric properties. In particular, allowing for the active role of electronic states in vibronic effects resulting in dipole instability and ferroelectricity, it can be shown that the temperature of ferroelectric phase transitions, as well as other ferroelectric properties, change under the influence of an external magnetic field, even if the crystal is nonmagnetic and has no unpaired electrons.[390,391] This effect has been observed experimentally.[392–394]

Other details of the vibronic theory of ferroelectricity and its applications are dealt with elsewhere.[12–14]

5.3. Vibronic Crystal Chemistry. Plasticity and Distortion Isomerism

Plasticity of the Coordination Sphere around Vibronic Centers. An interesting consequence follows from the vibronic stereochemistry rules discussed above. It relates to the "softness" or "plasticity" of the coordination spheres of vibronic centers having degenerate or pseudodegenerate electronic terms, in the sense that such a center allows for (even requires) a nontotally symmetric (distorted) configuration of the ligands with several (or an infinite number of) equally probable (equivalent) directions of distortion. For instance, for an $E-e$ problem without quadratic interactions (when G_E is small, see Section 2.1) the coordination sphere allows any tetragonal distortion along the symmetrized Q_θ and Q_ϵ displacements, but within the limits $Q_\theta^2 + Q_\epsilon^2 = \text{const}$ (see Figure 2.5). If quadratic effects dominate, only three directions of tetragonal distortion along the fourth-order axes remain equally probable. There may be other possibilities of three tetragonal, four trigonal, six orthorhombic, etc., equivalent directions of distortion (see Chapter 2).

In the absence of external perturbations the softness of the coordination sphere results in continuous transitions between equivalent directions of distortions. In the case of crystal structures, the crystal environment of a given coordination center may stabilize one of its equivalent distortions, reducing the dynamic distortions to static ones. The external factor stabilizing one of many equivalent JT distorted configurations, according to the above results, can be: (1) any weak low-symmetry perturbation, such as the influence of the next (subsequent) coordination sphere, intermolecular (van der Waals) interactions, or hydrogen bonds; (2) cooperative interaction of dynamically distorted complexes leading to a phase transition (CJTE and CPJTE, Section 5.2).

Different static distortions of the vibronic coordination sphere can be obtained by means of an appropriate selection of stabilizing factors. The selection is at present limited by experimental difficulties in chemical synthesis and crystal growing. But *the empirical data* already available suffice to *confirm the great variety of possible distortions originating from the vibronic properties of JT centers and determined by the crystal environment (plasticity)*. To illustrate this statement the six-coordinate compounds of Cu(II), Mn(III), and Cr(II) are taken as examples. Their ground states are twofold orbitally degenerate E terms. From ESR (Section 4.1), as well as from other direct and indirect data, it follows that there is a strong JTE in these systems. X-ray analysis explicitly shows that the six-coordinated cluster about these metals is not a regular octahedron even when all the ligands are identical, and in the majority of known cases the octahedron is tetragonally distorted. Tables 5.2–5.5 give the crystallographically determined distances to the two axial (R_L) and four equatorial (R_S) ligands in a series of CuO_6 (Table 5.2), CuN_6 (Table 5.3), and other CuX_6 (Table 5.4) clusters, as well as in similar octohedral Mn(III) and Cr(II) ones (Table 5.5) in different compounds.[395,396,53]

It follows from these tables that the six-atom polyhedra of atoms around the Cu(II), Mn(III), and Cr(II) centers in different crystals are mainly elongated octahedra, $R_L > R_S$, with two ligands on the long axis and four on the short axes. Although for some of the tabulated compounds the ligands out of the first coordination sphere atoms are considerably different (and in the crystal the interatomic distances also depend on the packing of the molecules in the lattice), the large number of such compounds confirms statistically that the deformation of the coordination sphere around Cu(II), Mn(III), and Cr(II) is due to internal forces, i.e., to the *E* term JTE. The fact that elongated octahedra are observed confirms the assumption of strong vibronic coupling and strong quadratic vibronic interaction in the $E-e$ problem for the systems under consideration: one of the three AP minima is stabilized by the crystal environment. For every case listed in Tables 5.2–5.5 the origin of the stabilizing factor can, in principle, be determined.

The cluster octahedron is regular for several compounds, which means that distortions are not stabilized by the crystal environment or cooperative effects. In other words, the phase transition to lower-symmetry structures due to cooperative effects has not taken place at the temperature of the X-ray measurements in question but should be expected at lower temperatures. For some of these compounds (such as $CuSiF_6 \cdot 6H_2O$ and $K_2PbCu(NO_2)_6$) this point of view has been confirmed by ESR and other direct measurements, partly discussed above.

Reported cases of tetragonally compressed octahedra in $(MH_4)_2Cu(SO_4)_2 6H_2O$, $Cu(PCP)_3(ClO_4)_2$, $Cu(dien)_2(NO_3)_2$, $Cu(en)_3SO_4$,[484] Mn $(acac)_3$, etc., need additional thorough investigation. In similar cases (e.g., in nitrate of bis(terpyridin)Cu(II),[485] $Cu(en)_3Cl_2$,[484] and also in the above K_2CuF_4 and $K_2PbCu(NO_2)_6$), a more careful study shows that in fact there are no tetragonally compressed octahedra but rather tetragonally elongated octahedra which, being antiferrodistortive ordered, give an average picture of the diffraction similar to that for ferrodistortive ordered compressed octahedra. A similar conclusion about elongated instead of compressed octahedra emerges from an ESR investigation.[373] Recently[486] the origin of elongated octahedra in Cu(II) compounds was explained by the contribution of configuration interaction with the nearest excited state, which includes the 4*s* orbital of the copper atom. This negative contribu-

Table 5.2. Equatorial (R_S) and Axial (R_L) Interatomic Distances Cu–O in Cu(II) Compounds Containing CuO_6 Clusters

Compound	R_S, Å	R_L, Å	Reference
$Cu(C_6H_4OHCOO)_2 \cdot 4H_2O$	1.88	3.00	397
Cu(glycollate)$_2$	1.92	2.54	398
Cu(acac)$_2$	1.92	3.08	399
Cu(salicylaldehydate)$_2$	1.92	3.15	400
Cu(ω-nitroacetophenate)$_2$	1.93	2.60	401
$Na_2Cu(CO_3)_2$	1.93	2.77	402
$Cu(C_6H_5COO)_2 \cdot 3H_2O$	1.94	2.51	403
$Na_2Cu(PO_3)_4$	1.94	2.52	404
Cu(meso-tartrate) $\cdot 3H_2O$	1.94	2.54	405
Cu(OMPA)$_2$(ClO_4)$_2$	1.94	2.55	406
Cu(phtalate)$_2 \cdot H_2O$	1.94	2.58	407
$Cu(OH)_2$	1.94	2.63	408
CuB_2O_4	1.94	3.07	409
$CuSO_4$	1.95	2.37	410
$Cu_2P_4O_{12}$	1.95	2.38	411
Cu(d-tartrate) $\cdot 3H_2O$	1.95	2.40	405
Cu(meso-tartrate) $\cdot 3H_2O$	1.95	2.48	405
Cu(2-hydroxy-2-methylpropionate)$_2 \cdot 2H_2O$	1.95	2.56	398
$CuUO_4$	1.95	2.56	412
$Cu(C_8H_5O_4)_2 \cdot 2H_2O$	1.95	2.68	413
Cu(H-maleate)$_2 \cdot 4H_2O$	1.95	2.68	414
CuO	1.95	2.78	415
$CuSO_4 \cdot 3H_2O$	1.96	2.42	416
$Ca(Cu,Zn)_4(OH)_6(SO_4)_2 \cdot 3H_2O$	1.96	2.43	417
Cu(phtalate)$_2 \cdot H_2O$	1.96	2.46	407
Cu(phenoxyacetate)$_2 \cdot 2H_2O$	1.96	2.50	398
$Ca(Cu,Zn)_4(OH)_6(SO_4)_2 \cdot 3H_2O$	1.96	2.52	417
$PbCuSO_4(OH)_2$	1.96	2.53	418
$CuSO_4 \cdot 5H_2O$	1.97	2.41	419
$Cu(NaSO_4)_2 \cdot 2H_2O$	1.97	2.41	420
$Cu_2O(SO_4)$	1.97	2.52	421
β-$Cu_2P_2O_7$	1.97	2.58	422
$Cu_2(OH)_2CO_3$	1.97	2.58	423
$Cu_6(Si_6O_{19}) \cdot 6H_2O$	1.97	2.68	424
$Cu(C_2H_5OCH_2COO)_2 \cdot 2H_2O$	1.98	2.38	425
$CuWO_4$	1.98	2.40	426
$CuIO_3(OH)$	1.98	2.59	427
$Cu_5(PO)_2(OH)_4$	1.98	2.69	428
Cu(crotonate)$_2 \cdot 2H_2O$	1.99	2.33	429
$Tl_2[Cu(SO_3)_2]$	1.99	2.44	430
$Ba_2Cu(HCOO)_6 \cdot 4H_2O$	2.00	2.18	431
$Cu(HCOO)_2 \cdot 2H_2O$	2.00	2.30	432
Cu(1,2-ethanediol)$_3$(SO_4)	2.00	2.33	433
$Cu(CH_3C_6H_4SO_2) \cdot 4H_2O$	2.00	2.34	434
$Cu_2(OH)_2CO_3$	2.00	2.36	423
$Cu(HCOO)_2 \cdot 2H_2O$	2.01	2.37	432

continued

Table 5.2 — *continued*

Compound	R_S, Å	R_L, Å	Reference
$[C_{14}H_{19}N_2]Cu(hfacac)_3$	2.02	2.18	435
$Cu(CH_3OCH_2COO)_2 \cdot 2H_2O$	2.03	2.13	398
$CdCu_3(OH)_6(NO_3)_2 \cdot H_2O$	2.03	2.43	436
$K_2BaCu(NO_2)_6$	2.04	2.29	437
$Cu_2(OH)AsO_4$	2.04	2.34	438
$Cu_4(NO_3)_2(OH)_6$	2.04	2.34	439
$CuCrO_4$	2.05	2.15	440
$Cu_4(NO_3)_2(OH)_6$	2.05	2.23	439
$(NH_4)_2Cu(C_2O_4)_2 \cdot 2H_2O$	2.05	2.49	441
$(NH_4)_2Cu(C_2O_4)_2 \cdot 2H_2O$	2.05	2.74	441
$Ca(Cu, Zn)_4(OH)_6(SO_4)_2 \cdot 3H_2O$	2.06	2.23	417
$Cu(OMPA)_3(ClO_4)_2$	2.07	2.07	443
$Cu(IPCP)_3(ClO_4)_2$	2.07	2.11	444
$Cu_2(OH)PO_4$	2.07	2.28	445
$Cu(NO_3)_2HgO \cdot 3H_2O$	2.10	2.10	446
$Cu(PCP)_3(ClO_4)_2$	2.11	2.04	447
$Ca(Cu, Zn)_4(OH)_6(SO_4)_2 \cdot 3H_2O$	2.11	2.11	417
$Cu(ClO_4)_2 \cdot 6H_2O$	2.13	2.28	448
$Cu(H_2O)_8(UO_2)_4(AsO_2)_2$	2.14	2.58	449
$(NH_4)_2Cu(SO_4)_2 \cdot 6H_2O$	2.15	1.97	450
$Cu(H_2O)_6(ClO_4)_2$	209–216	2.28	451

Abbreviations: PCD = N-phenyl-2-carbamoyl-5,5′-dimethylcyclohexane-1,3-dienate; acac = acetoacetate; OMPA = octamethylpyrophosphoramide; hfacac = hexafluoracetylacetonate; IPCP = tetraisopropyl-methylenediphosphonate; PCP = octa methylmethylenediphosphonic diamide.

Table 5.3. Equatorial (R_S) and Axial (R_L) Interatomic Cu–N Distances in Cu(II) Compounds Containing CuN_6 Clusters

Compound	R_S, Å	R_L, Å	Reference
β-Cu(phtalocyanine)$_2$	1.94	3.28	452
Cu(ethylenebidiguanide)$Cl_2 \cdot H_2O$	1.96	3.17	453
$Cu(C_4H_7N_5O)_2(ClO_4)_2$	1.97	3.14	454
$Cu(NH_3)_4(NO_2)_2$	1.99	2.65	455
$Na_4Cu(NH_3)_4Cu(S_2O_3)_2 \cdot NH_3$	2.01	2.88	457
Cu(phen)$_3(ClO_4)_2$	2.05	2.33	458
Cu(dien)$_2Br_2 \cdot H_2O$	2.04	2.43	459
Cu(N,N′-$(CH_3)_2$en)$_2$(NCS)$_2$	2.06	2.52	460
[Cuen$_2$]Hg(SCN)$_4$	2.08	2.58	461
Cu(l-pn)$_3Br_2 \cdot 2H_2O$	2.09	2.31	462
$K_2PbCu(NO_2)_6$	2.11	2.11	463
Cuen$_3SO_4$	2.15	2.15	464
Cu(dien)$_2(NO_3)_2$	2.22	2.01	465

Abbreviations: phen = o-phenanthroline; en = ethylenediamine; pn = 1,2-propanediamine; dien = diethylenetriamine; pn = 1,2-propandiamine.

Table 5.4. Equatorial (R_S) and Axial (R_L) Distances Cu–X in Cu(II) Compounds Containing CuX_6 Clusters, Where X = F, Cl, Br

Compounds	R_S, Å	R_L, Å	References
Ba_2CuF_6	1.85	2.08	466
Na_2CuF_4	1.91	2.37	467
CuF_2	1.93	2.27	468
K_2CuF_4	1.95	2.08	469
$KCuF_3$	1.96	2.07	470
$CuCl_2$	2.30	2.95	471
$CsCuCl_3$	2.28; 2.36	2.78	472
$CuBr_2$	2.40	3.18	473

Table 5.5. Equatorial (R_S) and Axial (R_L) Distances Me–X in Some Compounds Containing MeX_6 Clusters, Where Me = Mn(III), Cr(II) and X = O, S, F, Cl, I

Compounds	R_S, Å	R_L, Å	Reference
$K_2MnF_5H_2O$	1.83	2.07	474
$(NH_4)_2MnF_5$	1.85	2.10	475
K_2NaMnF_6	1.86	2.06	476
MnF_3	1.79–1.91	2.09	477
Cs_2KMnF_6	1.92	2.07	478
$Mn(trop)_3$, I	1.94	2.13	479, 480
$Mn(trop)_3$, II	1.94–1.99	2.09	479, 480
CrF_2	1.99	2.43	481
$Mn(acac)_3$	2.00	1.95	396
$KCrF_3$	2.14	2.00	470
$Mn(Et_2dtc)_3$	2.38–2.43	2.55	482
$CrCl_2$	2.40	2.92	482
CrI_2	2.74	3.24	483

tion to the energy in the $^2B_{1g}$ state (elongated octahedron) has been shown to be twice that in the $^2A_{1g}$ state (tetragonally compressed octahedron). A similar comprehensive understanding of the question has not yet been achieved for Mn(III), Cr(II), and low-spin Co(II) compounds.[480,487,396]

Another conclusion derived from the data in Tables 5.2–5.5 concerns the diversity of Cu–O and Cu–N distances, which vary greatly from one system to another while preserving some relationship between R_L and R_S. It can be easily seen that R_L decreases when R_S increases, and vice versa. The expressions of Section 2.1 can be employed to obtain some useful relationships between the interatomic distances in question. If we transform from symmetrized to Cartesian coordinates (Table 1.2) and note that all coordinates $Q^0_\alpha = 0$ except $Q^0_{E\theta} = \rho_0$ and $Q^0_A = \rho_{A_0}$ (equation (2.14)), we derive

$$R_L - \bar{R}_0 = \frac{1}{\sqrt{3}}(1 - 2\eta)\rho_0$$
$$R_S - \bar{R}_0 = \frac{1}{\sqrt{3}}(-\tfrac{1}{2} - 2\eta)\rho_0 \qquad (5.31)$$

where $\eta = G_E/K_A$, G_E is the quadratic vibronic constant, K_A is the force constant for totally symmetric displacements A, $\bar{R}_0 = R_0 + \Delta R_0$ is that part of the average metal–ligand distance independent of ρ_0, R_0 is the initial distance, and $\Delta R_0 = (1/\sqrt{6})\rho^0_{A_0} = (1/\sqrt{6})F_A/K_A$ is the variation in R_0 as a result of linear vibronic coupling to the totally symmetric displacements A, F_A being its vibronic constant. It can be assumed that the value of $\bar{R}_0$ determined by the main σ bonds between the metal and nearest-neighbor atoms does not change essentially from one system to another, provided the nearest-neighbor environment remains the same.

It is seen from equation (5.31) that the differences in the distances R_L and R_S for the same cluster in different compounds are due to changes in parameters ρ_0, ρ_{A_0}, and η (disregarding cooperative effects). The main contribution comes from ρ_0, which depends on F_E, G_E, and K_E (equation (2.14)). Most important seem to be changes in the force constant K_E, which is mostly affected by the second coordination sphere causing an inhomogeneous distortion of the cluster in question (strictly speaking, when taking into account the next coordination sphere the problem becomes a multimode one, Section 2.4).

Eliminating ρ_0 from the two equations (5.31), we obtain the following relationship between R_L and R_S:

$$R_L(1 - 4\eta) + R_S(2 + 4\eta) = 3\bar{R}_0 \qquad (5.32)$$

Since η is small (see below), this relationship is linear in R_L and R_S for any series of compounds, for which the averaged metal–ligand distances $\bar{R}_0$ do not change drastically (and the harmonic approximation employed above is valid). Figure 5.30 shows that the linear relationship (5.32) is approximately valid: all points (R_L, R_S) fall within a narrow band which approximates a line to within about 10%. The points with extremely large distortions to the left and right in Figure 5.30 clearly lie outside the assumed harmonic approximation (large distortions acquire cubic or higher-order displacements). This result confirms the vibronic nature of the observed distortions.

If a_L and a_S are the values at which the line (5.32) intersects the R_L and R_S axes, respectively, then

$$\eta = (2 - C)/4(1 + C) \qquad (5.33)$$

where $C = a_L/a_S$ is the tangent of the angle of line slope. For the lines in Figure 5.30, $C = 1.4$ and $\eta = 0.063$. Note that this small value of η does not mean that G_E is small, in general. Indeed $\eta = G_E/K_A$, and K_A, the force constant of totally symmetric (breathing) displacements, can be very large. (The ratio G_E/K_E seems to be more informative for G_E.) Since K_A is usually determined from IR absorption data, the value of G_E can be calculated if η is known. Then $\rho_0 = 2\sqrt{3}(R_L - R_S)$ can be calculated from equation (5.31) and, if the value of K_E is also known, the vibronic constant F_E can be estimated from equations (2.14). If $R_L + 2R_S$ is taken as the averaged distance for a series of compounds, the value of ρ_{A_0} can be evaluated and $\rho^0_{A_0} = F_A/K_A$ can be calculated from equations (2.14).

There may be many other ways of applying the plasticity properties. In particular, electronic lability effects in orbitally near degenerate states[488] and small perturbations (such as solute–solvent interaction[489] and mutual influence of ligands[490]) in JT systems are closely related to the plasticity phenomena discussed in this section.

Distortion Isomers. A possible situation arises when the crystal stabilizes not one, but two or several coordination sphere configurations which, being close in energy, differ by the distortion magnitude and direction. Such configurations of the same compound may exhibit themselves as different crystal isomers. An interesting example of this kind are the distortion isomers of Cu(II), first synthesized and studied by Gažo et al.[394,395] These isomers possess the same total composition and the same Cu(II) ligand environment composition but differ in the interatomic metal–ligand distances (in the distorted coordination sphere). These distortion isomers also differ in their properties, such as color, appearance, crystal form, chemical behavior, solubility, and spectroscopic properties. The isomers pass from one to another under the influence of pressure, heating, or long-term storage. In some cases, besides the two principal isomers (usually called α and β), a series of intermediate species have also been obtained.

One of the simplest compounds having distortion isomers is $Cu(NH_3)_2X_2$, where X = Cl, Br. Their possible JT origin was suggested when the isomers were first discovered.[491] However, the differences in the ligands NH_3 and X remove any degeneracy, and therefore the distortion cannot be of pure JT origin.[492] Liehr[39] suggested that the PJTE may be operative under conditions similar to those in question.

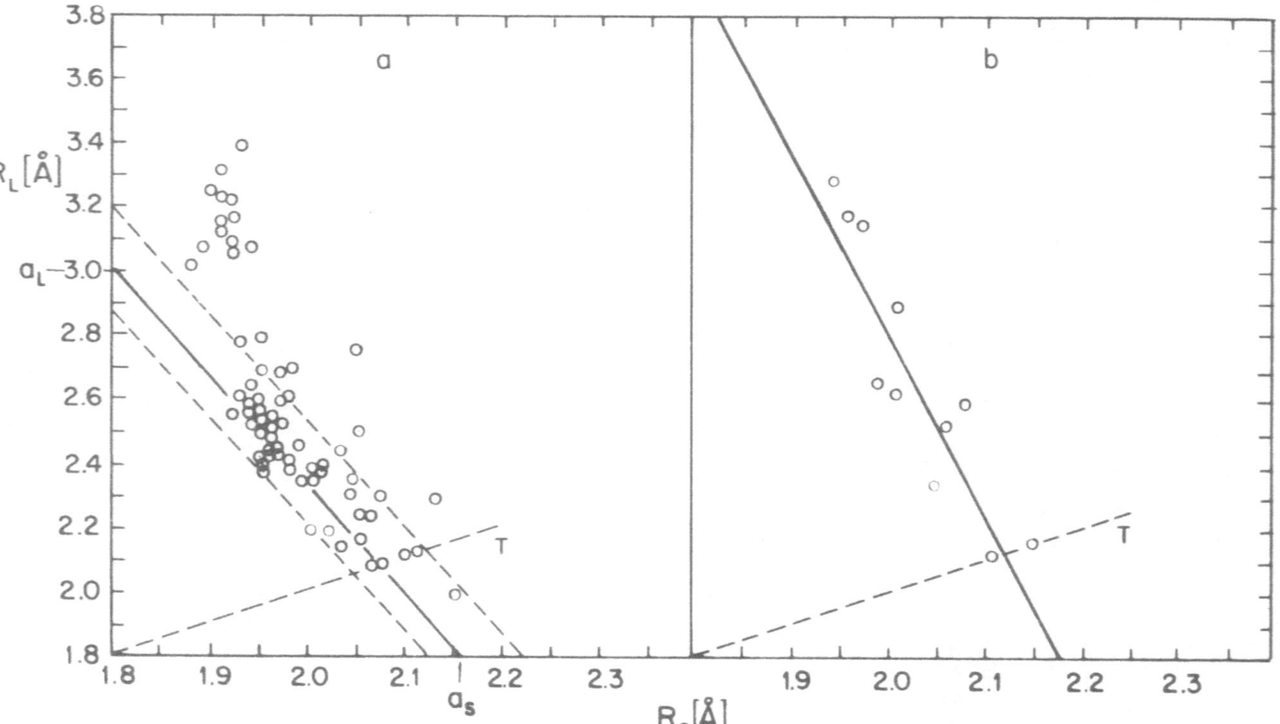

Figure 5.30. Axial (R_L) versus equatorial (R_S) metal–ligand distances in octahedral clusters of divalent copper: (a) CuO_6; (b) CuN_6. With the exception of very large values of R_L, for which the harmonic approximation is invalid, all points (R_L, R_S) lie within a narrow band near a straight line. T denotes the line $R_L = R_S$.

The above results yield a quite natural explanation of the origin of distortion isomers due to the vibronic properties of the Cu(II) center accompanied by the stabilizing influence of, and cooperative effects in the crystal. Consider the crystal $Cu(NH_3)_2X_2$ as an example.[493,395] This crystal comprises mutually parallel chains each of which is arranged as illustrated in Figure 5.31a, where all the X atoms occupy equivalent bridge positions. There is a strong interaction between the Cu(II) centers through bridging atoms inside the chain, while bonding between chains is weak (van der Waals and/or hydrogen bonds). Each copper atom is surrounded by four X atoms in the plane of the square and by two NH_3 groups in positions 5 and 6. The degeneracy of the ground Cu(II) state in the octahedron is removed as a consequence of the difference between X and NH_3, and the ^2E_g term splits into $^2A_{1g}$ and $^2B_{1g}$. Let this splitting be 2Δ.

If we assume that the X atoms in the plane form a regular square, the polyhedron $Cu(NH_3)_2X_4$ is a tetragonally distorted octahedron belonging to D_{4h} symmetry. Consider the PJTE on the A_{1g} and B_{1g} terms of such a complex. Noting that this case corresponds completely to the two-level case examined in Section 2.3, the results obtained there may be applied directly. In particular, the normal coordinate Q which mixes the states A_{1g} and B_{1g} transforms according to B_{1g} (since $A_{1g} \times B_{1g} = B_{1g}$) and the corresponding B_{1g} displacements in the D_{4h} group coincide with the Q_ε displacements of the O_h group given in Figure 1.1. From equation (2.34), the vibronic constant of the coupling to these displacements is given by the expression

$$F = \left\langle \Psi_{A_{1g}} \middle| \left(\frac{\partial V}{\partial Q}\right)_0 \middle| \Psi_{B_{1g}} \right\rangle \tag{5.34}$$

where $\Psi_{A_{1g}}$ and $\Psi_{B_{1g}}$ are the wave functions of the mixing pseudodegenerate A_{1g} and B_{1g} states (in the crystal field approach, these two are $3d_{z^2}$ and $3d_{x^2-y^2}$ functions, repectively, for a d hole in the Cu^{2+} (d^9) configuration). The respective AP are given by expressions (2.36) (Figure 2.12b):

$$\varepsilon_\pm = \tfrac{1}{2}KQ^2 \pm [\Delta^2 + F^2Q^2]^{1/2} \tag{5.35}$$

where $K = M\omega^2$, ω being the corresponding B_{1g} vibrational frequency. If the instability condition $\Delta < F^2/K$ is satisfied, we obtain two minima on the lower sheet of the AP at $\pm Q_0$, determined by equation (2.39).

Thus the PJTE at each center distorts the bipyramid environment such that the square of its base with four atoms X at the apexes transforms to a rhombus with the major diagonal along Q_x (corresponding to the minimum I in Figure 5.32b) or Q_y (minimum II).

Due to strong interaction between the distortions of neighboring

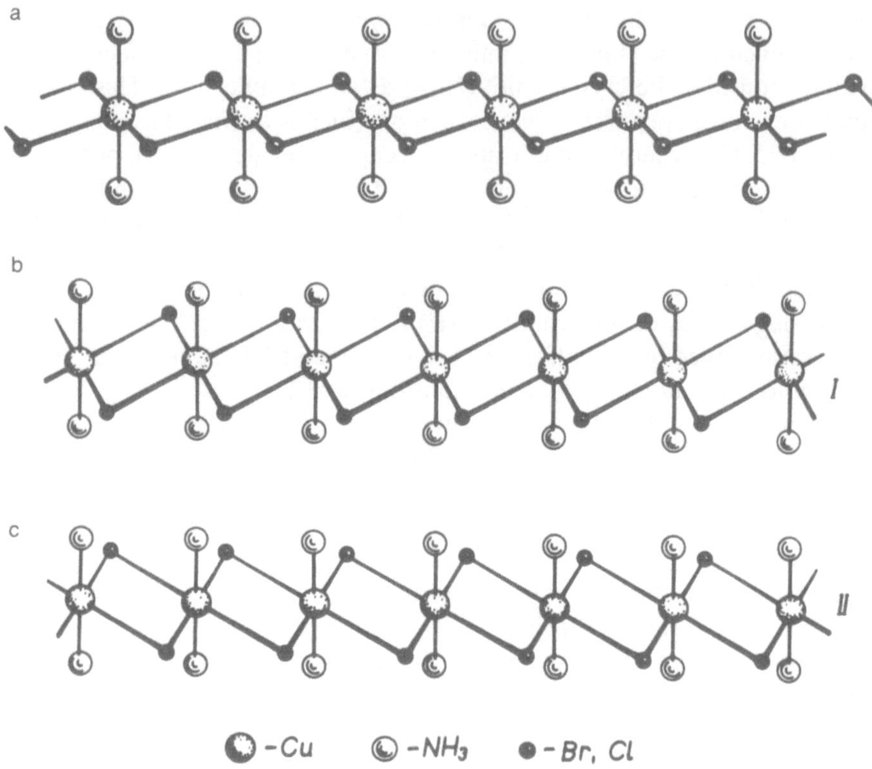

Figure 5.31. Chain structure of the crystal $Cu(NH_3)_2X_2$ in the cubic undistorted unstable β isomer (a), and in two equivalent configurations of the α isomer I and II (b and c) resulting from the in-chain cooperative PJTE.

centers in the chain through the ligands X in common, the CPJTE leads to ferrodistortive ordering of these distortions along the chain, and this ordering remains unchanged up to high temperatures. Here, the PJT distortions are stabilized by the cooperative phase transitions. As a result of this ordering it can be assumed that at room temperature each of the chains has two stable configurations, I and II (Figures 5.31b, c), corresponding to two minima for each center (I and II in Figure 5.32b). We can conclude that there is a possibility that two equivalent chain structures of $Cu(NH_3)_2X_2$ originate from the PJTE at each $Cu(NH_3)_2X_4$ center with ferrodistortive ordering of the distortions due to the CPJTE.

One more important crystal factor, interaction between the chains, remains to be considered. Analysis of the structure of the crystal composed of parallel chains indicates that intermolecular interaction between the chains is optimum when the chains are not distorted and the entire crystal is cubic. In this cubic state the intermolecular distances (between strongly interacting atoms of different chains) are minimal, and they increase with interchain distortions toward configuration I or II. The interchain interaction as a function of the interatomic distances can, in principle, be estimated semiempirically (for instance, by means of the atom–atom potential approach). In the absence of such calculations we assume that interchain interaction in the crystal has a minimum at $Q = 0$ (Figure 5.32a), where Q is the coordinate of the cooperative interchain distortion corresponding to B_{1g} distortion at each center. Adding the AP energy of the chain (Figure 5.32b) to the interchain interaction energy as a function of Q (Fig. 5.32a) we obtain the total energy of the crystal (Figure 5.32c). Under the above assumptions, the total energy is seen to have three minima: besides the two minima I and II for the stable distorted configurations of the chains, there is a third, shallow minimum for the undistorted crystal.

It has been assumed above that all the chains are at the same type of minimum, either I or II, and the intrachain distortions are ordered along the crystal. If different chains are at different types of minima (and the entire crystal is not ordered) the total energy may have additional minima which, however, are expected to be more shallow than minima I and II, but deeper than the minimum at $Q = 0$ (on the assumption that the intrachain PJTE stabilization energy is larger than that of interchain interactions).

It can be shown that the above description explains qualitatively the origin of all the main features of the distortion isomers in $Cu(NH_3)_2X_2$. The α isomer corresponds to the deepest minimum of the AP (I or II) with the structures illustrated in Figures 5.31b,c. The unstable β isomer with a cubic

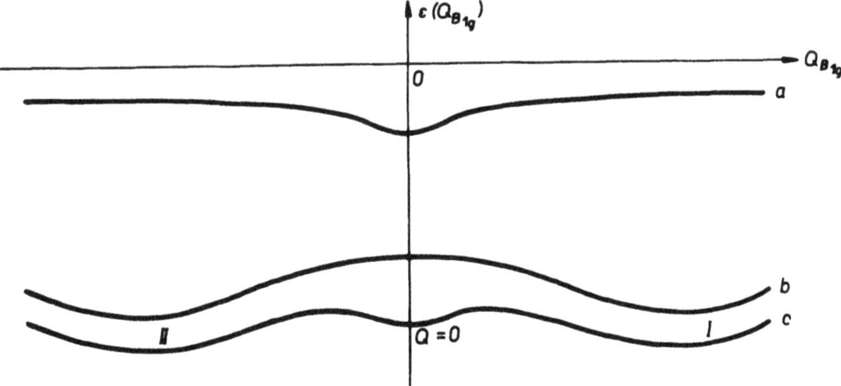

Figure 5.32. Energy of the $Cu(NH_3)_2X_2$ crystal as a function of the cooperative B_{1g} type in-chain distortion (Figure 5.31): (a) interchain interaction; (b) intrachain CPJT interaction; (c) summarized value.

structure corresponds to the more shallow minimum at $Q = 0$, and the intermediate preparations with noncubic structures correspond to the aforementioned additional minima for the uncorrelated chain distortions.

The interpretation agrees well with the experimental features of the isomers, including their behavior under stress and temperature, depending on the conditions of their preparation.[394,395]

It is noteworthy that good crystals (single crystals) can be obtained only for the β isomer,[394] i.e., when (as follows from the theory) the intermolecular (interchain) interactions, favoring crystal formation, are at a maximum. On the other hand, the α isomer and intermediate preparations cannot be obtained in the form of good crystals (but only as powders), in agreement with the assumption of the theory of poor interchain interactions in these situations. The fact that the β (and only the β) isomer contains as impurities the NH_4X compounds may also be important.[494] The latter crystallized in the cubic structure isomorphic to the β isomer of $Cu(NH_3)_2X_2$, and these crystals easily form mixed crystals. The cubic structure of the β isomer is apparently also stabilized by the large amount of NH_4X impurity molecules, which alternately occupy the place of the Cu(II) centers in the chains, strongly reducing the intrachain effect. It has been shown recently[495] that without the NH_4X admixture the β isomer rapidly converts to the α isomer. It can be prepared and preserved under pressure, however.

The PJT origin of the intrachain distortion was confirmed by approximate calculations of PJTE for each $Cu(NH_3)_2X_4$ polyhedron. In so doing, the electronic energy as a function of the $Q_{B_{1g}}$ distortion was estimated by means of the angular overlap method and, using empirical data on the value of K, the AP curve (Figure 5.32b), including its minima, was obtained.[496] The results confirm that the distortion is of PJT origin: $Q_{B_{1g}}$ (calculated) ≈ 0.5 Å, while $Q_{B_{1g}}$ (experimental) $\cong 0.4$ Å.

In principle, the same idea may be used to examine the origin of other distortion isomers, described elsewhere.[394] Moreover, the above method of investigating the origin of crystal structures has general significance and can be applied to other compounds and quasi-one-dimensional structures. The vibronic origin of crystal modifications can be applied to any class of chemical compound.

By way of example we shall briefly consider one more group of distortion isomers based on dioximes of trivalent cobalt and rhodium, mentioned in Section 5.1. It was shown that chemical compounds of the type $[Co(DH)_2APh_3Hal]$, where DH is dimethylglyoximate $[CH_3(HON)C{=}C(NO)CH_3]^-$; A = P, As, Sb; $Ph_3 = (C_6H_6)_3$, Hal = Cl, Br, I, crystallize in two modifications.[256,257] In both of them the dimethylglyoximate residues are *trans* to each other, but in one of them (rhombic space group D_2^4) the two DH residues form together with the central atom a planar configuration, whereas in the other (monoclinic, C_{2h}^5) configuration these two DH plane groups geniculate at an angle of about 15°. From another standpoint, the monoclinic configuration differs from the rhombic one in that the out-of-plane displacement of the Co atom entails an appropriate bending of the two DH plane groups. Two similar modifications were found in $[Rh(DH)_2PPh_3Cl]$: bent[258] and planar[259] (the planar configuration was interpreted for the case where the P atom is replaced by Sb, but these two compounds are isostructural). These two modifications (two isomers) of the crystal differ only in small internal distortions around the metal centers, which suggests their vibronic origin.[497]

In cobalt dioximes, as in some metal porphyrins (Section 5.1), there are near energy levels of electronic states (Figure 5.9) which result in a PJTE with respect to the out-of-plane displacement of the Co atom. Owing to this PJTE the position of the Co atom in the dioxime plane can be unstable. A more comprehensive PJTE treatment for the system as a whole may explain not only the displacement of the Co atom with respect to the DH plane, but also the bending of the latter (similar to bending of linear A–B–A molecules due to the Renner effect, discussed in Section 5.1). This explains the origin of the low-symmetry monoclinic modification.

Analysis of the intermolecular interaction in the crystal is necessary in order to understand the origin of the rhombic isomer. It can be shown that (as expected from general considerations) when passing from the rhombic to the monoclinic isomer the intermolecular contacts in the crystal deteriorate, and hence stabilization of the crystal as a whole decreases (complete quantitative estimates have not yet been performed). It follows that the two isomers, as in the above distortion isomers of Cu(II), occur due to the two types of minima of crystal energy as a function of distortions: one due to internal vibronic interaction and corresponding to a PJT (or JT) distortion of the local environment of the metal, the other due to intermolecular interactions which prefer nondistorted molecules.

The general requirement for two (or more) vibronic (distortion) isomers to be observed at the same (or in a narrow interval of) temperatures lies in the fact that the depths of the two (or more) minima of the AP of the crystal as a whole should have the same (or close) values.

General Validity of the Vibronic Model in Stereochemistry, Crystal Chemistry, and the Theory of Structural Phase Transitions. Summing up the contents of this chapter, one can see that vibronic effects are essential to describe almost every feature of the stereochemistry of free molecules, crystal chemistry, and structural phase transitions in a number of classes of chemical compounds. This number is increasing rapidly; the above systems may be regarded as examples (for more details see the Bibliographic Review[14]). In this connection the question arises: is the vibronic model of general use in the above type of problems or, if not, what are the limits of its application?

To answer this question, it is first necessary to note that the analysis of the origin of polyatomic nuclear configurations without the notion of vibronic coupling is impossible. A stable nuclear configuration requires repulsive forces between the nuclei to be compensated by the chemical bonding produced by a special electronic distribution. This mutual compensation of the repulsive and attractive forces provides for the static stability of the system. Besides, the conditions of dynamic stability of the system (a minimum, and not a maximum or a saddle point of the AP) must be provided. Static stability requires the linear vibronic constant to vanish ($F = 0$), while for dynamic stability a positive force constant ($K > 0$) is needed (see Chapter 1). The change of stability by electronic rearrangement is described by the orbital vibronic constant (Section 1.3). Moreover, it has been shown[28] (Section 1.4) that any dynamic instability of an orbital nondegenerate system is of a PJTE nature. For degenerate states the instability follows immediately from the JT theorem. Thus the instability (and hence the stability) of any molecular polyatomic system is determined by the vibronic coupling.

The proof of the vibronic nature of the dynamic instability of polyatomic systems has been extended to dielectric crystal structures.[498] It results in a statement that *any structural phase transition in these systems is of vibronic origin.* Even when the structural phase transition originates from an ordering of stable asymmetrical atomic groups in the crystal (order–disorder transitions), the existence of two or more equivalent stable positions of these groups (and hence the instability of the position of maximum symmetry) can be reduced to vibronic interactions.

For instance, in ferroelectrics with hydrogen bonds (such as KH_2PO_4) the structural phase transition to the ferroelectric state is assumed to result from the ordering of proton positions along the hydrogen bond (for which there are two equivalent minima). Since it has been established that dynamic instability in any polyatomic system is of vibronic origin (Section 1.4), the two minima of the proton along the hydrogen bond and the presence of a potential barrier between them (a maximum of the AP) result from vibronic mixing of the ground state with the near excited state.

It might appear that the vibronic origin of the dynamic instability of the crystal contradicts the existence of ionic crystals. Indeed, if the crystal consists of absolutely rigid ions, there are no electronic states and vibronic interactions. However, such a crystal is unstable in the whole of configurational space. The ionic crystal can be stable under certain assumptions based on the nonrigidity of the ions (polarizability, Born repulsion, etc.) which, from the microscopic point of view (i.e., starting from first principles), are possible only if one takes into account electronic states and vibronic interactions.[498]

Thus, in principle, the vibronic model has no limitations. *The origin of the spatial configuration of any polyatomic system, including the formation of the crystal structure and its change by phase transitions, as well as the change of these configurations by electronic rearrangements, can be considered* in terms of vibronic theory *as resulting from vibronic effects.* In practice, however, the vibronic approach is not always especially useful. In particular, vibronic interactions can be neglected for nondegenerate, nonpseudodegenerate, and hence highly stable molecules (having a large energy gap between the ground and excited states), and when considering properties not concerned with (and not essentially coupled to) nuclear displacements.

It is evident that these nonvibronic systems with nonvibronic properties are, with some exceptions, not concerned with the most interesting systems, phenomena, and properties of modern chemistry. The latter are intimately connected with and based on nuclear configuration formation, distortion, and dynamics. In a sense, the general importance and wide use of vibronic effects in stereochemistry and crystal chemistry, as demonstrated in this chapter, can be described as a new trend — *vibronic stereochemistry and crystal chemistry.*

[illegible] are possible only if one takes into account electronic states and electronic interactions.

Thus, in principle, the electronic theory has no limitations. The origin of the spatial configuration of any polyatomic system, including the formation [illegible]

6

Activation Mechanisms in Chemical Reactions and Catalysis

The use of vibronic effects in the elucidation of mechanisms underlying chemical reactions and for the prediction of chemical activation is one of the latest and perhaps most interesting applications of the concept of vibronic interactions.

6.1. Nature of Instability of the Activated State of a Chemical Reaction and the Possibility of Its Experimental Observation

Activated States of Chemical Reactions. The activated or transition states (activated complexes) of chemical reactions for the most part determines their more important characteristics — the mechanism and rate of the elementary act. This has led to increased interest in both theoretical and experimental research on the structure and properties of activated complexes, and the determination of the magnitude and shape of reaction potential barriers. Beginning with Eyring, Wigner, and Peltzer[499] many scientists have contributed considerably to a proper understanding of certain important aspects of the topic (see Eyring et al.[500] and Nikitin[501] and references therein). Important progress was achieved by means of qualitative and semiquantitative approaches based on perturbation theory and LCAO MO methods.[16,27,502–505] As a whole, however, the problem remains far from being completely solved, and the vibronic concept may contribute substantially to its solution.[17–21,28,506]

The chemical literature reports much work in which activated states are treated as real states and the problem of their possible experimental observation and investigation is discussed. However, using first principles of quantum mechanics, it can be easily shown that the activated state of a chemical reaction, understood literally (the activated complex in its ground unstable state), cannot in principle be observed experimentally.

Consider the potential energy surface for a chemical reaction with known mechanism and assume that the curve in Figure 6.1 represents the section of this surface along the line of its "steep slope" on the lowest barrier between the two minima of the initial reactants A and final products B (Q is the reaction coordinate). It is evident that point M of the maximum of the barrier along Q on the multidimensional surface is a saddle point, otherwise the reaction path through M will not correspond to the lowest barrier. The possibility (in principle) of observing the activated state of point M by means of spectroscopic methods requires this state to be stationary or quasistationary from a quantum-mechanical standpoint. Meanwhile, it is well known that stationary state solutions may be obtained only for potentials with a well (minimum) under the condition that its depth is not very small compared to the kinetic energy of the system in the well.[22]

There are no stationary solutions for the saddle point type potential near the saddle point M. It follows that spectroscopic observation of an activated state which corresponds to this point M is impossible.

Cases arise when there is an additional minimum of the adiabatic potential near the top of the barrier, shown in Figure 6.1 by a dashed line. If the depth of this additional well is sizeable[22] it may contain quasi-stationary states, which are real and can be observed experimentally. However, these states may be regarded as additional isomeric states of the system rather than activated states of chemical reactions, and they will be disregarded here.

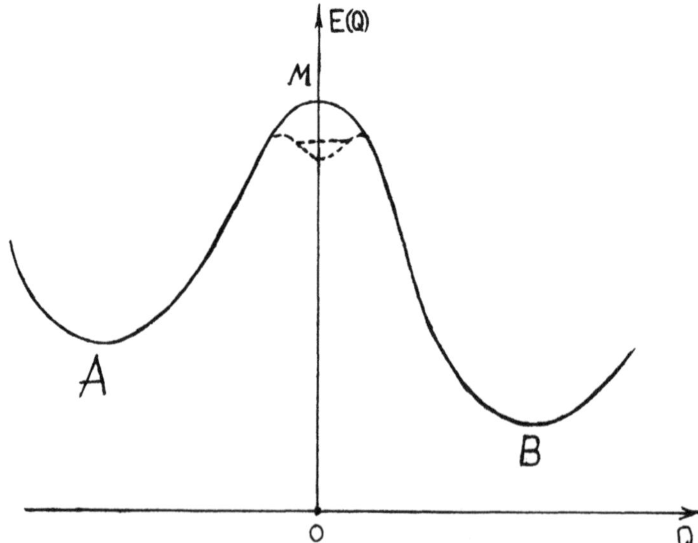

Figure 6.1. Schematic presentation of the potential energy curve of a classical chemical reaction along the reaction path Q. A, B, and M are the initial reagents, reaction products, and barrier maximum point, respectively.

In principle, point M can be assumed to be reachable through vibrational excitation at high temperatures. However, the corresponding excited vibrational state, which includes point M, would have an abnormally large amplitude between the reactants A and products B and is hardly possible as a stationary spectroscopic state (due to rapid relaxation); it cannot be used to reveal the characteristics of the system at point M.

Although the activated state in question cannot be observed directly, calculation of the AP (the chemical reaction barrier size and shape near saddle point M) is very important as it determines the rate of the elementary act and the mechanism of the reaction. (This is not the only case when important physical and chemical properties are determined by unobservable computed quantities.)

While it is impossible to observe experimentally the activated state of a chemical reaction, this does not mean that the activated complex as a molecular system with given nuclear configuration cannot be described in general. Indeed, all the above deals with the activated complex in its ground (unstable) state. This does not mean that there are no other (excited) electron distributions (for the same nuclear configuration) which are bonding. If such a bonding electronic state exists, then the appropriate adiabatic potential has a well (minimum) near the nuclear configuration of point M. Here, stationary states are quite possible and may be observed experimentally. It will be shown below that such stable excited states do exist, and that they are the source of dynamic instability of the ground state.

Stable Excited States of Activated Complexes. Consider the following important question: what is the origin of the instability of the ground state of an activated complex? This state is dynamically unstable, since it corresponds to the maximum of the AP along the coordinate Q (the electrostatic forces are equilibrated). It was shown in Section 1.4 that dynamic instability of any real molecular system along a direction Q is due to vibronic mixing of the electronic state in question with the near excited states by the nuclear displacements Q.

Consider the expression for the curvature of the AP in the Q direction — the force constant K^{Γ} of the system in state Γ from equation (1.27) (see also Bader[506]):

$$K^{\Gamma} = \left\langle \Gamma \left| \left(\frac{\partial^2 V}{\partial Q^2} \right)_0 \right| \Gamma \right\rangle - \sum_{\Gamma'}{}' \frac{\left| \left\langle \Gamma \left| \left(\frac{\partial V}{\partial Q} \right)_0 \right| \Gamma' \right\rangle \right|^2}{\varepsilon_{\Gamma'} - \varepsilon_{\Gamma}} \tag{6.1}$$

The leading term $K_0^{\Gamma} = \langle \Gamma | (\partial^2 V/\partial Q^2)_0 | \Gamma \rangle$, as shown elsewhere[28] (see also Section 1.4), is always positive. Therefore a negative value of K^{Γ}, and hence dynamic instability, only appears due to the second term in equation (6.1), which describes vibronic mixing of state Γ with excited states Γ'.

For further analysis equation (6.1) is employed in the form (1.18): $K^{\Gamma} = K_0^{\Gamma} - \Sigma'_{\Gamma'} |F^{(\Gamma\Gamma')}|^2/\Delta_{\Gamma'\Gamma}$, where $F^{(\Gamma\Gamma')}$ is the nondiagonal linear vibronic constant and $2\Delta_{\Gamma'\Gamma} = \varepsilon_{\Gamma'} - \varepsilon_{\Gamma}$. Consider the particular, but very important case when the greatest contribution to the vibronic term in equation (6.1) originates from only one term with the lowest value of $\Delta_{\Gamma'\Gamma}$ and highest $F^{(\Gamma\Gamma')}$. This means that there is a distinguishable near and very active excited state Γ', which is far from the other states. Then, the problem can be reduced to a two-level one. Neglecting the small contributions of the other terms and denoting $F^{(\Gamma\Gamma')}$ by F and $\Delta_{\Gamma'\Gamma}$ by Δ, we have

$$K^{\Gamma} = K_0^{\Gamma} - F^2/\Delta \tag{6.2}$$

If $F^2/\Delta > K_0^{\Gamma}$, then $K^{\Gamma} < 0$ and the system becomes unstable. This criterion agrees with that of equation (2.38) obtained from the PJTE for a two-level system. Unlike equation (6.1), which is valid only for small values of $(F^2/\Delta)Q^2$, the expressions for the two-level problem obtained in Section 2.3 (not based on perturbation theory) are of more general use. In particular, using equation (2.36) and taking into account that the curvatures K_0^{Γ} and $K_0^{\Gamma'}$ may differ, one obtains the following equations for the AP of the two mixing states Γ and Γ':

$$\varepsilon_{\Gamma}(Q) = \tfrac{1}{2}K_0^{\Gamma}Q^2 - (\Delta^2 + F^2Q^2)^{1/2} \tag{6.3}$$

$$\varepsilon_{\Gamma'}(Q) = \tfrac{1}{2}K_0^{\Gamma'}Q^2 + (\Delta^2 + F^2Q^2)^{1/2} \tag{6.4}$$

If the expressions under the square root are expanded in small values of Q, we derive

$$\varepsilon_{\Gamma}(Q) = \tfrac{1}{2}(K_0^{\Gamma} - F^2/\Delta)Q^2 + (F^4/\Delta^3)Q^4 - \cdots \tag{6.5}$$

$$\varepsilon_{\Gamma'}(Q) = \tfrac{1}{2}(K_0^{\Gamma'} + F^2/\Delta)Q^2 - (F^4/\Delta^3)Q^4 + \cdots \tag{6.6}$$

As a result of vibronic mixing, the force constants are now given by

$$K^{\Gamma} = K_0^{\Gamma} - F^2/\Delta \tag{6.7}$$

$$K^{\Gamma'} = K_0^{\Gamma'} + F^2/\Delta \tag{6.8}$$

For the ground state, K^{Γ} coincides with equation (6.2) obtained in the framework of simple perturbation theory. However, in addition to this result, we reach two new important conclusions. Instability occurring under the condition $F^2/\Delta > K_0$ is only local instability, since for larger Q the positive terms of the order of Q^4 in equation (6.5) become more pronounced (vibronic anharmonicity[12]). Therefore, in addition to the maximum point at $Q=0$ the curve $\varepsilon(Q)$ has two minima at $\pm Q_1 = (F^2/K_0^2 - \Delta^2/F^2)^{1/2}$ (Figure 2.12b). But it is more important that we have obtained here also the nearby excited state curvature $K^{I''}$. Taking into account that[28] $K_0^{I''} > 0$, we have $K^{I''} > 0$ (see equation (6.8)), and we conclude that the excited state, which induces instability of the ground state, is stable.

Thus the two-level model, in addition to illustrating the origin of the instability of the ground state as due to vibronic mixing with excited states, also indicates the nearby bonding excited state which causes instability of the ground state. This excited state can be found by means of symmetry rules from the requirement that Δ is small and the vibronic constant F is large. The force constant (6.8) and other characteristics are correlated with those of the unstable ground state by equations (6.3)–(6.8).

If there are more than one near vibronically active excited state, the two-level scheme becomes invalid and a more complicated many-parameter calculation is necessary in order to solve the problem quantitatively.

Existence of Bonding States of Arbitrary (Neutral) Polyatomic Groups. The above results suggest the possible existence of bonding states for arbitrary configurations of any polyatomic group. Such bonding states have been shown to exist as excited states for any unstable system, at least in the two-level model. The following general statement can be formulated about the existence of bonding states for polyatomic groups: *for any nuclear configuration of a neutral polyatomic group there are bonding electronic (ground or excited) states.*

The electronic state is regarded as bonding if its adiabatic potential surface has a well (minimum), and the nuclear configuration under consideration corresponds to any point of the well area (not necessarily to the minimum point itself). If the minimum well depth is not very small and it contains some vibrational states,[22] the bonding electronic state means that the pertinent nuclear configuration lies within the amplitude of stationary (or quasistationary) vibrations of the system.

A simple LCAO MO scheme can be used to prove this statement. Consider first diatomic molecules. Using the well known LCAO MO qualitative energy level scheme for diatomic molecules with identical light atoms (for different atoms the situation is similar; the conclusions deduced below are independent of the energy level ordering), it is seen that, in the ground state, the bonding $\sigma_g 1s$ orbital is populated first, then the antibonding $\sigma_u 1s$ orbital follows, and so on, the electronic configuration of the molecule being dependent on the overall number of electrons. For instance, the He_2 molecule has four electrons, and in its ground state both the bonding $\sigma_g 1s$ and antibonding $\sigma_u 1s$ orbitals are occupied. It follows that the He_2 ground state is unstable, but the transfer of an electron from the antibonding $\sigma_u 1s$ orbital to any bonding, say $\sigma_g 2s$, orbital results in an excited bonding electronic state (with respect to the corresponding, including excited, atomic states).

Analysis of the qualitative LCAO MO schemes leads to the conclusion that all neutral and not very high ionic diatomic systems have bonding electronic states, the number of which is very large (infinite). The physical reason is that such excited atomic states always exist for which the molecular states will be bonding. Exceptions are highly charged systems for which both atomic and molecular states may have continuous energy spectra with no bonding states.

The proof of the theorem for polyatomic molecules may be given similarly, using the diatomic picture as a base. For any nuclear configuration a general LCAO MO energy level scheme can be used to illustrate the existence of excited bonding states. However, a complication arises here due to the multidimensionality of the potential energy surface, which means that the sought electronic states must be bonding with regard to the decomposition of the molecule not only into atoms (equivalent to a minimum on the surface along the totally symmetric coordinate), but into any group of atoms, or any molecular fragments (equivalent to a minimum along all the nontotally symmetric coordinates).

Consider a triatomic molecule ABC. Using the above results one can easily find the bonding states of any of its diatomic fragments. Taking the bonding states of, say, AB as a basis, one can use them as atomic states and consider the system $AB - C$ as a "diatomic" one, for which, as shown above, there are (at least excited) bonding electronic states. These latter will also be bonding with regard to the decomposition of ABC into atoms, $ABC = A + B + C$, since the initial orbitals of the AB fragment, used in the scheme, are bonding. But there is no evidence that this state will be bonding with regard to the decomposition $ABC = A + BC$ (or $ABC = AC + B$), too, which means that the electronic state under consideration is not bonding with regard to the "diatomic" system $A - BC$. In this case the more stable BC bonding states should be used as a basis to construct new bonding states of the molecule. The latter will be bonding with regard both to the decomposition $ABC = A + BC$ and to $ABC = AB + C$. The nature of the bonding state, the depth of the local minimum of the adiabatic potential and its shape, will also depend on the reciprocal positions of the fragment and the atom (there may be several different local minima for different nuclear configurations of the interacting fragments).

For four or more atoms the reasoning is similar to that given above. For instance, for a tetraatomic system the bonding states of two diatomic fragments may be taken as a basis for constructing bonding states of the system as a whole. If the fragments with the most stable bonding orbitals are chosen as initial states, the bonding orbitals obtained will be bonding also with regard to decomposition into any possible fragments. With an increase in the number of atoms, the number of possible fragments and ways of decomposition also increase, but since the number of initial fragment bonding states is large enough (infinite) the bonding states sought for the system as a whole exist.

The reasoning in this section remains incomplete without a discussion of the important question about the depths of the adiabatic potential minima corresponding to the LCAO MO bonding states under consideration. If the minimum depth ΔE is much smaller than the kinetic energy of the system in the bonding state well (which for small wells is of the order of $\hbar^2/Md^2$ where d and M are its linear dimensions and mass, respectively), i.e. $\Delta E \ll \hbar^2/Md^2$, there are no localized vibrational states in the well,[22] and such "bonding" electronic states have no physical meaning as real bonding. In fact the requirement $\Delta E \gtrsim \hbar^2/Md^2$ is not difficult to obtain, and it is satisfied in most cases, even for shallow minima. However, there is no evidence that it will always be satisfied. Therefore, in the cases of small and very shallow minima the proof of existence of LCAO MO bonding states for any group of atoms, carried out in this section, must be supplemented by additional estimates of the depth and linear dimensions of the potential well.

Possible Experimental Information about Activated Complexes. The existence of stationary excited bonding states in activated complexes of chemical reactions gives rise to assumptions of possible experimental detection and investigation. While a discussion of experimental procedure is beyond the scope of this book it seems that, in principle, all spectroscopic methods from electron and photoelectron spectroscopy in the range of X-ray, UV, visible, as well as vibrational and rotational IR and radio spectroscopy, including ESR, NMR, etc., may be used for this purpose.

The most important information to be derived from these experiments regards the nuclear configuration of the activated complex which determines the reaction mechanism. It can be obtained, in principle, by any of the aforementioned experimental methods. Along with the nuclear configuration, important information about the reaction potential barrier can be derived using the earlier-obtained correlation between the excited and ground states. By way of example we consider here experimental data on the optical absorption band for the two-level approximation when the excited state minimum position coincides with the activated complex nuclear configuration.

Suppose one has succeeded in obtaining experimentally the shape of the absorption band due to electron transitions from the minimum A of the reactants (or B of products) to the bonding excited state of the activated complex (Figure 6.2) and the frequencies of the band maximum Ω_{max}, pure electron transition Ω_0, Q type vibrations of the reactants ω_A, and the activated complex ω_e. Using these spectral parameters the following relationships can be deduced from Figure 6.2 (for more details see, e.g., Bersuker[53]):

$$D = \hbar[\Omega_0 + \tfrac{1}{2}(\omega_A - \omega_e)] - 2\Delta \tag{6.9}$$

$$\tfrac{1}{2}K_e d^2 = \hbar(\Omega_{max} - \Omega_0 + \omega_e/2) \tag{6.10}$$

where $K_e = M\omega_e^2$, D is the barrier height, and d is the distance between the coordinates of the A minimum and activated complex maximum M,

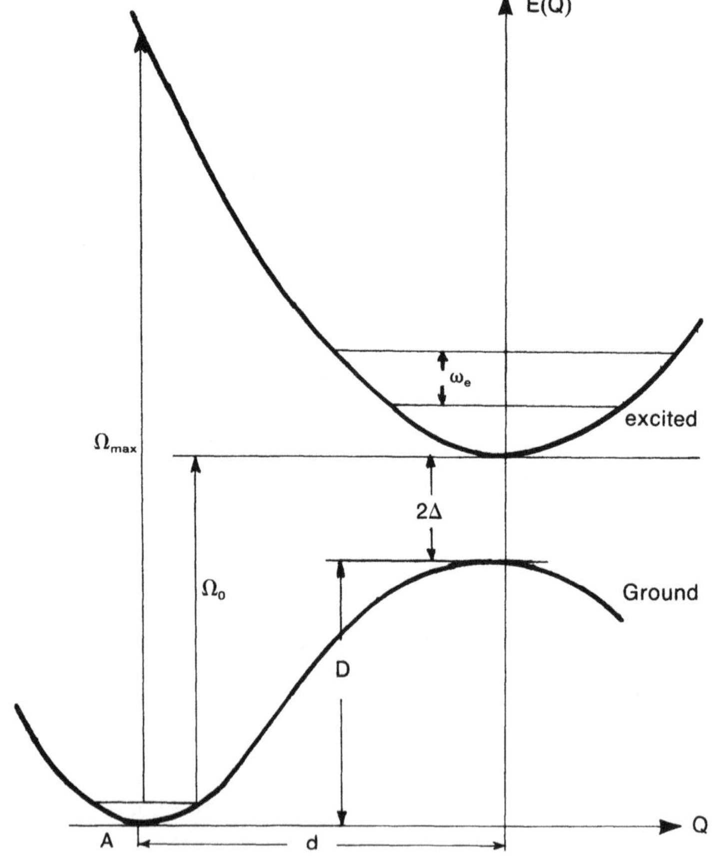

Figure 6.2. Parameters linking the unstable transition state of the chemical reaction with the stable excited state which, through vibronic coupling, causes the instability of the ground state of the activated complex.

determining the barrier width. In fact, the vibrational frequencies are much smaller than the electron transition frequencies and, in accord with the above, Δ is also much smaller than Ω_0. Therefore formulas (6.9) and (6.10) may be simplified to

$$D \cong \hbar\Omega_0 - 2\Delta \approx \hbar\Omega_0 \tag{6.11}$$

$$\tfrac{1}{2}K_e d^2 \cong \hbar(\Omega_{max} - \Omega_0) \tag{6.12}$$

It follows from the discussion in Section 1.5 that the positive fixed electron ("nuclear") part of the force constant is almost independent of the electronic state, and therefore the relationship $K_0^{\Gamma'} \approx K_0^{\Gamma''}$ seems to be valid. Hence

$$K^{\Gamma'} \cong K^{\Gamma''} - 2F^2/\Delta \tag{6.13}$$

Thus, by investigating the light absorption band shape measured in the process of the chemical reaction, one can estimate the barrier height D and its width d and, using additional estimates for the values of the vibronic constant F and energy separation Δ, the curvature of the barrier near its maximum can also be obtained. These parameters, together with the nuclear configuration at the point M, almost completely characterize the activated state of the chemical reaction.

Difficulties associated with the investigation of the electronic absorption band under consideration may naturally arise due to its expected large width and the unusual experimental conditions, but these difficulties do not seem to be fundamental. For simple systems, such investigations are known to have been successfully conducted and even used in excimeric lasers.[507] (Activated complex excited bonding states may also be used in excimeric lasers.) Difficulties can arise in the determination of constants F and Δ, but they can be calculated approximately by means of standard quantum-chemical methods.

The light absorption experiment discussed here is taken as an illustration. Quite a number of similar studies of activated complexes of a chemical reaction may be carried out when other physical methods are incorporated.

6.2. Orbital Symmetry Rules in Mechanisms of Chemical Reactions

The Vibronic Approach. Knowledge of the mechanisms governing chemical reactions and the possibility that new chemical reactions can be predicted based on this knowledge, is one of the most important problems of modern theoretical chemistry. Major achievements in this area have been attained in the last two decades stimulated by the general development of quantum chemistry. The orbital symmetry rules determining allowed and forbidden reaction mechanisms and locating the most favorable pathway with the lowest activation energy, should be related to these results. The principles of allowed and forbidden reactions based on symmetry properties of interacting orbitals were first formulated by Woodward and Hoffmann.[16] Further developments and somewhat different formulations based on the work of Bader[506] have been given by Pearson and others (see Pearson[17] and references cited therein).

We shall demonstrate how the basic features of these orbital symmetry rules can be derived from the PJTE for the activated state Γ of the chemical reaction.[53] From the results obtained in Sections 6.1 and 2.3 we know that if there is only one near active excited state Γ', the nuclear configuration of the unstable ground state of the activated complex softens in a certain direction $Q_{\bar{\Gamma}}$, and the latter is determined by the condition that the vibronic constant

$$F^{(\Gamma\Gamma')} = \left\langle \Gamma \middle| \left(\frac{\partial V}{\partial Q_{\bar{\Gamma}}}\right)_0 \middle| \Gamma' \right\rangle \tag{6.14}$$

is nonzero. The symmetry $\bar{\Gamma}$ of this displacement Q is determined by the symmetries of the wave functions of states Γ and Γ': $\bar{\Gamma} = \Gamma \times \Gamma'$. Therefore, if the symmetries of the ground and near excited states of the activated complex are known, the direction of its labilization (corresponding to a certain mechanism of the reaction) can be predicted.

Consider two approaching molecular closed shell systems A and B, each of which is stable separately, e.g., owing to the large gaps Δ between the ground and excited states. Interaction gives rise to the united system (activated complex) AB with its own symmetry and energy intervals Δ. What are the directions of the lowest (or even negative) curvature of the AP of the system AB? To answer this question, as shown above, the symmetry of the wave functions of the ground and low-lying excited states must be known.

The MO of the united system AB may be easily found approximately, provided the MO of the separate molecules A and B are known. Assume that the energy levels of the one-electron MO of molecules A and B and their occupation numbers are as shown in Figure 6.3a, with total wave functions Ψ_1^A and Ψ_1^B and one-electron functions ψ_1^A and ψ_2^A, ψ_1^B and ψ_2^B for HOMO and LUMO, respectively. It can further be assumed that at large distances where interaction between A and B is weak, the total wave function is $\Psi^{AB} \approx \Psi^A \Psi^B$, and the MO energy levels are as shown in Figure 6.3b. In this scheme, the first excited state resulting from an electron transition from the HOMO ψ_1^A to the LUMO ψ_2^B (Figure 6.3c) is equivalent

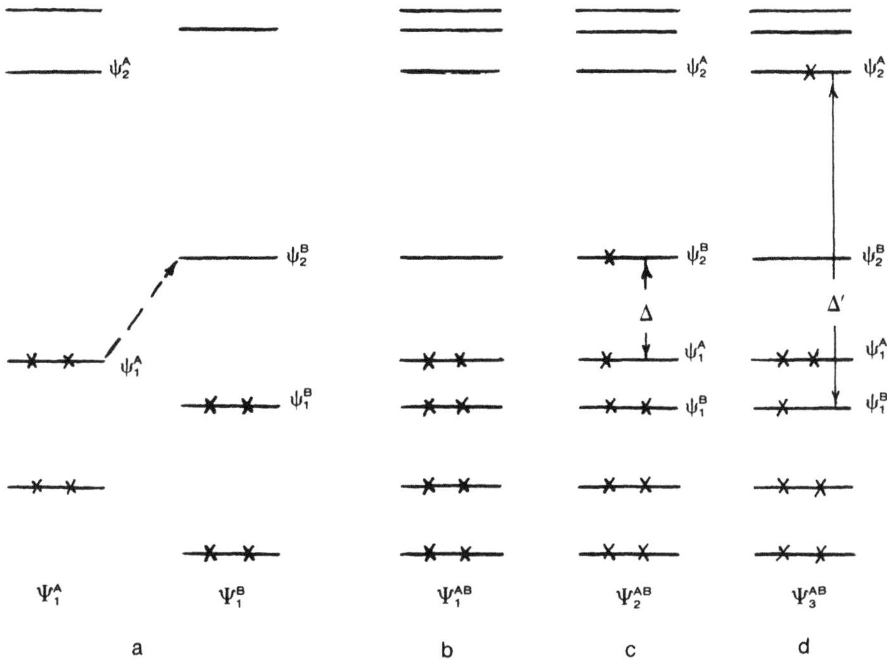

Figure 6.3. One-electron MO energy level schemes of two reagents entering a chemical reaction: (a) noninteracting systems A and B; (b) ground state of the joint system AB in the case of weak interaction not changing the MO positions; (c) first excited state AB; (d) higher excited state.

to an electron transfer from A to B, shown in Figure 6.3a by a dashed arrow.

The symmetry of the fully occupied MO forming closed shells (and hence the ground state Ψ_1^{AB}) belongs to the totally symmetric representation A_1. Therefore, in general the symmetry of the total wave function is determined by the symmetry of the HOMO, which are not fully occupied by electrons. Consequently, the symmetry of the excited state Ψ_2^{AB} is determined by the symmetry Γ' of the product $\psi_1^A\psi_2^B$. Since $A_1 \times \Gamma' = \Gamma'$, the symmetry Γ' of the product $\psi_1^A\psi_2^B$, in accord with the above, determines the symmetrized direction Q of the labilization of the activated state. The next excited state with wave function $\Psi_3^{AB} \approx \psi_1^B\psi_2^A$ results from electron transition from the HOMO of B to the LUMO of A (Figure 6.3d). However, for this excited state $\Delta' > \Delta$ and, *ceteris paribus*, it should be less effective in determining the labilization direction Q. The excitations inside the A or B molecules do not contribute in this sense to the labilization (instability).

We shall illustrate these statements by examples. Consider the exchange reaction $H_2+D_2 \rightarrow 2HD$ and assume that the H_2 and D_2 molecules approach each other so forming a rectangular transition state (Figure 6.4a). The HOMO $\psi_1^A = \sigma(1s)$ is bonding, whereas the LUMO $\psi_2^B = \sigma^*(1s)$ is antibonding. Figure 6.4b indicates that these two functions have opposite parity with respect to the reflection in the plane O–O, and therefore the function $\Psi_2^{AB} \approx \psi_1^A \psi_2^B$ changes sign as a result of such a reflection. Consequently, the instability coordinate Q also changes sign by the reflection. The odd displacements corresponding to Q are shown in Figure 6.4c. It is seen that, starting from the rectangular activated state, the two molecules H_2 and D_2 following the instability coordinate Q will not continue their rectangular approach, but will turn along the arrows shown in the figure.

The parallel (rectangular) approach of the H_2 and D_2 molecules corresponds to the coordinate Q' of the synchronous reaction $H_2+D_2=2HD$ (Figure 6.4d). The coordinate Q' does not change sign by reflection in the O–O plane. For Q' the vibronic constant $F^{(\Gamma\Gamma')}$ equals zero according to equation (6.14), and hence the system is not labilized (softened) in the Q' direction by vibronic mixing with the lowest excited state Γ'. When moving from the minimum of the initial reactants along the Q' coordinate through the rectangular activated complex to the reaction products, the curvature of the AP from equation (6.2) is not lowered by the vibronic contribution of the first excited state with the smallest value of Δ.

It can be shown that the next excited state $\Psi_3^{AB} \approx \psi_1^A \psi_2^B$ also makes no vibronic contribution. Only the highly excited state, for which both molecules are in their antibonding $\sigma^*(1s)$ states (and hence Δ is large), makes a small contribution to lowering of the curvature. This means that the potential barrier for the reaction under consideration will be very high. Therefore the mechanism of the reaction of direct exchange $H_2+D_2=2HD$ is forbidden from the viewpoint of the orbital symmetries. This forbiddenness is removed for the axial approach of the molecules (or at any angle above zero and below 180°). Hence the reaction passes through an intermediate dissociation of the molecules.

For a more complicated example, such as the reaction $N_2+O_2=2NO$, the treatment may be carried out similarly. The essential difference lies in the MO wave functions: $\psi_1^A = \sigma(2p)$, $\psi_2^B = \pi^*(2p)$, $\psi_1^B = \pi^*(2p)$, and $\psi_2^A = \pi^*(2p)$ (Figure 6.5). Taking into account vibronic mixing with the lowest excited state $\Psi_2^{AB} \approx \psi_1^A \psi_2^B$, the labilized coordinate is seen to be the same odd one Q, as in the case of the H_2+D_2 reaction. For a parallel approach of two molecules N_2+O_2, corresponding to the Q' displacements (Figure 6.4d), the integral (6.4) is equal to zero, the reaction of direct exchange thus also being forbidden. However, unlike the case of H_2+D_2, here the excited state $\Psi_3^{AB} \approx \psi_1^B \psi_2^A$ is seen from Figure 6.5 to be of even

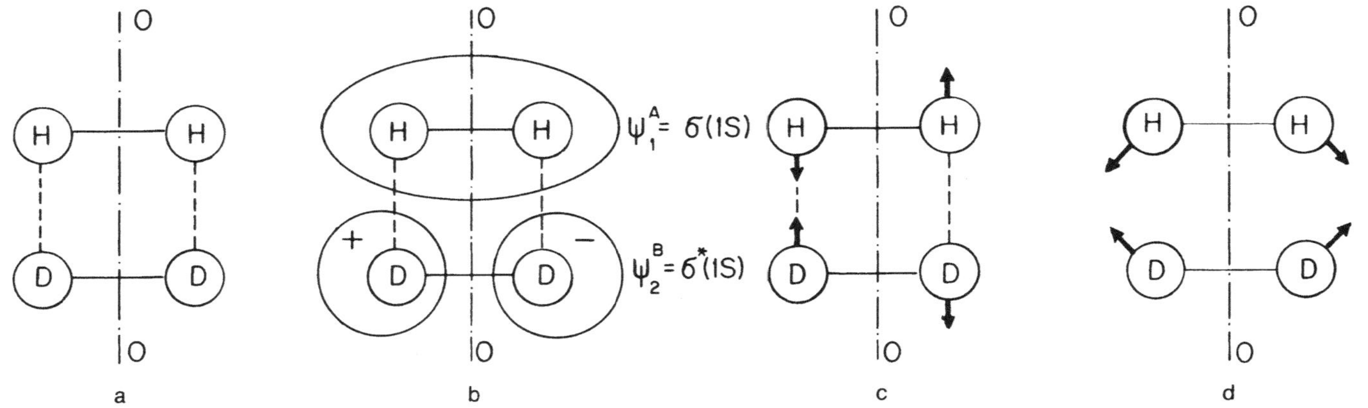

Figure 6.4. Mechanism of the exchange reaction $H_2 + D_2 = 2HD$; (a) rectangular transition state; (b) symmetry of the ground, ψ_1^A, and first excited, ψ_2^B, MO of Figure 6.3; (c) coordinate Q of labilization (softening or instability) of the transition state resulting from vibronic PJT mixing of the above two states; (d) coordinate Q' of in-parallel rapprochement leading to synchronous exchange reaction.

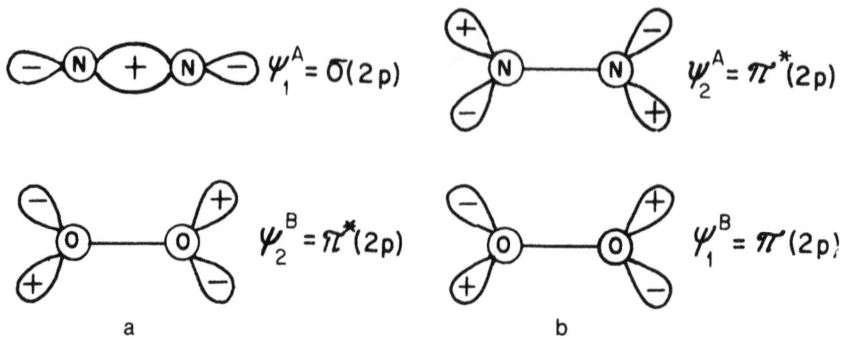

Figure 6.5. Mechanism of the exchange reaction $N_2 + O_2 = 2NO$: (a) ground, ψ_1^A, and first excited, ψ_2^B, MO of Figure 6.3; (b) MO ψ_2^A and ψ_1^B, the vibronic interaction between which allows the reaction of synchronous exchange.

parity, and its contribution labilizes the Q' coordinate which allows the direct exchange $N_2 + O_2 = 2NO$. However, the value of Δ' for this state is so great (the Ψ_3^{AB} state corresponds to electron transfer from the acceptor oxygen molecule to the donor nitrogen one) that this contribution, in the presence of a much stronger one labilizing the Q coordinate, may be neglected. In other words, vibronic mixing with the Ψ_3^{AB} state reduces relatively weakly the AP curvature in the Q' direction and the barrier remains high (although it may be lower than in the case of $H_2 + D_2$), i.e., the mechanism of the direct exchange $N_2 + O_2 = 2NO$ is forbidden. However, if the N_2 molecule is first excited to the $\pi^*(2p)$ state, the forbiddenness is removed. Thus the reaction of the direct exchange $N_2 + O_2 = 2NO$ is allowed as a photochemical reaction.

In general, if the reaction mechanism is considered by approaching the two molecules A and B without changing the symmetry of the united system AB, the coordinate of the reaction will be the totally symmetric coordinate of the activated complex AB. Therefore, the reaction mechanism under consideration will be allowed if and only if the wave functions of the HOMO of A, ψ_1^A, and the LUMO of B, ψ_2^B (or ψ_1^B and ψ_2^A), have the same symmetry (a nonzero overlap integral). Otherwise the reaction mechanism is forbidden. This explains the origin of the term "preservation of orbital symmetry in chemical reactions" which is often met in the literature. This term is also concerned with the general preservation rules which form its foundation.[508]

If there are several near HOMO and/or LUMO, the contribution of each pair should be examined and the results summarized (strictly speaking, the aforementioned multilevel problem must be solved).

Comparison with the Classical Woodward–Hoffmann Description. The above vibronic approach to the problem states that the possibility for the reaction to proceed under a given mechanism with a low activation barrier depends on the presence of low-lying excited electronic states with appropriate symmetry which lower the curvature of the AP in the given direction. With allowance for the visual interpretation of the PJTE (Section 2.3, Figure 2.13), one can say that for the reaction to proceed with a low activation energy, it is necessary that the energy gained by the formation of new bonds (in the process of the reaction) at an early stage compensates the energy lost in the breaking of bonds. Since the ground states are occupied, the formation of new bonds may proceed only by the involvement of reactant excited states. The latter should have appropriate symmetry and low-lying energy levels, otherwise the newly formed bonds are very weak.

It is just this picture which appears in the Woodward–Hoffmann rules in the way they were first formulated.[16] In Figure 6.6, the visual treatment of these rules is reproduced for the formation of cyclobutane from two ethylene molecules through the rectangular activated complex. It is seen that the formation of σ bonds between the carbon atoms of the two

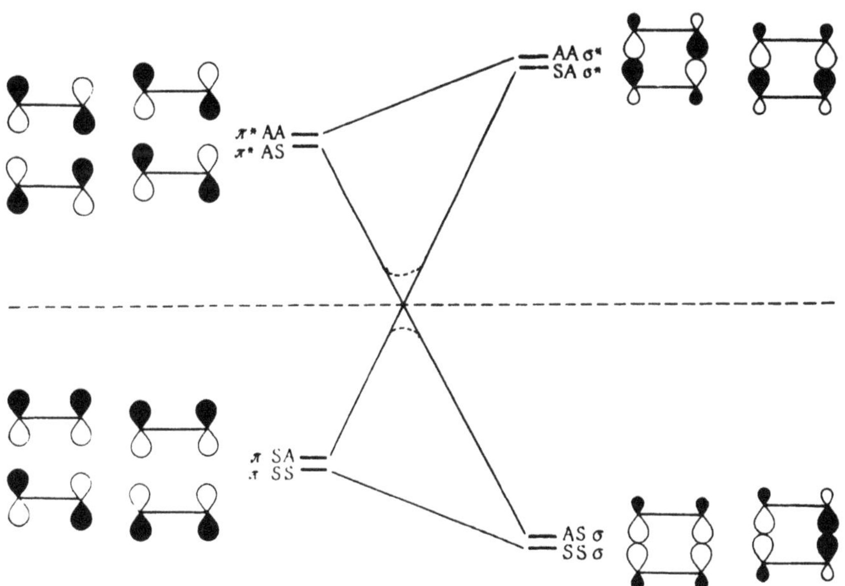

Figure 6.6. MO energy level correlation diagram, which illustrates the Woodward–Hoffmann orbital symmetry rule in the formation of the cyclobutane molecule from two ethylene molecules.[16] S and A indicate the symmetry properties (symmetric and antisymmetric, respectively) with respect to reflections in the two symmetry planes of the rectangular transition state.

molecules (before the break of the π bonds in each molecule) can take place only by involving excited MO having the same symmetry as the ground state. If the corresponding excited state, as in the case of ethylene, is too high in energy, the formation of σ bonds does not compensate for the breakdown of the π bonds. This is easily seen from the one-electron MO correlation diagram (Figure 6.6), in which the energy levels of the newly formed MO and old MO broken in the process of the reaction are connected by straight lines. As a result of the high position of the excited MO, the corresponding lines in Figure 6.6 increase steeply, and in the intermediate area corresponding to the activated complex the energy is rather high — the reaction barriers assume high values. The latter can be estimated approximately by remembering that the above excited state (Figure 6.6) corresponds to a transition of the bonding electrons to the antibonding MO in both molecules resulting in an excitation energy of ~5 eV (115 kcal/mol).

A more detailed energy level diagram of the system as a whole is given in Figure 6.7. One can see that, along with the mentioned excited states, there are others much lower in energy but of different symmetry. In particular, starting from the excited state AA the two ethylene molecules can form the cyclobutane molecule (also in its excited state) without any activation barrier. Consequently, this reaction is allowed photochemically.

Turning to the above vibronic approach one can see that, first of all, the latter contains all the features of the phenomenon. If we start from two ethylene molecules approaching in parallel and forming a rectangular activated state, the curvature of the AP K^{Γ} (equation (6.2)) in the direction Q of these motions does not decrease: the energy gap Δ to the excited state, for which $F \neq 0$, is very large. When moving along Q the energy of the ground (bonding) state increases and the energy of the excited (antibonding) state decreases, with the result that there is a lowering of the value of Δ and a consequent decrease in the curvature K^{Γ} until $K^{\Gamma} = 0$ (shown in Figures 6.6 and 6.7 by a dashed line). After that, K^{Γ} increases again due to the formation of σ bonds. If the initial value of Δ is large, there is a great increase in the energy required to mix states before Δ becomes small enough so that the negative vibronic contribution to the curvature F^2/Δ turns down the AP. As a result the reaction barrier is high.

Repeating the main point of the Woodward–Hoffmann orbital symmetry rules, the vibronic approach gives them a foundation based on the first principles of quantum mechanics, stemming from equation (6.2) (or the more general equation (6.1)). Also of great importance is that, as distinct from the qualitative Woodward–Hoffmann approach, *the vibronic approach allows one to deduce quantitative criteria for the favorable mechanisms of the chemical reaction.* For a given reaction one can inspect all the low-lying excited states of the activated complex and determine the

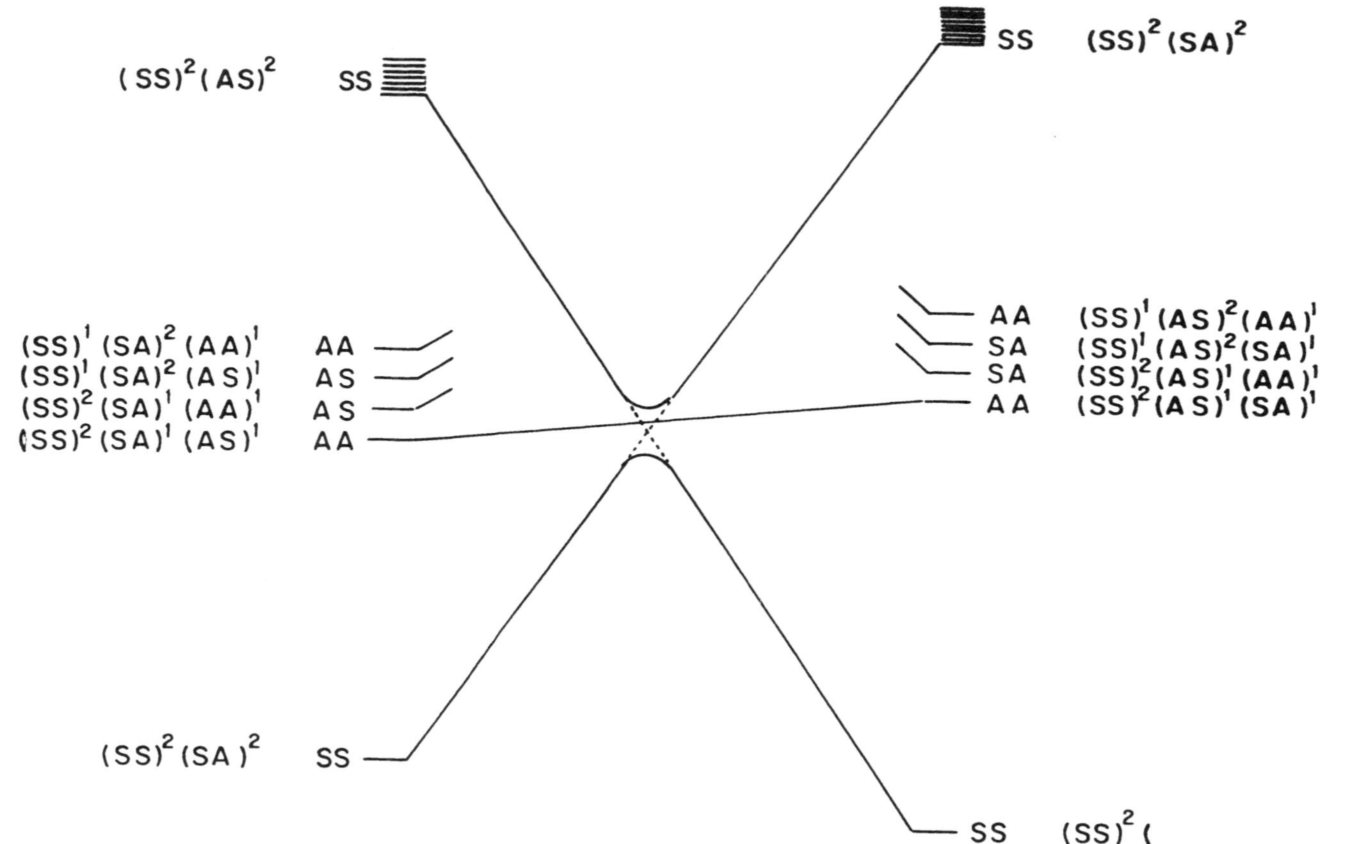

Figure 6.7. More detailed correlation diagram for the case in Figure 6.6, illustrating the behavior of the multielectron energy (singlet) terms. The electronic configuration of the outer four electrons and the symmetry of the resulting term are also indicated.

coordinate Q for which the vibronic contribution F^2/Δ (more precisely, its summarized value) is a maximum. For the exchange reaction $H_2 + D_2 = 2HD$, the most favorable mechanism in the vibronic approach is that of the coordinate Q' in Figure 6.4c.

Example with Catalyst Participation. The orbital symmetry rules in their qualitative version are widely used in the study and prediction of chemical reaction mechanisms. Examples of such applications can be found in monographs,[16,17] reviews, and original papers (see also Bibliographic Review[14]).

The example given below is especially important as it reveals the role of a third molecular system (the catalyst) in creating a symmetry forbidden reaction between the two given molecules allowed. As forbidden reaction we take the previously considered formation of cyclobutane from two ethylene molecules through a rectangular activated complex. Its MO energy level correlation diagram is given in Figure 6.6. Following Mango and Schachtschneider,[509] who were the first to consider this case, we assume that both molecules are *cis*-coordinated to the transition metal atom M with the two C=C bonds perpendicular to the plane of the metal–ligand bonds.

In the reaction under consideration the two parallel coordinated ethylene molecules move along the above described Q coordinate to produce the coordinated cyclobutane molecule:

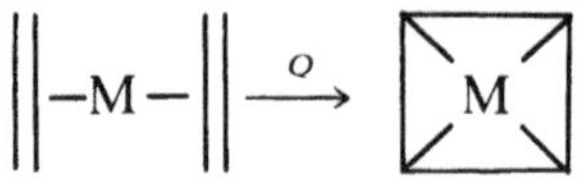

$$\| -\mathrm{M}- \| \xrightarrow{Q} \boxed{\mathrm{M}} \qquad (6.15)$$

One can see that with the participation of the metal atom, the symmetry of the system and the symmetry of the reaction coordinate Q are the same as for the reaction without catalyst participation. In particular, the two planes of symmetry are preserved with respect to reflections, and hence the classification of the MO states remains as shown in Figure 6.6. However, *the positions of the MO energy levels under the influence of the metal atom vary considerably, and this is the main effect of the catalyst in the scheme under consideration.*

The MO correlation diagram when M is a transition metal with a d^8 electronic configuration is given in Figure 6.8; only the π bonds of the ethylene molecules (transforming to σ bonds in cyclobutane) and the d states of the metal atom are shown (the s and p states of the metal being disregarded). These energy levels must be populated by six electrons (two electrons from the d^8 metal and two from each of the two ethylene molecules). Compared with the MO correlation diagram without the metal participation (Figure 6.6), the energy levels of the cyclobutane molecule

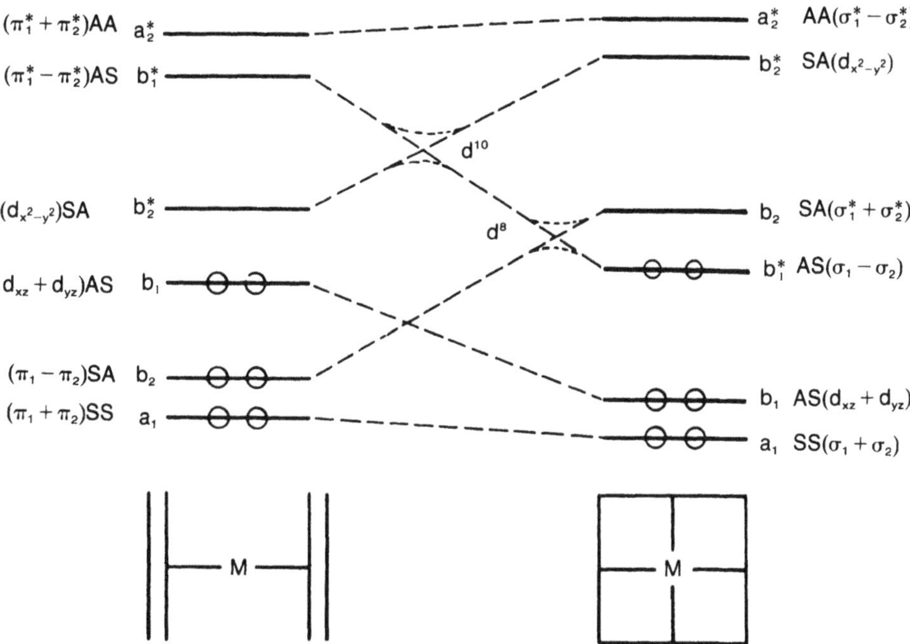

Figure 6.8. Correlation diagram for cyclobutane formation from two ethylene molecules with catalyst (transition metal atom M) participation. The crossings d^8 and d^{10} illustrate the reaction barrier formation for corresponding metal d^n configurations.

vary substantially: the energy level of the antibonding σ MO $(\sigma_1^* + \sigma_2^*)$ with SA symmetry is lowered, while the AS $(\sigma_1 - \sigma_2)$ MO energy increases. As a result, the energy gap Δ between the ground and excited states decreases and the vibronic contribution to lowering of the AP curvature in the Q direction increases, the reaction barrier (shown in Figure 6.8 as the d^8 crossing) becoming essentially smaller. Note that for d^{10} metals, these (or somewhat changed) energy levels have to be populated by eight electrons resulting in occupation of the b_2^* $(d_{x^2-y^2})$ and b_2 $(\sigma_1^* + \sigma_2^*)$ levels (see Figure 6.8). This does not substantially increase the reaction barrier (shown in Figure 6.8 as the d^{10} crossing). These results agree well with empirical data on d^8 and d^{10} metal activity as catalyst in the reactions in question.[509] Quantitatively, the influence of the catalyst depends critically on the magnitude of the energy level shifts produced by coordination.

6.3. Vibronic Activation in Elementary Acts of Chemical Reactions and Catalysis

Vibronic Structure as a Basis for a New Approach to the Problem of Chemical Transformations. At the end of Section 6.2 it was shown how the participation of a third molecular (polyatomic) system, approaching some of the energy levels of two interacting molecules, amplifies vibronic interaction in the initial reagents and final products and lowers the activation energy of the reaction. The description, however, was mainly qualitative.

The orbital vibronic constants (OVC) introduced in Section 1.3 and the picture of vibronic structure, obtained there by means of these constants, may serve as a good basis for a new approach to the problem of chemical transformations. This new approach, in particular, leads to a deeper understanding and a quantitative (or semiquantitative) description of the influence of a particular molecular system on the reactivity of another in the process of their interaction.[18–21] This is especially important in catalysis, where the coordination of the molecule to the active center of the catalyst (homogeneous or heterogeneous) changes considerably the reactivity of a molecule with respect to a chemical reaction with another atom or molecule (or the unimolecular reaction of decomposition).

Because of the importance of catalysis, this presentation is given in terms of activation of molecular systems by coordination to the catalyst, although the approach suggested here can also be used in other aspects of the problem of chemical transformations. In particular, at the end of this section it will be shown that the JTE determines the direction (mechanism) of the chemical reaction without catalyst participation. Because of the novelty of this approach, many of its possibilities still have to be clarified. Qualitative consideration of the changes in molecular geometry by coordination due to electronic population of excited states may often be met in the literature (e.g., see Ugo[510] and Řeřicha[510a]).

We shall use here some of the results obtained in Section 1.3. In the main this is the introduction of the orbital vibronic constants $f_{\bar{\Gamma}\bar{\gamma}}^{(ij)}$ and orbital force constants $k_{0\bar{\Gamma}}^{i}$ and $k_{\bar{\Gamma}}^{i}$, which are the dynamic parameters of molecular structure (cf. the integral constants $F_{\bar{\Gamma}\bar{\gamma}}^{(ij)}$ and $K_{\bar{\Gamma}}^{\Gamma}$). In particular, the diagonal linear OVC $f_{\bar{\Gamma}\bar{\gamma}}^{i}$ equals the force with which the electron of the ith MO distorts the nuclear framework in the direction of the symmetrized displacements $Q_{\bar{\Gamma}\bar{\gamma}}$ minus the proportion of the nuclear repulsion force in this direction per electron. Consequently, the total force distorting the molecule in this direction (the integral vibronic constant $F_{\bar{\Gamma}\bar{\gamma}}^{\Gamma}$) is (the "addition rule")

$$F_{\bar{\Gamma}\bar{\gamma}}^{\Gamma} = \sum_{i} q_i^{\Gamma} f_{\bar{\Gamma}\bar{\gamma}}^{i} \tag{6.16}$$

where q_i^{Γ} is the electron occupation number for the ith MO in the electronic state Γ under consideration. If the system is (statically) stable with respect to the displacement Q, then

$$F^{\Gamma}_{\bar{\Gamma}\bar{\gamma}} = \sum_i q^{\Gamma}_i f^i_{\bar{\Gamma}\bar{\gamma}} = 0 \tag{6.17}$$

Note that the OVC differ for different orbitals: the nuclear repulsion per electron is independent of the MO, whereas the electron distribution changes considerably from one MO to another. In particular, in diatomics the bonding influence of the electron of the bonding MO is stronger than the nuclear repulsion per electron, $f^i_R > 0$ (R is the interatomic distance), whereas for the antibonding orbitals the opposite is true: $f^i_R < 0$. At the point of stability these different values of OVC exactly compensate each other, as follows from equation (6.17).

Similar relationships can be obtained for the nondiagonal OVC and orbital force constants. In Koopmans approximation (Section 1.3), if two states Γ and Γ' differ in the replacement of the ith by the jth MO, we have

$$F^{(\Gamma\Gamma')}_{\bar{\Gamma}\bar{\gamma}} = f^{(ij)}_{\bar{\Gamma}\bar{\gamma}} \tag{6.18}$$

and

$$\begin{aligned} K^{\Gamma}_{\bar{\Gamma}} &= \sum_i q^{\Gamma}_i k^i_{\bar{\Gamma}} \\ k^i_{\bar{\Gamma}} &= k^i_{0\bar{\Gamma}} - \sum_j{}' \left(|f^{(ij)}|^2/\Delta_{ij}\right)(m_j - q_j) \end{aligned} \tag{6.19}$$

where $\Delta_{ji} = \frac{1}{2}(\varepsilon_j - \varepsilon_i)$. Unlike equation (1.26), here j numbers only the free and not fully occupied MO, m_j being the maximum occupation number of the jth MO (equations (1.26) and (6.19) are equivalent).

It is seen that, for the orbital force constants, there is also formally an addition rule: the curvature of the AP in a given direction $K^{\Gamma}_{\bar{\Gamma}}$ equals the sum of orbital contributions $k^i_{\bar{\Gamma}}$. However, as distinct from OVC, the orbital force constant $k^i_{\bar{\Gamma}}$ is determined not only by the ith MO, but also by the contribution of vibronic mixing with other MO. From the latter, only the free or not fully occupied MO should be taken into account, since the contribution of mixing of the fully occupied MO enters into the sum for $K^{\Gamma}_{\bar{\Gamma}}$ twice with opposite signs, and compensate each other.

The introduction of orbital vibronic and force constants, as indicated in Section 1.3, considerably amplifies the description of the electronic structure of molecules by means of the one-electron MO scheme. Knowledge of the energy levels and the electron distribution at each MO presents the static electronic structure and properties of the molecular system in a given state. Supplemented with orbital vibronic and force constants these MO give much more complete information, adding to the static electronic structure new parameters of dynamic structure: the contribution of each MO to the distorting forces and force constants in a given direction (Figure 1.3).

The supplementary parameters of molecular structure characterize the MO in more detail. They give a quantitative estimate of the bonding or antibonding nature of the MO, including even polyatomic systems when, owing to the complicated genealogy of the orbitals, examination of this question by means of energy characteristics is difficult or even impossible. Even when the energy characteristics (bonding, antibonding, nonbonding) are effective, the orbital vibronic and force constants supplement this information by indicating the direction and magnitude of the distortion and/or softening (or hardening) of the nuclear framework by the MO electron.

The additional (orbital vibronic) parameters of molecular structure are of special importance in analyzing the origin of molecular transformations, in particular, analysis of the origin of the activation of one given molecule by another. Any influence of a given molecular system upon another (van der Waals interaction, hydrogen bonding, static polarization, chemical bonding with electron collectivization, oxidation, reduction, excitation, ionization, coordination) starts with a change in the mobile electron distribution which, in turn, may lead to a change in the much more inertial nuclear configuration.

The variation in electronic structure, if small enough, in the first approximation may be described by the changes in the MO electronic population numbers — orbital charge transfer Δq_i. As long as these Δq_i are not very large (of the order of one electron), one can use them, following the Koopmans approximation (the approximation of "frozen orbitals"), as the characteristics of the intermolecular influence. In fact, the MO wave functions are also changed, but this is a second-order effect and may be neglected, provided intermolecular interaction is small. The second-order terms can be taken into account in a more sophisticated consideration.

The electronic redistribution presented by the new orbital population numbers $q_i^{\prime} + \Delta q_i$ generally results in noncompensating orbital distorting forces and force constants. If the initial system is stable, $F_{\bar{\Gamma}} = 0$, $K_{\Gamma} > 0$ (subsequently, subscript Γ of the electronic state of the system as a whole is omitted), the substitution $q_i \to q_i + \Delta q_i$ in equations (6.17) and (6.19) results in the following relationships:

$$F_{\bar{\Gamma}} = \Delta F_{\bar{\Gamma}} = \sum_i \Delta q_i f_{\bar{\Gamma}}^i \tag{6.20}$$

$$\Delta K_{\bar{\Gamma}} = K_{\bar{\Gamma}}^{\prime} - K_{\Gamma} = \sum \Delta q_i k_{\bar{\Gamma}}^i + \sum_{i,j}{}^{\prime} q_i \Delta q_j \, |f_{\bar{\Gamma}}^{(ij)}|^2 / \Delta_{ji} \tag{6.21}$$

Hence the electronic rearrangement, taking into account the orbital population number changes Δq_i, results in a nonzero distorting force $F_{\bar{\Gamma}} \neq 0$

and a change of force constant $\Delta K_{\bar{\Gamma}}$ in the direction $Q_{\bar{\Gamma}}$, for which $f^{i}_{\bar{\Gamma}} \neq 0$ (where i is the index of the MO for which $\Delta q_i \neq 0$).

The direction of the distorting force $\bar{\Gamma}$ depends on the symmetry Γ_i of the ith MO. As discussed above (Chapter 1), $f^{i}_{\bar{\Gamma}}$ is nonzero if the symmetric product $[\Gamma_i \times \Gamma_i]$ contains $\bar{\Gamma}$. If Γ_i is nondegenerate, $\bar{\Gamma} = A_1$ is totally symmetric, i.e., electrons of nondegenerate MO distort the molecule in the direction of totally symmetric displacements A_1, which do not change the symmetry of the system. If Γ_i is degenerate, $\bar{\Gamma}$ can also be nontotally symmetric, but it should be JT active (Table 1.3).

As for the change in force constant $\Delta K_{\bar{\Gamma}}$, $\bar{\Gamma}$ can be of any allowed symmetry in the appropriate point group. This is seen visually from the second term in equation (6.21), in which $f^{(ij)}_{\bar{\Gamma}}$ is nonzero if $\bar{\Gamma} = \Gamma_i \times \Gamma_j$, and Γ_i and Γ_j (for the ground and excited states, respectively) may belong to any symmetry representation. Similar expressions can be obtained for the change in the anharmonicity constants, etc.

The distorting forces, and changes in the force constants and anharmonicity constants due to electronic rearrangements, directly explain the change in reactivity of the molecule — its chemical activation. An approximate quantitative method for estimating these variations as a function of the electronic rearrangement is proposed below. However, some general conclusions about the conditions for vibronic activation by coordination may be drawn at this stage using equations (6.20) and (6.21), from which it follows that the distorting and softening (or hardening) influence of the coordination center on a given molecule is linearly dependent on the orbital charge transfers Δq_i. On the other hand, the total transfer $\Delta q = \Sigma_i \Delta q_i$ cannot be very large (cf. the electroneutrality principle proposed by Pauling[511]). Therefore, Δq_i may be large, in principle, only if more than one orbital is involved in the bonding and if the values of Δq_i have different signs.

In the often considered case of only two orbitals involved in the bonding — HOMO and LUMO (see below) — the two values of Δq_i have opposite signs: $\Delta q_1 < 0$ and $\Delta q_2 > 0$. If the HOMO is bonding (i.e., $f^{(1)} > 0$) and the LUMO is antibonding ($f^{(2)} < 0$), then, according to equation (6.20), the resulting distorting force is given by

$$F = \Delta q_1 f^{(1)} + \Delta q_2 f^{(2)} = -(|\Delta q_1|\,|f^{(1)}| + |\Delta q_2|\,|f^{(2)}|)$$

In other words, the contribution of the two orbitals to the distorting force is determined here by their absolute values (a similar conclusion is valid for the contribution to the change in the force constant). Consequently, the greater the absolute values of two charge transfers of opposite sign, the greater the mutual vibronic influence of the interacting molecular systems. If there are more than two orbitals active in the bond formation, the possibility increases for a favorable combination of charge transfers which would increase the vibronic influence (but preserve the required small total charge transfer Δq).

Hence follows the conclusion concerning *the role of multiorbital bonding in the vibronic influence* of one molecular system upon another, as, for example, in catalysis. *The special role of transition and rare earth elements* and their compounds *in catalysis is due to their ability to form multiorbital bonds* with various molecular groups.

Change of Reactivity as a Result of Electronic Rearrangement. A quantitative definition of chemical reactivity must first be given in order to analyze the origin of molecular reactivity and changes in reactivity under electronic rearrangement. This is a difficult task. It is evident that reactivity should characterize the rate of the elementary act of a chemical reaction. Therefore, it is natural to try to relate the reactivity to the reaction activation energy. The latter, however, in general depends not only on a given molecule, but also on the molecule with which the given molecule reacts.

Consider a simple reaction AB + C = A + BC. Assume that activation of this reaction is due to activation of the A–B bond (described by the increasing portion of the reaction curve, Figure 6.1). When a specific activation is reached binding with the C atom begins, so lowering the energy (the decreasing portion of the reaction curve). For a more rigorous description of this process, it is necessary to introduce two reaction coordinates — interatomic distances R_1 (A–B) and R_2 (B–C).

The potential energy surface of the reaction in terms of coordinates R_1 and R_2 is shown schematically in Figure 6.9. It is clear that atom C is more active (reactive) the earlier turning point M is reached (M is the barrier maximum), and the lower the activation energy. This is illustrated in Figure 6.10, which shows the potential energy along the reaction trajectory on the plane perpendicular to Figure 6.9 and passing through the R_1 axis. The activation energy D varies for different C atoms and reaches the value of the dissociation energy if the reaction is unimolecular.

In order to investigate changes in the reactivity of a given molecule A–B due to a change in electronic structure, *the activation energy D may be taken as a quantitative measure of the reactivity of this molecule in the reaction under consideration.* The value of D can be determined if the increasing portion of the reaction curve from the initial minimum to the maximum M is known. One can approximate this curve $\varepsilon(Q)$ by, say, a cubic polynomial (another simple possibility is the Morse approximation[20]):

$$\varepsilon(Q) = \tfrac{1}{2}KQ^2 - \gamma Q^3 \tag{6.22}$$

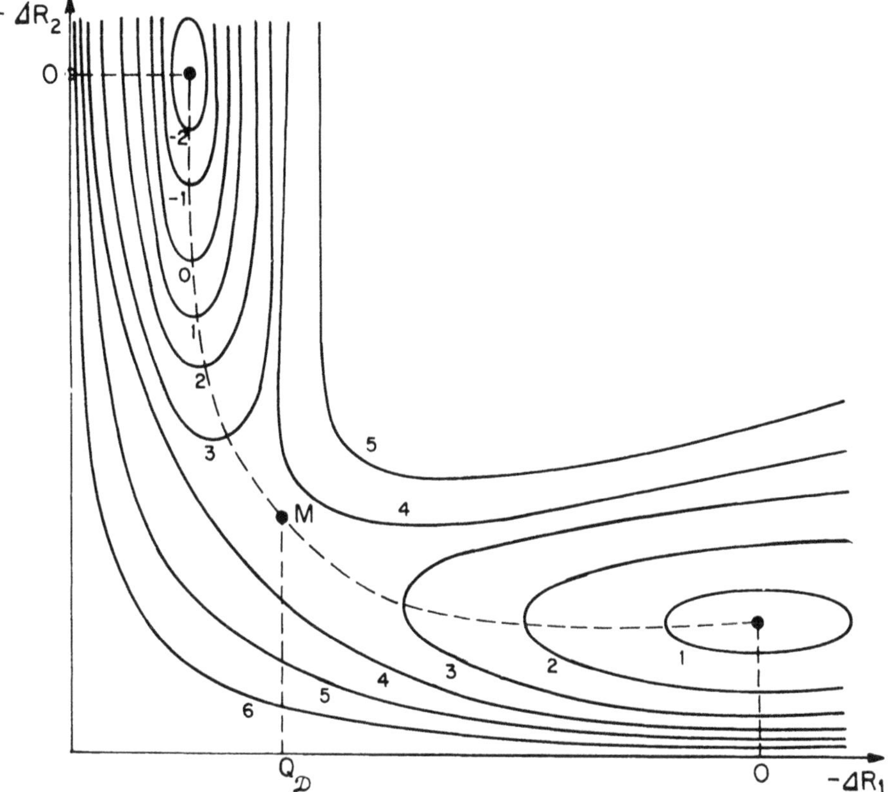

Figure 6.9. Schematic representation of the potential energy surface of the AB + C = A + BC reaction by means of equipotential curves. Energy increases with the number of the curve. The reaction trajectory is indicated by a dashed line. M is the turning point at the barrier maximum.

According to equations (6.20) and (6.21), a nonzero distorting force F, as well as changes in the force constant $\Delta K = K' - K$ and anharmonicity constant $\Delta\gamma = \gamma' - \gamma$ appear after electronic rearrangement. As a result curve (6.22) takes the form

$$\varepsilon'(Q) = FQ + \tfrac{1}{2}K'Q^2 - \gamma' Q^3 \tag{6.23}$$

Equations (6.22) and (6.23) easily yield the following relations (see also Figure 6.11; Q_0 is the shift of the minimum position due to the electronic rearrangement):

$$D = K^3/54\gamma^3 \qquad Q_D = K/3\gamma = (6D/K)^{1/2} \qquad Q_{D'} = K'/3\gamma' - Q_0$$
$$Q_0 = (K'/6\gamma')[1 - (1 + 12\gamma' F/K'^2)^{1/2}] \tag{6.24}$$

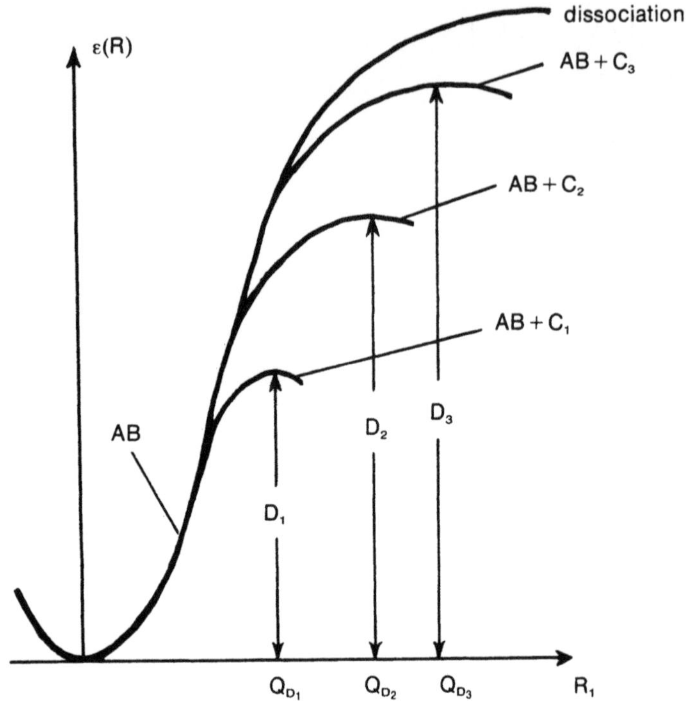

Figure 6.10. Increasing portion of the reaction potential energy curve along the reaction path in Figure 6.9, characterizing activation of the AB bond. The turning point and the reaction barrier height D depend on the reactivity of the C atom.

$$
\begin{aligned}
D' &= (K' - 6\gamma' Q_0)^3/54\gamma'^2 \\
&= (K'^3/54\gamma'^2) - (K'^2 Q_0^2/3\gamma') + 2K'Q_0^2 - 4\gamma' Q_0^3
\end{aligned}
$$

$$
\begin{aligned}
\Delta D &= D' - D \\
&= -D[1 - (K'/K)^3(\gamma/\gamma')^2] - (K'^2 Q_0/3\gamma') + 2K'Q_0^2 - 4\gamma' Q_0^3 \qquad (6.25)
\end{aligned}
$$

The last formula can be simplified. The coefficient of the cubic term can be obtained by means of spectroscopic anharmonicity corrections ωx,[22] known in most cases from empirical data.

We have $\gamma/\gamma' = \beta^{1/2}(K/K')^{3/2}(\hbar\omega'/\hbar\omega) = \beta^{1/2}(K/K')$, where $\beta = \omega x/\omega' x'$ (for small electronic rearrangements, $\beta \approx 1$). Finally we obtain

$$
\begin{aligned}
D'/D &= \beta(K'/K)(1 - 2KQ_0/\sqrt{6\beta K D})^3 \\
\Delta D &= -D[1 - \beta(K'/K)(1 - 2KQ_0/\sqrt{6\beta K D})^3]
\end{aligned}
\qquad (6.26)
$$

or

$$\Delta D = -D[1-\beta(K'/K)] - K'Q_0\sqrt{6\beta KD}/K + 2K'Q_0^2$$
$$-4K'KQ_0^3/\sqrt{54\beta KD}$$

The four terms in this equation, as in Equation (6.25), can be given the following physical interpretation. The first term is due to the change in the force constant upon coordination (softening, or second-order effect). The second and third terms together result from the occurrence of the distorting force (linear effect) and equal the work performed by this force along the path $Q_D - Q_0$. The last term is a third-order (anharmonicity) correction, which has to be small. The largest contribution is expected from the linear terms (see Table 6.1 below).

Equation (6.26) allows one to calculate the change in activation energy ΔD for the reaction with given D, provided the vibronic constants and the changes in the MO population numbers determining parameters K', Q_0, and β are known. Equation (6.26) can also be regarded as an empirical formula relating ΔD to the molecular characteristics K, K', Q_0, and β treated as empirical parameters.

For a complicated molecule with many degrees of freedom the above consideration and conclusions apply to each normal coordinate, in particular, to the coordinate of the reaction (which may be a linear combination of normal coordinates). In these multidimensional cases the expressions derived above can also be used to determine the change in the molecular geometry caused by an electronic rearrangement.

Formula (6.26) is approximate, and is valid only for small changes in the electronic structure compared with the initial structure. Such electronic

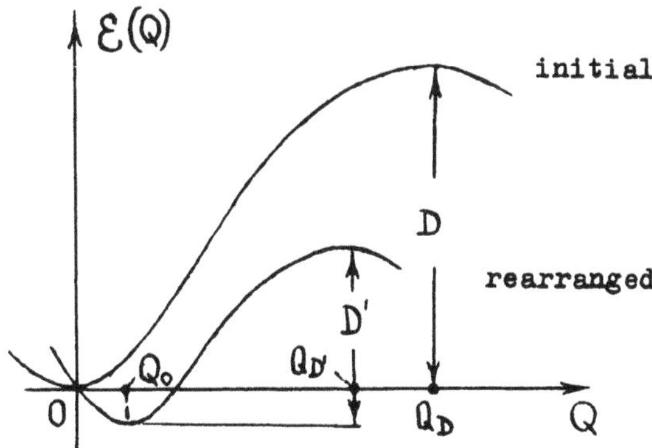

Figure 6.11. Change in the reaction potential energy curve resulting from the influence of the catalyst.

rearrangements occur in the above-mentioned processes of oxidation, reduction, excitation, ionization, and coordination of one molecular system to another in the process of the chemical reaction, or to a coordination center as a ligand, on a solid surface during adsorption, etc.

The latter cases, which can be joined together under the general name "activation by coordination", are of special importance. Here, the changes in MO population numbers as a result of charge transfer to the coordination center and back are fractional. While cases of integer charge transfer (oxidation, reduction, ionization, excitation, etc.) can be treated by other methods (such as quantum-chemical calculations of the electronic structure of the initial and final systems), an analysis of the properties of molecular systems with fractional charge changes can apparently be carried out only by this approach involving the OVC.

The assumption of weak changes in the electronic structure by coordination is valid in most cases. This is confirmed by spectroscopic investigations which indicate that the molecule preserves its principle structural features, not very much changed by coordination. Certainly, there are cases when this statement is not true. For example, a hydrogen molecule may dissociate completely by coordination, thus changing its structural features. Even in such cases, if the process is evaluated according to the above approach at an early stage when the coordination is still weak, the direction of the reaction, as well as some other features, may be predicted qualitatively.

Parametrization. Convenient methods for determining the parameters in the above formulas are needed for quantitative estimates of the reactivity changes due to electronic rearrangement. It follows from equation (6.26) that the change in the activation energy of the reaction depends on the molecular and orbital parameters of the activated molecule, including the OVC and orbital charge transfers Δq_i. The latter are determined by the electronic structure of both the coordinated molecule and the coordination center, as well as by the geometry of coordination. The mode of coordination determines which orbitals of the coordinated molecule that overlap those of the coordination center have the largest Δq_i.

From general considerations confirmed by computation, it follows that (disregarding some special cases) when simple molecules coordinate, only two outer orbitals are influenced substantially: the HOMO and the LUMO. Neglecting the changes in the MO population number for the inner orbitals, one can considerably simplify equations (6.20) and (6.21), which can be written in the following form (superscripts 1 and 2 refer to HOMO and LUMO, respectively):

$$F = f^{(1)}\Delta q_1 + f^{(2)}\Delta q_2 \qquad (6.27)$$

$$\Delta K = k^{(1)}\Delta q_1 + k^{(2)}\Delta q_2 \tag{6.28}$$

$$\beta - 1 = C^{(1)}\Delta q_1 + C^{(2)}\Delta q_2 \tag{6.29}$$

where we introduce force constant coefficients $k^{(1)}$ and $k^{(2)}$ (mentioned in Section 1.4), which are combinations of VC and indicate how the force constant changes by adding one electron to the corresponding MO. Equation (6.29), like equations (6.27) and (6.28), describes the linear dependence of the ratio of the anharmonicity constants β on the charge transfer Δq_i, coefficients $C^{(1)}$ and $C^{(2)}$ being complicated combinations of cubic and lower-order OVC.

All six coefficients $f^{(1)}$, $f^{(2)}$, $k^{(1)}$, $k^{(2)}$, $C^{(1)}$, and $C^{(2)}$ can be easily determined if the values of F, ΔK, and β are known for any two independent processes of electronic rearrangement (for two pairs of values of Δq_1 and Δq_2 excluding the trivial value $\Delta q_1 = \Delta q_2 = 0$). These processes may be ionization ($\Delta q_1 = -1$, $\Delta q_2 = 0$), reduction ($\Delta q_1 = 0$, $\Delta q_2 = 1$), or excitation ($\Delta q_1 = -1$, $\Delta q_2 = 1$), provided the above empirical parameters are available. For the distorting force according to equation (6.24), we have

$$F = -K'Q_0[1 - KQ_0(6\beta K'D_0)^{-1/2}] \tag{6.30}$$

where K and D_0 are constants for the unperturbed molecule (D_0 is the dissociation energy in the direction under consideration), whereas K', Q_0, and β characterize the electronically rearranged system.

Force constants K and K' can be obtained from the appropriate vibrational frequencies. The magnitude of distortion Q_0 in the Q direction is in general available from X-ray (or other diffraction method) analysis of the structure of the electronically rearranged system as compared with the unperturbed one. For simple electronic rearrangement, such as ionization or excitation, Q_0 can be obtained from spectroscopic data. In some cases, especially for simple molecules, Q_0 can also be determined from the known value of K' using empirical relationships between K' and Q_0 (see, e.g., Charitonov and Bazileva[512]).

If the coefficients in equations (6.27)–(6.29) are known, the distorting force F and the variations in the force constant ΔK and anharmonicity constant β can be estimated for given charge transfers Δq_1 and Δq_2, after which the change in the activation energy ΔD can be evaluated from equation (6.26) as a function of D. On the other hand, if F and ΔK are known, Δq_1 and Δq_2 can be obtained for the HOMO and LUMO.

This empirical parametrization may be inconvenient if the estimated vibronic constants are very sensitive to the processes used for their estimation. In this case the processes closest to the expected charge transfer must be used in the empirical scheme. Calculations of the orbital vibronic and force constant coefficients are devoid of this fault. For the activated state, the value of Q_0, which is the small difference in the interatomic distance in the two states, free and activated by coordination, is most difficult to obtain by direct X-ray measurements with the high accuracy needed. This difficulty can be overcome if one uses empirical correlations between interatomic distances and vibrational frequencies (see below). These correlations yield values of Q_0 which correspond more to the true values in the activated molecule than those obtained by X-ray measurements. The latter are obtained from crystal structures in which the interatomic distances are influenced by the lattice structures.

Equation (6.26) can also be used as an empirical relation between ΔD and D:

$$\Delta D = aD + bD^{1/2} + c + dD^{-1/2} \tag{6.31}$$

where coefficients a, b, c, and d are determined from empirical parameters K, K', Q_0, and β. The value of D, i.e., the activation energy for the reaction under consideration in the absence of the catalyst, is often unknown. In these cases the curves $\Delta D = f(D)$ can be plotted for each coordination center or other source of electronic rearrangement in the molecule. By comparing these curves, the relative efficiency of different coordination centers in lowering the activation energy can be established for each value of D.

Examples: Activation of Carbon Monoxide, Nitrogen, and Nitrogen Oxide by Coordination.[18,19,21,512a] The electronic structure[513] and spectroscopic parameters of the CO molecule and its ions have been well studied. This makes it possible to obtain direct estimates of the main orbital vibronic constant coefficients. For CO the HOMO is 5σ, while the LUMO is 2π (Figure 6.12). From empirical data on the force constant K for the free molecule, and K', Q_0, and β for the CO^+ ion[514,512] ($\Delta q_1 = -1, \Delta q_2 = 0$) and for the excited state CO ($5\sigma^2 2\pi^0 \rightarrow 5\sigma^1 2\pi^1$) ($\Delta q_1 = -1, \Delta q_2 = 1$),[515,512] equations (6.27)–(6.29) (average values pertaining to the two states $A\,^1\Pi$ and $a\,^3\Pi$ are taken for the $5\sigma^1 2\pi^1$ configuration; another parametrization was used elsewhere[21]) yield

$$\begin{aligned} f^{(1)} &= -4.54 \cdot 10^{-4}\ \text{dyn}, & f^{(2)} &= -12.1 \cdot 10^{-4}\ \text{dyn} \\ k^{(1)} &= -0.080 \cdot 10^{6}\ \text{dyn/cm}, & k^{(2)} &= -0.83 \cdot 10^{6}\ \text{dyn/cm} \\ C^{(1)} &= 0.134, & C^{(2)} &= 0.0104 \end{aligned} \tag{6.32}$$

It is evident both from the OVC values and from the force constant coefficients that both orbitals 5σ and 2π are antibonding, the latter being much more antibonding than the former. Unlike the qualitative statement

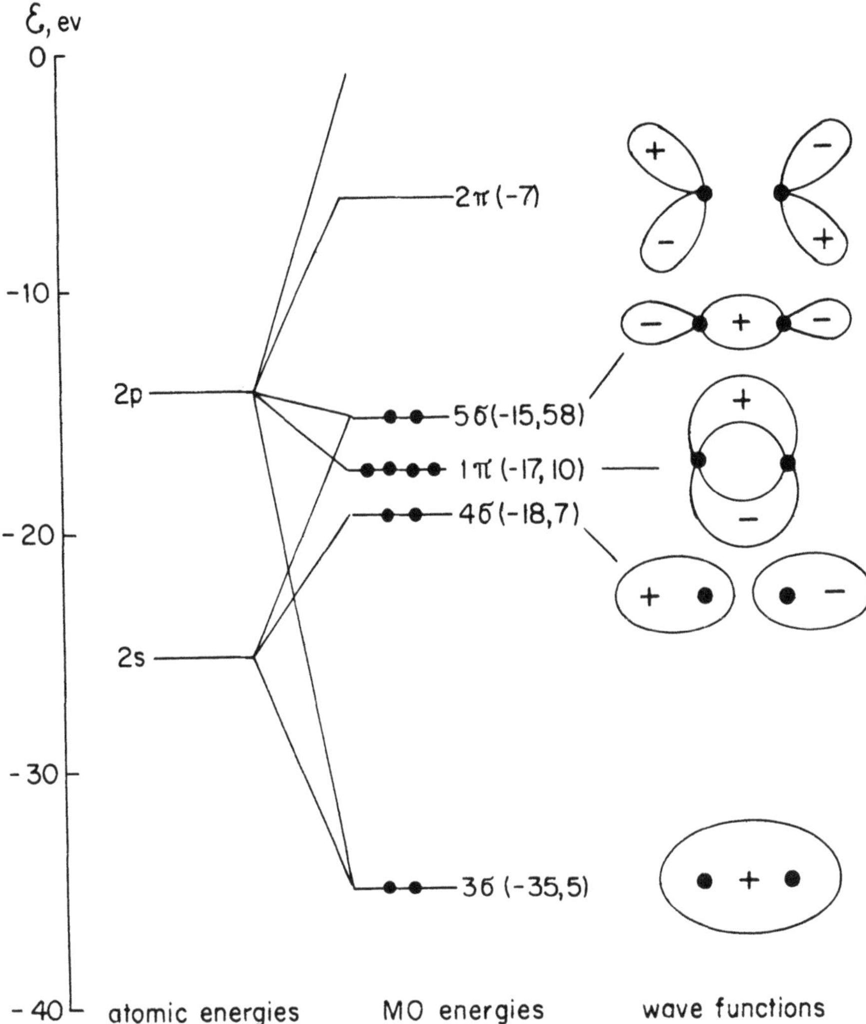

Figure 6.12. Energy level scheme and MO symmetries for two isoelectronic molecules CO and N_2. The figures in parentheses are the N_2 energy levels in eV.

about the nature of the corresponding MO (bonding or antibonding), which can often (but not always) be made without using the vibronic approach, the latter can be used to determine the quantitative degree of MO participation in the chemical bonding, and separately in the distorting force, force constant, anharmonicity correction, etc.

With knowledge of constants (6.32), one can analyze the behavior of the CO molecule in other situations, in particular, when it is coordinated to another molecular system or solid surface. If the charge transfers Δq_1 and Δq_2 are known, then F, ΔK, β, and hence ΔD (for given D) can be obtained, and vice versa. It is seen from these constants that the greater the positive values of Δq_1 and Δq_2 (i.e., the greater the increase in the population numbers of HOMO and LUMO), the larger the negative force F (which acts as an antibonding factor to lower the activation energy) and the negative ΔK (acting in the same way). However, since the HOMO is fully occupied by electrons in the free molecule, Δq_1 can only be negative or zero. Consequently, the activation of the CO molecule is greater, the larger the positive value of Δq_2 and the smaller the negative value of Δq_1, the former being much more important. Hence (for reactions in which the activation energy is determined by the activation of the CO molecule), say for a linear coordination, a catalyst is most efficient, the greater its π donor properties and the lower its σ acceptor ability.

Consider the linear coordination of carbon monoxide to the Ni atom on an NiO surface. The value $K' = 1.710 \cdot 10^6$ dyn/cm is known from empirical data.[516] Using the above-mentioned relationship between K' and Q_0 for the CO molecule,[512]

$$K' = (1.7957\text{–}8.2926\ Q_0\ [\text{Å}]) \cdot 10^6\ \text{dyn/cm}$$

one can determine $Q_0 = 0.0126$ Å, $F = -2.119 \cdot 10^{-4}$ dyn and $\Delta K = K' - K = -0.192 \cdot 10^6$ dyn/cm, and then estimate $\Delta q_1 = -0.32$, $\Delta q_2 = 0.26$, and $\beta = 0.93$ from equations (6.27)–(6.29). These data provide in detail a mechanism for the CO molecule activation when coordinated to the NiO surface. As a result of coordination there is a 0.3 electron charge transfer to the metal from the weak antibonding 5σ orbital and about a 0.3 electron charge transfer from the metal to the strong antibonding 2π orbital of CO.

Using the relationship between K and Q_0, a general correlation between the CO vibrational frequency and charge transfers Δq_1 and Δq_2 from the HOMO 5σ $(-\Delta q_1)$ and to the LUMO 2π (Δq_2) by coordination has been derived. It is shown in Figure 6.13 in the form of a correlation diagram. Note, however, that the empirical function $K' = f(Q_0)$ is valid only in the region where the coordination is strong and does not apply to very small charge transfers and CO frequency changes.

The variation in CO reactivity on the NiO surface can be estimated from equation (6.26) (D and ΔD in kcal/mol):

$$\Delta D = -0.17D - 1.40D^{1/2} + 0.78 - 0.15D^{-1/2} \tag{6.33}$$

This curve $\Delta D = f(D)$ together with similar curves for CO activation in a series of metal carbonyl, is shown in Figure 6.14.[516a]

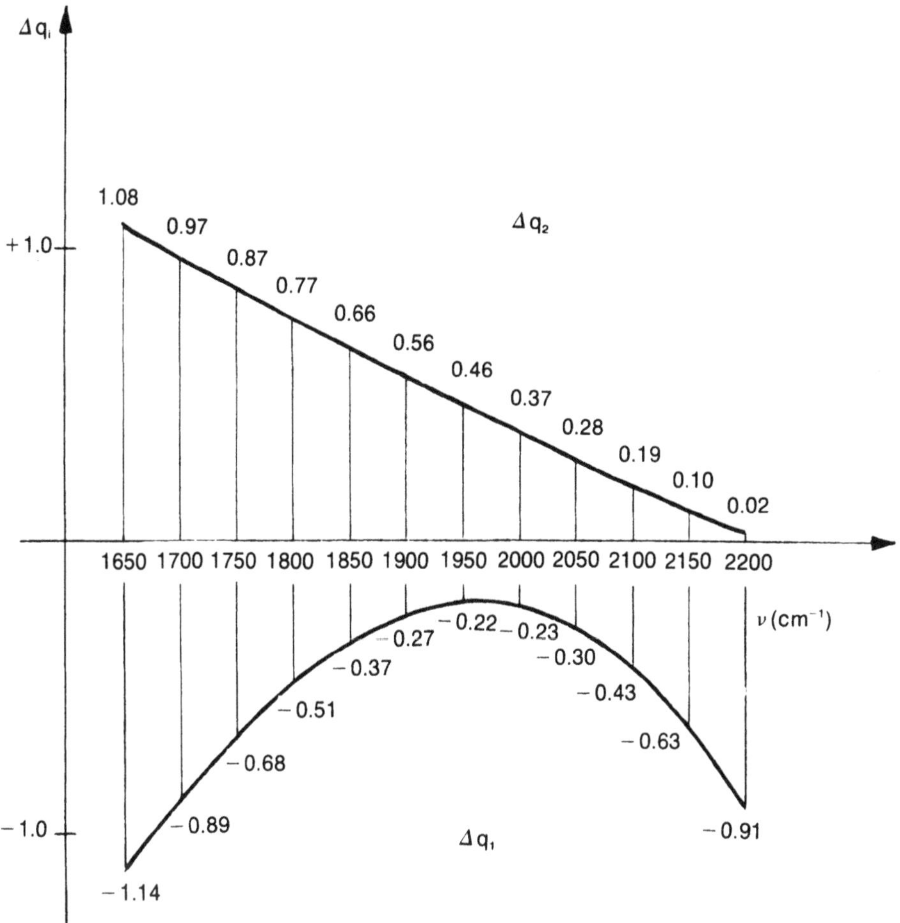

Figure 6.13. Correlation diagram between vibration frequency $\nu = \omega/2\pi$ of the CO molecule and charge transfers Δq_1 and Δq_2 from the HOMO 5σ $(-\Delta q_1)$ and to the LUMO 2π (Δq_2) by coordination to another molecule or catalyst active center.

An example with a known value of D must be considered in order to obtain a numerical estimate of the value of ΔD. If we assume that the activation energy of the reaction $CO + O_2 = CO_2 + O$ in flames is $D = 48$ kcal/mol,[517] and this is associated with the activation of the CO bond only, then by linear coordination to the NiO surface, the activation energy, equation (6.33), is lowered by $\Delta D \cong -17$ kcal/mol and becomes $D' = D + \Delta D \cong 31$ kcal/mol. The experimental value for this reaction on an NiO catalyst (in excess CO) is $D_{\text{exp}} = 25.4$ kcal/mol.[518]

For the nitrogen molecule N_2, like CO, the HOMO is 5σ and the LUMO is 2π. Using data on N_2^+ and the excited states $a\,^1\Pi_g$ and $B\,^3\Pi_g$,[519]

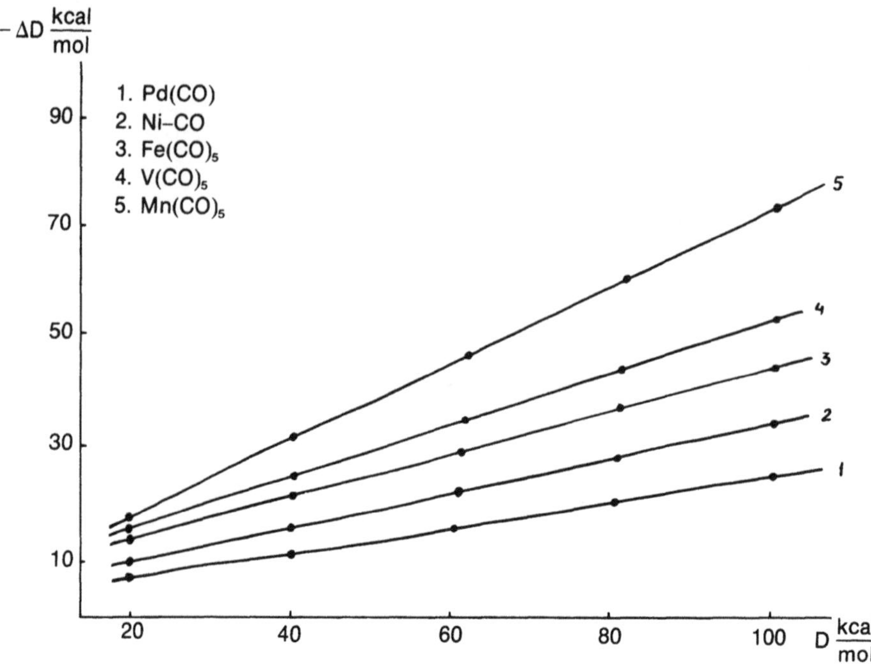

Figure 6.14. Vibronic activation of carbon monoxide: curves $-\Delta D = f(D)$ for CO coordination with some transition metal systems.

which emerge from the $(5\sigma)^2 \rightarrow (5\sigma)^1(2\pi)^1$ excitation ($\Delta q_1 = -1$, $\Delta q_2 = 1$) and equations (6.27)–(6.29), we have

$$f^{(1)} = 3.51 \cdot 10^{-4} \text{ dyn}, \qquad f^{(2)} = -8.18 \cdot 10^{-4} \text{ dyn}$$
$$k^{(1)} = 0.286 \cdot 10^6 \text{ dyn/cm}, \qquad k^{(2)} = -0.785 \cdot 10^6 \text{ dyn/cm} \tag{6.34}$$
$$C^{(1)} = 0.112 \qquad C^{(2)} = 0.127$$

It is seen that, in contrast to the CO molecule, the HOMO 5σ is bonding while the LUMO 2π, like CO, is antibonding. Therefore an activation center with high π donor and σ acceptor properties is needed for activation of linearly coordinated dinitrogen. If the coordination center is a π acceptor, a charge transfer results either from the inner bonding π_u orbital, when the coordination is linear (longitudinal), or from the antibonding σ_u orbital for transversal coordination. It follows that activation in this case depends on the geometry of coordination (more details are given elsewhere[18–20]). Gagarin[520], too, has estimated the OVC.

The constants (6.34) and empirical data on K' and Q_0 pertaining to the coordinated nitrogen molecule can be employed to estimate the charge

Table 6.1. Semiempirical Calculated Charge Transfers Δq_1 and Δq_2 and Vibronic Reduction in Activation Energy ΔD by Coordination to Different Complexes for the Nitrogen Molecule (D and ΔD in kcal/mol)

Coordination system[a]	ν', cm^{-1}	K' 10^6 dyn/cm	$\Delta K = K' - K$, 10^6 dyn/cm	Q_0, Å	F, 10^{-4} dyn	Δq_1	Δq_2	$\Delta D = aD + bD^{1/2} + c + dD^{-1/2}$			
								a	b	c	d
$[RuH_2(N_2)(PPh_3)_3]$	2147	1.900	−0.394	0.0229	−4.19	−0.16	0.44	−0.14	−2.72	2.87	−1.01
$[\{Ru(NH_3)_5\}_2N_2]^{4+}$	2100	1.821	−0.479	0.0303	−5.25	−0.57	0.41	−0.21	−3.37	4.81	−2.28
$[Os(NH_3)_5N_2]Br_2$	2028	1.695	−0.599	0.0428	−6.77	−1.00	0.40	−0.31	−4.32	8.94	−6.17
$[(tol)(PPh_3)_2Mo{-}N_2{-}Fe(C_5H_5)(dmpe)]^+$	1930	1.536	−0.758	0.0619	−8.54	−1.21	0.52	−0.37	−5.62	16.9	−17.0

[a] tol = toluol; dmpe = $Me_2PCH_2CH_2PMe_2$.

transfers Δq_1 and Δq_2 and the reduction in activation energy ΔD as a function of D, using equations (6.27)–(6.29). Table 6.1 presents some examples of such calculations (data from Ref.[521]). The value of Q_0 is estimated from the empirical formula (ν' in cm^{-1})

$$Q_0[\text{Å}] = 0.9482 - 0.7105\,(\nu' \cdot 10^{-3}) + 0.1302\,(\nu' \cdot 10^{-3})^2.$$

It is seen that, as above, the biggest contribution to ΔD originates from the linear effect, i.e., the distorting force (terms b and c), the contribution from the softening (second-order) effect being smaller while the third-order anharmonicity correction, d, is smallest.

As distinct from the above two examples, the HOMO and LUMO in the free NO molecules are from the same antibonding 2π orbital occupied by one electron. Calculations[522] show that, in addition to this orbital, the next lower-in-energy fully occupied 5σ orbital takes part in the charge transfers by coordination (the remaining inner orbitals are not involved). Coefficients a, b, c, and d which determine the activation energy reduction ΔD (equation (6.31)) are given in Table 6.2 for several systems. Appropriate curves are illustrated in Figure 6.15. In contrast to the previous examples, it is seen that there is a case (the iron complex) where $\Delta D > 0$, which shows an increase in activation energy as a result of coordination (the iron complex is an anticatalyst for appropriate reactions with the NO molecule).

Mechanisms of Chemical Reactions in Systems with the JTE. Consider a molecular system in an orbitally degenerate (or pseudodegenerate) electronic state, in which the JTE (or PJTE) results in instability of the maximum symmetry configuration. Minima with lower symmetry are formed and dynamic transitions occur between them (Chapters 2 and 3). Assume that such a system enters into a chemical reaction with another

Table 6.2. Coefficients a, b, c, and d in the Relationship $\Delta D = aD + bD^{1/2} + c + dD^{-1/2}$ for NO Molecules Activated by Different Coordination Systems

Coordination system	a	b	c	d
$Fe(CN)_5NO^{2-}$	0.03	0.78	0.19	0.02
$Mn(CN)_5NO^{2-}$	−0.02	−0.23	0.02	$-4.7 \cdot 10^{-4}$
$Mn(CN)_5NO^{3-}$	−0.18	−2.6	2.83	−1.01
$Cr(CN)_5NO^{3-}$	−0.27	−3.56	5.77	−3.11
$Cr(CN)_5NO^{4-}$	−0.37	−4.26	9.58	−7.17
$V(CN)_5NO^{3-}$	−0.35	−4.19	9.07	−6.54

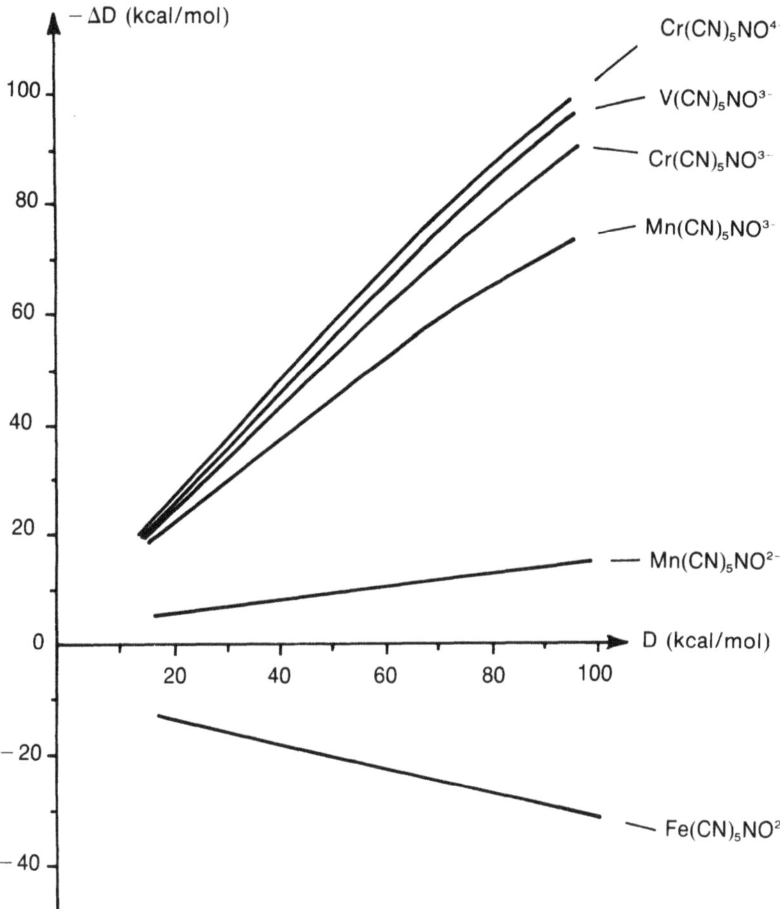

Figure 6.15. Vibronic activation of nitrogen oxide: curves $-\Delta D = f(D)$ for NO coordination with some transition metal cyanides. The iron complex is an anticatalyst for the NO molecule.

molecule. (The conclusions reached below are also true for the limiting case of a unimolecular reaction.) Neglecting vibronic effects (or, more precisely, considering only the directions for which there are no vibronic effects), the reaction curve has the usual shape given in Figure 6.1 (see also Figure 6.16). Note that the reaction barrier at the point Q_D, in accord with Section 6.2, is formed due to vibronic mixing with the excited electronic state of the other reagent molecule (while in the unimolecular reaction it is due to mixing with its own upper-lying dissociation states).

In the direction of JT active displacements Q (Table 1.3) or PJT active ones (Section 2.3), this reaction curve changes due to the distorting force $F \neq 0$ (in the case of the JTE) or a negative force constant $K < 0$ (for the

PJTE) which arises at the point $Q = 0$. The former case is illustrated in Figure 6.16. Comparison of this figure with Figure 6.11 indicates that the two curves with and without the JTE are quite similar to the two reaction curves with and without the influence of a third system (catalyst). In other words, the influence of the JTE on the chemical reaction proceeding in the JT active directions is qualitatively the same as the influence of the catalyst (or another appropriate source of electronic rearrangement).

This result is quite understandable if one takes into account that both perturbation influences result in distorting forces, softening, etc. This analogy and the similarities in the reaction curves allow for direct use of the above equations to estimate the changes in the reaction parameters due to the JTE and PJTE. In particular, the change in activation energy ΔD can be calculated directly from equation (6.26).

There are also essential differences together with this important similarity between the two sources of chemical activation. They concern, first, the direction in which the activation energy is lowered, i.e., the allowed mechanism of the reaction. In systems with JTE and PJTE, this direction is predetermined by the electronic structure of the molecule itself (by the JT and PJT active displacements), whereas in the case of a catalyst influence the direction of the reaction depends on the nature of the catalyst and the electronic rearrangement produced by the latter on a given molecule. Therefore, different catalysts in principle can cause different reactions to dominate. The catalyst can also change the mechanism of

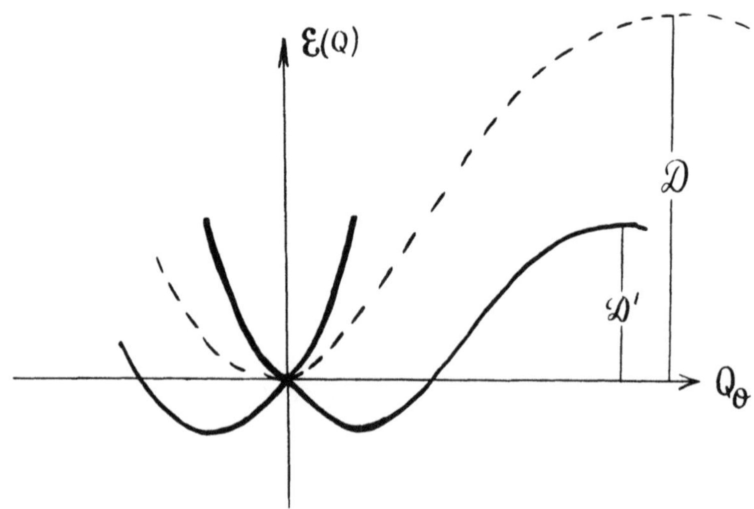

Figure 6.16. Shape of the AP in the direction of the JT active coordinate Q at large distances with (solid line) and without (dashed line) the JTE, illustrating lowering of the activation energy of corresponding chemical reactions induced by the JTE.

reaction which occurs without the catalyst. On the other hand, for reactions of free molecules proceeding under the influence of vibronic effects, the mechanism is known *a priori*, and this allows us to predict the reaction course and its products.

By way of example consider the reaction of substitution in, or unimolecular decomposition of, an octahedral transition metal complex in an orbitally twofold degenerate electronic E state, e.g., for bivalent copper known as having a strong JTE. As a result of the JTE, the AP of the system has three equivalent minima at each of which the octahedron is elongated along the tetragonal axis, two ligands being further away from the central atom than the other four (see Section 2.1, Figure 2.3, and Tables 5.2–5.4 which illustrate this statement). In Q_θ and Q_ε coordinates (Figure 1.1) the distortion takes place along the Q_θ direction; Q_θ is thus the coordinate along which the reaction activation energy is lowered compared with other directions *ceteris paribus*.

As a result, the above reactions should proceed in the direction of the Q_θ displacements for which the two axial ligands move away from (while the other four approach) the complex. For the unimolecular reaction this implies dissociation of the ligands in the *trans*-positions. For the substitution reaction, it means the formation of *trans*-substituted complexes. Visually, this conclusion corresponds to the following picture: when elongated octahedra are formed during the pulsating motions, the two ligands in *trans*-positions are weakly bonded and more easily reached by the attacking reagents.[523]

This result is in complete agreement with well-known empirical data on the behavior of octahedral complexes of Cu(II) in solution or gas phase. It is known that these complexes lose two ligands in the *trans*-position forming square-planar structures, while substitution reactions produce only *trans*-substituted complexes (except for bidendate ligands which can only be *cis*-coordinated). On the other hand, if the JTE in such a system is quenched, the potential barrier increases and the reaction rate decreases substantially.

This conclusion also agrees with experimental data.[524] Theory predicts similar effects in octahedral complexes of Cr(II), Mn(III), Ag(II), etc., having E ground terms.

The expected vibronically allowed mechanisms of reactions are different for other nonoctahedral molecular configurations with E terms and for other ground terms. They can be determined qualitatively from known distortions in the JTE and PJTE. In particular, for a tetrahedral system with a T term (of CH_4^+ type), in which the formation of four minima of the AP is expected, each system has one long and three short bonds. Here, the dissociation and substitution with a lower activation energy is expected, *ceteris paribus*, to take place along the elongated bond.

Another feature of activation energy lowering caused by the JTE, distinguishing it from that of an electronic rearrangement, lies in the relative contributions of distortion and softening. In the JTE the largest contribution to ΔD comes from the distorting force F, the change in force constant ΔK due to quadratic terms of the vibronic interactions being small. In contrast, the electronic rearrangement causes the contribution from ΔK to be much larger (the contribution due to anharmonicity variations is small in both cases). For the PJTE, $F = 0$, but $\Delta K < 0$ and $|\Delta K| > K$. Hence a simple comparison here with electronic rearrangement is difficult.

If only the main contribution to ΔD arising from the distorting force (in the approximation of linear vibronic coupling) is taken into account in the case of the JTE, the following simple equation for ΔD is obtained instead of equation (6.26):

$$\Delta D = -(AB)^{1/2}[1-(A/B)^{1/2} + A/3B] \tag{6.38}$$

where $A = 4E_{JT}$, E_{JT} being the stabilization energy in the JTE, and $B = 3D$. Hence the reduction in activation energy due to the JTE for reactions proceeding in JT active directions may be very substantial.

There are also other suggested uses of the JTE in determining chemical reaction rates.[525–527] In particular, *ab initio* calculations[525a] have shown that the photolysis reaction $CH_4 \rightarrow CH_2 + H_2$ proceeds along the JT coordinate of the excited T term. In another study,[526] slow electron transfer between two low spin coordination systems Co(II)→Co(III) is attributed to the small Franck–Condon factor due to the JTE on the E term of the low spin Co(II); the pertinent measurements were performed on the system Co([4]diene $N_4)(OH_2)_2^{2+}$ → Co([14]diene $N_4)(OH_2)_2^{3+}$). Vekhter and Rafalovich[527] examined the probability of electron transfer between two transition metal complexes, one of which is in a degenerate state (E or T), and showed that the JTE is most essential in determining the reaction channels, reaction rate, and their dependence on temperature, as well as the energy gap between the ground states of the two complexes. Influence of JT rapid inversion of distortions on substitution reaction kinetics has also been observed.[528]

References*

1. M. Born and R. Oppenheimer, *Ann. Phys.* **84**, 457 (1927).
2. M. Born and K. Huang, *Dynamical Theory of Crystal Lattices*, Clarendon Press, Oxford (1951).
3. E. Teller, An historical note, in *The Jahn–Teller Effect in Molecules and Crystals* (R. Englman, ed.), Wiley–Interscience, London (1972).
4. H. A. Jahn and E. Teller, *Proc. R. Soc. London, Ser. A* **161**, 220 (1937).
5. R. Englman, *The Jahn–Teller Effect in Molecules and Crystals*, Wiley–Interscience, London (1972).
6. F. S. Ham, in: *Electron Paramagnetic Resonance* (S. Geschwind, ed.), p. 1, Plenum Press, New York (1972).
7. A. Abragam and B. Bleaney, *Electron Paramagnetic Resonance of Transition Ions*, Ch. 21, Clarendon Press, Oxford (1970).
8. I. B. Bersuker, *Coord. Chem. Rev.* **14**, 357 (1975).
9. I. B. Bersuker, B. G. Vekhter, and I. Ya. Ogurtsov, *Usp. Fiz. Nauk* **116**, 605 (1975).
10. G. A. Gehring and K. A. Gehring, *Rep. Prog. Phys.* **38**, 1 (1975).
11. C. A. Bates, *Phys. Rep.* **35C**, 187 (1978).
12. I. B. Bersuker and B. G. Vekhter, *Ferroelectrics* **19**, 137 (1978).
13. I. B. Bersuker and V. Z. Polinger, *Vibronnie Vzaimodeistvya v Molekulakh i Kristallakh*, Nauka, Moscow (1982).
14. I. B. Bersuker, *The Jahn-Teller Effect: A Bibliographic Review*, IFI/Plenum, New York, 1984.
15. U. Öpik and M. H. L. Pryce, *Proc. R. Soc. London, Ser. A* **238**, 425 (1957).
16. R. B. Woodward and R. Hoffmann, *The Conservation of Orbital Symmetry*, Verlag Chemie, GmbH, Weinheim/Bergstrasse (1970).
17. R. G. Pearson, *Symmetry Rules for Chemical Reactions*, Wiley–Interscience, New York (1976).
18. I. B. Bersuker, *Kinet. Katal.* **18**, 1268 (1977).
19. I. B. Bersuker, *Chem. Phys.* **31**, 85 (1978).
20. I. B. Bersuker, *Teor. Eksp. Khim.* **14**, 3 (1978).
21. I. B. Bersuker, in: (IUPAC) Coordination Chemistry-20 (D. Banerjea, ed.), p. 201, Pergamon Press, Oxford and New York (1980).

* Most of the references in Russian are available also in English translation: JETP (Sov. Phys. — JETP), Fizika Tverdogo Tela (Sov. Phys. — Solid State), Uspekhi Fiz. Nauk (Sov. Phys. — Uspekhi), Zh. Struct. Khim., Teoret. i Eksp. Khim., Optika i Spektroskopiya, etc. All the references are given in their original version.

22. L. D. Landau and E. M. Liphshitz, *Kvantovaya Mekhanika. Nerelativistskaya Teoriya*, Nauka, Moscow (1974).
22a. R. J. Buenker, *Gazz. Chim. Ital.* **108**, 245 (1978).
23. R. M. Hochstrasser, *Molecular Aspects of Symmetry*, Benjamin, New York, Amsterdam (1966).
23a. J. P. Fackler, Jr., *Symmetry in Coordination Chemistry*, Academic Press, New York and London (1979); J. P. Fackler, Jr., ed., *Symmetry in Chemical Theory*, Wiley, New York (1973).
24. P. Roman, *Advanced Quantum Theory*, pp. 576–590, Addison-Wesley, Reading, Massachusetts (1965).
25. J. S. Griffith, *The Irreducible Tensor Method for Molecular Symmetry Groups*, Prentice-Hall, Englewood Cliffs, New Jersey (1962).
26. J. S. Slater, *Electronic Structure of Molecules*, McGraw-Hill Book Company, New York (1963).
27. L. Salem, *Chem. Phys. Lett.* **3**, 99 (1969).
28. I. B. Bersuker, *Nouveau J. Chimie* **4**, 139 (1980); *Teoret. Eksp. Khim.* **16**, 291 (1980).
28a. I. B. Bersuker, N. N. Gorinchoi, and V. Z. Polinger, to be published.
29. R. F. W. Bader and A. D. Bandrauk, *J. Chem. Phys.* **49** 1666 (1968).
30. H. A. Jahn, *Proc. R. Soc. London, Ser. A* **164**, 117 (1938).
31. R. Renner, *Z. Phys.* **92**, 172 (1934); H. Sponer and E. Teller, *Rev. Mod. Phys.* **13**, 75 (1941).
32. G. Herzberg, *Electronic Spectra and Electronic Structure of Polyatomic Molecules*, Van Nostrand, Toronto, New York, London (1966).
32a. A. N. Petelin and A. A. Kiselev, *Int. J. Quantum Chem.* **6**, 701 (1972).
32b. M. Köppel, W. Domcke, and L. S. Cederbaum, *J. Chem. Phys.* **74**, 2945 (1981).
33. I. B. Bersuker, *Teor. Eksp. Khim.* **2**, 518 (1966).
34. E. Ruch and A. Schonhofer, *Theor. Chim. Acta* **3**, 291 (1965); E. L. Blount, *Math. Phys.* **12**, 1890 (1971).
35. A. D. Liehr and C. J. Ballhausen, *Ann. Phys. (NY)* **3**, 304 (1958).
36. I. B. Bersuker, Doctoral dissertation, Leningrad State University (1964); *Doklady AN SSSR* **132**, 587 (1960); *Zh. Strukt. Khim.* **2**, 350, 734 (1961); I. B. Bersuker and Yu. G. Titova, *Zh. Strukt. Khim.* **5**, 121 (1964).
37. R. E. Coffman, *J. Chem. Phys.* **44**, 2305 (1966).
38. J. H. Van Vleck, *J. Chem. Phys.* **7**, 61 (1939).
39. A. D. Liehr, *J. Phys. Chem.* **67**, 389 (1963); *Prog. Inorg. Chem.* **3**, 281 (1962); **4**, 455 (1962); **5**, 385 (1963).
40. M. Bacci, *Phys. Rev.* **B17**, 4495 (1978).
41. M. C. M. O'Brien, *Phys. Rev.* **187**, 407 (1969).
42. I. B. Bersuker and V. Z. Polinger, *Phys. Lett.* **44A**, 495 (1973).
43. I. B. Bersuker and V. Z. Polinger, *JETP* **66**, 2078 (1974).
44. S. Muramatsu and T. Iida, *J. Phys. Chem. Solids* **31**, 2209 (1970).
45. M. Bacci, M. P. Fontana, A. Ranfagni, and G. Viliani, *Phys. Lett.* **50A**, 405 (1975); *Phys. Rev.* **B11**, 3052 (1975).
46. M. Bacci, A. Ranfagni, M. Cetica, and G. Viliani, *Phys. Rev.* **B12**, 5907 (1975).
47. R. S. Dagis and I. B. Levinson, in: *Optika i Spektroskopiya. III. Molekul Spektroskopiya*, p. 3, Nauka, Leningrad (1967).
48. B. R. Judd, *J. Chem. Phys.* **68**, 5643 (1978).
49. B. R. Judd, *Colloques Internat. CNRS* **255**, 127 (1978).
50. V. P. Khlopin, V. Z. Polinger, and I. B. Bersuker, *Fiz. Mol.* **3**, 70 (1977).
51. V. P. Khlopin, V. Z. Polinger, and I. B. Bersuker, *Theor. Chim. Acta* **48**, 87 (1978).
52. I. B. Bersuker, *Phys. Lett.* **20**, 589 (1966).
53. I. B. Bersuker, *Elektronnoe Stroenie i Svoistva Koordinatsionnikh Soedinenii*, 2nd edition, Khimiya, Leningrad (1976).

54. I. B. Bersuker, B. G. Vekhter, G. S. Danil'chuk, L. S. Kremenchugskii, A. A. Muzalevskii, and M. L. Rafalovich, *Fiz. Tverd. Tela* **11**, 2452 (1969).
55. I. B. Bersuker and S. S. Stavrov, *Chem. Phys.* **54**, 331 (1981).
56. D. K. Pooler, *Phys. Rev.* **B21**, 4804 (1980).
56a. S. A. Borshch, I. Ya. Ogurtsov, and I. B. Bersuker, in: *Fiz. i Mat. Metody v Koord. Khimii*, p. 191, Shtiintsa, Kishinev (1980); *Zh. Strukt. Khim.* **23**, 7 (1982).
57. I. B. Bersuker, B. G. Vekhter, and M. L. Rafalovich, in: *Tezisi Dokladov Vsesoyuzn. Konf. po Dipolnim Momentam i Stroeniu Molekul*, p. 9, Rostov/Don (1967); *Teoret. Eksp. Khim.* **5**, 293 (1969).
58. J. H. Van Vleck, *Phys. Rev.* **57**, 426 (1940).
59. Yu. E. Perlin and B. S. Tsukerblat, *Effekty Elektronno-kolebatel'nogo Vzaimodeistviya v Opticheskikh Spektrakh Primesnikh Paramagnitnikh Ionov*, Shtiintsa, Kishinev (1975).
60. H. G. Longuet-Higgins, *Advances in Spectroscopy*, Vol. 2, p. 420, Interscience, New York (1961).
61. W. Moffitt and A. D. Liehr, *Phys. Rev.* **106**, 1195 (1956).
62. W. Moffitt and W. Thorson, *Phys. Rev.* **108**, 1251 (1957).
63. I. B. Bersuker and V. Z. Polinger, *Phys. Status Solidi B* **60**, 85 (1973).
63a. M. Wagner, in: *The Jahn–Teller Effect in Localized States in Crystals* (Yu. E. Perlin and M. Wagner, eds.), North-Holland, Amsterdam (1984) (to be published).
64. M. C. M. O'Brien, *Proc. R. Soc. London, Ser. A* **281**, 323 (1964).
65. I. B. Bersuker, *Opt. Spectrosk.* **11**, 319 (1961).
66. I. B. Bersuker, *JETP* **43**, 1315 (1962).
67. V. Z. Polinger, *Fiz. Tverd. Tela* **16**, 2578 (1974).
68. I. B. Bersuker, in: *Physics of Impurity Centers in Crystals*, p. 479, Acad. Sci. Est. SSR, Tallin (1972).
69. J. C. Slonczewski, *Phys. Rev.* **131**, 1596 (1963).
70. M. C. M. O'Brien, *J. Phys. C* **5**, 2045 (1972).
71. R. Englman and B. Halperin, *Ann. Phys.* **3**, 453 (1978).
72. Yu. B. Rozenfeld and V. Z. Polinger, *JETP* **70**, 597 (1976).
73. V. Z. Polinger and G. I. Bersuker, *Phys. Status Solidi B* **96**, 153 (1979).
73a. V. Z. Polinger and G. I. Bersuker, *Phys. Status Solidi B* **95**, 403 (1979).
74. *Offitsialnii Byulleten' Gosudarstvennogo Komiteta po Delam Izobretenii i Otkrytii, N40: Publikatsiya o Nauchnom Otkrytii* N 202 (I. B. Bersuker), October 1978.
75. I. B. Bersuker and B. G. Vekhter, *Fiz. Tverd. Tela* **5**, 2432 (1963).
76. F. S. Ham, *Phys. Rev.* **138**, 1727 (1965).
77. M. S. Child and H. C. Longuet-Higgins, *Phil. Trans. R. Soc. London, Ser. A* **254**, 259 (1962).
78. M. Caner and R. Englman, *J. Chem. Phys.* **44**, 4054 (1966).
79. B. S. Tsukerblat and B. G. Vekhter, *Fiz. Tverd. Tela* **14**, 2544 (1972).
80. B. G. Vekhter, *Phys. Lett., Ser. A* **45**, 133 (1973).
81. D. K. Faddeev and V. N. Faddeeva, *Vichislitelnie Metody Lineinoi Algebry*, Fizmatgiz, Moscow (1963).
82. H. C. Longuet-Higgins, U. Öpik, M. H. L. Pryce, and R. A. Sack, *Proc. R. Soc. London, Ser. A* **244**, 1 (1958).
83. W. Moffitt and W. Thorson, in: *Calcul des fonctions d'onde moleculaires*, p. 141, CNRS, Paris (1958).
84. S. Muramatsu and N. Sakamoto, *J. Phys. Soc. Jpn.* **44**, 1640 (1978).
85. C. W. Struck and F. Herzfeld, *J. Chem. Phys.* **44**, 464 (1966).
86. H. Uehara, *J. Chem. Phys.* **45**, 4536 (1966).
87. M. C. M. O'Brien, *J. Phys. C* **4**, 2524 (1971).
88. M. C. M. O'Brien, *J. Phys. C* **9**, 3153 (1976).
89. N. Sakamoto and S. Muramatsu, *Phys. Rev.* **B17**, 868 (1978).

90. S. I. Boldyrev, V. Z. Polinger, and I. B. Bersuker, *Fiz. Tverd. Tela* **23**, 746 (1981).
91. W. Thorson and W. Moffitt, *Phys. Rev.* **168**, 362 (1968).
92. B. R. Judd, *J. Chem. Phys.* **67**, 1174 (1977).
93. D. R. Pooler and M. C. M. O'Brien, *J. Phys. C* **10**, 3769, (1977).
94. B. Bleaney and D. J. E. Ingram, *Proc. Phys. Soc., Ser. A* **63**, 408 (1950).
95. B. Bleaney and K. D. Bowers, *Proc. Phys. Soc., Ser. A* **65**, 667 (1952).
96. A. Abragam and M. H. L. Pryce, *Proc. R. Soc. London Ser. A.* **63**, 409 (1950).
97. I. B. Bersuker, *JETP* **44**, 1239 (1963).
98. I. B. Bersuker, B. G. Vekhter, and V. Z. Polinger, in: *Problemy Magnitnogo Rezonansa*, p. 31, Nauka, Moscow (1978).
99. K. A. Müller, in: *Magnetic Resonance and Relaxation. Proc. XIV Colloque Ampère, Ljubljana, 1966* (R. Blinc, ed.), p. 192, North-Holland, Amsterdam (1967).
100. R. E. Coffman, *Phys. Lett.* **21**, 381 (1966).
101. R. E. Coffman, *J. Chem. Phys.* **48**, 609 (1968).
102. L. N. Shen and T. L. Estle, *J. Phys. C* **12**, 2103, 2129 (1979).
103. G. W. Ludwig and H. H. Woodbury, *Phys. Rev.* **113**, 1014 (1959); *Solid State Phys.* **13**, 1 (1962); H. H. Woodbury and G. W. Ludwig, *Phys. Rev.* **126**, 466 (1962).
104. F. S. Ham, *Phys. Rev.* **166**, 307 (1968).
105. I. V. Aleksandrov, *Teoriya Magnetnoi Relaksatsii*, Nauka, Moscow (1975).
106. A. Abragam, *The Principles of Nuclear Magnetism*, Clarendon Press, Oxford (1961).
107. N. Gauthier and M. B. Walker, *Can. J. Phys.* **54**, 9 (1976).
108. V. Z. Polinger and G. I. Bersuker, *Fiz. Tverd. Tela* **22**, 2545 (1980); G. I. Bersuker and V. Z. Polinger *JETP* **80**, 1788 (1981).
108a. C. A. Bates and P. Steggles, *J. Phys. C* **8**, 2283 (1975).
109. R. W. Reynolds, L. A. Boatner, M. M. Abraham, and Y. Chen, *Phys. Rev. B* **10**, 3802 (1974).
110. R. W. Reynolds and L. A. Boatner, *Phys. Rev. B* **12**, 4735 (1975).
111. L. A. Boatner, R. W. Reynolds, Y. Chen, and M. M. Abraham, *Phys. Rev. B* **16**, 86 (1977).
112. L. L. Chase, *Phys. Rev. Lett.* **23**, 275 (1969).
113. B. Bleaney, K. D. Bowers, and R. S. Trenam, *Proc. Soc. London, Ser. A* **228**, 157 (1955).
114. D. Bijl and A. C. Rose-Innes, *Proc. Phys. Soc. A* **66**, 954 (1953).
115. R. T. Tucker, *Phys. Rev.* **112**, 725 (1958); D. C. Burnham, *Bull. Ann. Phys. Soc.* **11**, 186 (1966).
116. R. E. Coffman, D. L. Lyle, and D. R. Mattison, *J. Phys. Chem.* **72**, 1392 (1968).
117. W. Hayes and J. Wilkens, *Proc. R. Soc. London, Ser. A* **281**, 340 (1964).
118. J. R. O'Connor and J. H. Chen, *Appl. Phys. Lett.* **5**, 100 (1964).
119. W. Hayes and J. R. Twidell, *Proc. Phys. Soc.* **82**, 330 (1963).
120. Y. T. Hochli, *Phys. Rev.* **162**, 262 (1967).
121. Y. T. Hochli and T. L. Estle, *Phys. Rev. Lett.* **18**, 128 (1967).
122. S. Geschwind and J. R. Remeika, *J. Appl. Phys.* **33**, 370 (1962).
123. W. Low and J. T. Suss, *Phys. Lett.* **7**, 310 (1963).
124. Y. Hochli, K. A. Muller, and P. Wysling, *Phys. Lett.* **15**, 5 (1965).
125. A. M. Ziatdinov, M. M. Zaripov, Yu. V. Yablokov, and R. L. Davidovich, *Phys. Status Solidi B* **78**, K69 (1976).
126. V. E. Petrashen, Yu. V. Yablokov, and R. L. Davidovich, *Phys. Status Solidi B* **88**, 439 (1978).
127. V. E. Petrashen, Yu. V. Yablokov, and R. L. Davidovich, *Phys. Status Solidi B* **101**, 117 (1980).
128. J. H. Ammeter and J. D. Swalen, *J. Chem. Phys.* **57**, 678 (1972).
129. J. H. Ammeter, *J. Magn. Res.* **30**, 299 (1978).

130. J. H. Ammeter, H. B. Bürgi, E. Gamp, V. Meyer-Sandrin, and W. P. Jensen, *Inorg. Chem.* **18**, 733 (1979).
131. I. B. Bersuker, S. A. Borshch, and I. Ya. Ogurtsov, *Phys. Status Solidi B* **59**, 707 (1973).
132. T. C. Gibb and N. H. Greenwood, *Tech. Rept. I. A. E. A.* **50**, 143 (1966).
133. M. Tanaka, T. Tokoro, and Y. Aijama, *J. Phys. Soc. Jpn.* **21**, 262 (1966).
134. I. B. Bersuker and I. Ya. Ogurtsov, *Fiz. Tverd. Tela* **10**, 3651 (1968); XXII IUPAC Congress, Abstracts, Sidney, p. 92 (1969); I. B. Bersuker, I. Ya. Ogurtsov, and E. I. Polinkovskii, *Izv. Akad. Nauk. Mold. SSR* **4**, 70 (1969).
135. F. S. Ham, *Phys. Rev.* **160**, 328 (1967); F. S. Ham, W. H. Schwarz, and M. C. M. O'Brien, *Phys. Rev.* **185**, 548 (1969).
136. I. B. Bersuker and B. G. Vekhter, in: *Trudy Komissii po Spektroskopii. Materialy XV Soyushchaniya po Spektroskopii*, Vol. 3(1), p. 520, Izd. VINITI (1965).
137. E. M. Gyorgy, R. C. LeCraw, and M. D. Sturge, *J. Appl. Phys.* **37**, 1303 (1966).
138. I. B. Bersuker, *JETP* **44**, 1577 (1963).
139. I. B. Bersuker, *Fiz. Tverd. Tela* **6**, 436 (1964).
140. S. A. Altshuller, V. I. Kochelaev, and A. M. Leushin, *Usp. Fiz. Nauk* **75**, 459 (1961).
141. M. D. Sturge, J. T. Krause, E. M. Gyorgy, R. C. LeCraw, and F. R. Merritt, *Phys. Rev.* **155**, 218 (1967).
142. M. D. Sturge, *Solid State Phys.* **20**, 91 (1967).
143. J. R. Fletcher and K. W. H. Stevens, *J. Phys. C* **2**, 444 (1969).
144. C. A. Bates, *Postepy Fiz.* **24**, 613 (1973); M. Abou-Ghantous, C. A. Bates, J. P. Fletcher, and P. C. Jaussand, *J. Phys. C* **8**, 3641 (1975).
145. J. Lange, *Phys. Rev.* **B14**, 4791 (1976); S. Guha and J. Lange, *Phys. Rev.* **B15**, 4157 (1977).
146. W. R. Thorson, *J. Chem. Phys.* **29**, 938 (1958).
147. M. S. Child, *J. Mol. Spectrosc.* **10**, 357 (1963).
148. M. S. Child, *Phil. Trans. R. Soc. London, Ser. A* **255**, 31 (1963).
149. R. L. Fulton and M. Gouterman, *J. Chem. Phys.* **41**, 2280 (1964).
150. I. B. Bersuker, *Opt. Spektrosk.* **12**, 528 (1962).
151. I. B. Bersuker, Yu. V. Shaparev, and I. Ya. Ogurtsov, *Opt. Spektrosk.* **36**, 315 (1974).
152. I. Ya. Ogurtsov, Yu. V. Shaparev, and I. B. Bersuker, *Opt. Spectrosk.* **45**, 672 (1978).
153. M. S. Child, *Mol. Phys.* **5**, 391 (1962).
154. D. S. Gemmel, E. P. Kanter, and W. J. Pietsch, *J. Chem. Phys.* **72**, 1402, 6818 (1980).
155. K. Fox, *Phys. Lett.* **27**, 233 (1971); *Phys. Rev.* **A6**, 907 (1972).
156. G. Plachek, *Releevskoe Rasseyanie i Raman Effekt*, GNTI, Kharkov–Kiev (1935).
157. R. Englman, B. Halperin, and S. Mukamel, in: *Light Scattering in Solids* (M. Balkanski ed.), p. 557, Hammarion, Paris (1976).
158. E. Mulazzi and N. Terzi, *Solid State Commun.* **18**, 721 (1976).
159. E. Mulazzi and N. Terzi, *Phys. Rev.* **B19**, 2332 (1979).
160. B. Weinstock and G. L. Goodman, *Adv. Chem. Phys.* **9**, 169 (1965).
161. S. Guha and L. L. Chase, *Phys. Rev.* **B12**, 1658 (1975).
162. S. Guha and L. L. Chase, *Phys. Rev. Lett.* **32**, 869 (1974).
163. M. Pawlikowski and M. Z. Zgierski, *Chem. Phys. Lett.* **48**, 201 (1977).
163a. V. V. Hizhnyakov, in: *Light Scattering in Solids* (M. Balkanski, ed.), Hammarion, Paris (1976).
164. M. Pawlikowski and M. Z. Zgierski, *J. Raman Spectr.* **7**, 106 (1978).
164a. W. H. Henneker, A. P. Fenner, W. Siebrand, and M. Z. Zgierski, *J. Chem. Phys.* **69**, 1704 (1978).
165. I. Ya. Ogurtsov and L. A. Kazantseva, *Mol. Strukt.* **55**, 301 (1979); *Zh. Strukt. Khim.* **20**, 414 (1979); I. Ya. Ogurtsov, L. A. Kazantseva, and A. A. Ischenko, *J. Mol. Phys.* **41**, 243 (1977).

166. S. Muramatsu, K. Nasu, M. Takahashi, and K. Kaya, *Chem. Phys. Lett.* **50**, 284 (1977).
167. R. S. Mulliken and E. Teller, *Phys. Rev.* **61**, 283 (1942).
168. G. Herzberg, *Discuss. Faraday Soc.* **35**, 7 (1963).
169. G. Herzberg, *Molecular Spectra and Molecular Structure III. Electronic Spectra and Electronic Structure of Poyatomic Molecules*, Van Nostrand, Toronto, New York, London (1966).
170. G. Fisher and G. Small, *J. Chem. Phys.* **56**, 5934 (1972).
171. C. Cossart-Magos, D. Cossart, and S. Leach, *J. Chem. Phys.* **69**, 4313 (1978).
172. C. Cossart-Magos, D. Cossart, and S. Leach, *Mol. Phys.* **37**, 793 (1979).
173. C. Cossart-Magos, D. Cossart, and S. Leach, *Chem. Phys.* **41**, 345 (1979).
174. C. Cossart-Magos, D. Cossart, and S. Leach, *Chem. Phys.* **41**, 363 (1979).
175. C. Cossart-Magos, S. Leach, *Chem. Phys.* **48**, 329 (1980).
176. C. Cossart-Magos and S. Leach, *Chem. Phys.* **48**, 349 (1980).
177. T. A. Miller and V. E. Bondybey, *Chem. Phys. Lett.* **58**, 454 (1978).
178. V. E. Bondybey, J. H. English, and T. A. Miller, *J. Am. Chem. Soc.* **160**, 5251 (1978).
179. V. E. Bondybey, T. A. Miller, and J. H. English, *J. Am. Chem. Soc.* **161**, 1248 (1979); *J. Chem. Phys.* **70**, 138 (1979); **71**, 1088 (1979).
180. T. A. Miller, V. E. Bondybey, and J. H. English, *J. Am. Chem. Soc.*, **70**, 2919 (1979).
181. T. Sears, T. A. Miller, and V. E. Bondybey, *J. Chem. Phys.* **72**, 6070 (1980).
182. C. Cossart-Magos and S. Leach, *J. Chem. Phys.* **64**, 4006 (1976).
183. A. Despres, V. Lejeune, E. Migirdicyan, and W. Siebrand, *Chem. Phys.* **36**, 41 (1979).
183a. L. S. Cederbaum, W. Domcke, H. Köppel and W. Von Niessen, *Chem. Phys.* **26**, 169 (1977).
184. B. S. Tsukerblat, Yu. B. Rosenfeld, V. Z. Polinger, and B. G. Vekhter, *JETP* **68**, 1117 (1975).
185. A. A. Kaplianskii and A. K. Przevusskii, *Opt. Spektrosk.* **19**, 597 (1965).
186. A. A. Kaplianskii and A. K. Przevusskii, *Opt. Spektrosk.* **20**, 1045 (1966).
187. L. L. Chase, *Phys. Rev.* **B2**, 2308 (1970).
188. A. Hjortsberg, B. Nygren, J. Vallin, and F. S. Ham, *Phys. Rev. Lett.* **39**, 1233 (1977).
189. J. E. Lowther, *J. Phys. C* **8**, 3448 (1975).
190. A. M. Stoneham, *Solid State Commun.* **21**, 339 (1977).
191. M. D. Sturge, *Phys. Rev.* **A140**, 880 (1965).
192. G. Viliani, M. Montagna, O. Pilla, A. Fontana, M. Bacci, and A. Ranfagni, *J. Phys. C* **11**, L 439 (1978).
192a. I. B. Bersuker and G. B. Vekhter, *Phys. Status Solidi B* **16**, 63 (1966); F. S. Ham, in: *Optical Properties of ions in Crystals* (H. M. Crosswhite and H. W. Moos, eds.), p. 357, Wiley, New York (1967); L. Cianchi, M. Mancini, and P. Moretti, *Rev. Nuovo Cimento* **5**, 187 (1975); *Lett. Nuovo Cimento*, **19**, 381 (1977); M. Bacci, *Lett. Nuovo Cimento* **16**, 340 (1976); A. Ranfagni and G. Viliani, *Solid State Commun.* **20**, 1005 (1976); *Phys. Status Solidi B* **84**, 393 (1977).
193. S. Sugano, Y. Tanabe, and H. Kamimura, *Multiplets of Transition-Metal Ions in Crystals*, Academic Press, New York and London (1970).
194. M. C. M. O'Brien, *Proc. Phys. Soc.* **86**, 847 (1965).
195. P. R. Moran, *Phys. Rev. A* **137**, 1016 (1965).
196. Y. Toyozawa and M. Inoue, *J. Phys. Soc. Jpn.* **20**, 1289 (1965); **21**, 1663 (1966).
197. B. G. Vekhter, Yu. E. Perlin, V. Z. Polinger, Yu. B. Rozenfeld, and B. S. Tsukerblat, *Cryst. Lattice Defects* **3**, 61, 69 (1972).
198. S. Muramatsu and N. Sakamoto, *J. Phys. Soc. Jpn.* **36**, 839 (1974).
199. P. Habitz and W. H. E. Schwarz, *Theor. Chim. Acta* **28**, 267 (1973).
199a. V. Loorits, *Izvestia AN Est. SSR, Fizika i Matematika* **29**, 208 (1980); H. Köppel, E. Haller, L. S. Cederbaum, and W. Domcke, *Mol. Phys.* **41**, 669 (1980).

200. K. Cho, *J. Phys. Soc. Jpn.* **25**, 1372 (1968).
201. R. Englman, E. Caner, and S. Toaff, *J. Phys. Soc. Jpn.* **29**, 306 (1970).
202. I. B. Bersuker, V. P. Khlopin, Yu. E. Perlin, V. Z. Polinger, Yu. B. Rozenfeld, B. S. Tsukerblat, and B. G. Vekhter, in: *Proc. XVI Int. Conf. Coord. Chem., Toronto*, p. 357 (1972).
203. V. P. Khlopin, B. S. Tsukerblat, Yu. B. Rozenfeld, and I. B. Bersuker, *Fiz. Tverd. Tela* **14**, 1060 (1972).
204. F. T. Chau and L. Karlson, *Phys. Ser.* **16**, 248 (1977).
205. R. H. Fetterman and D. B. Fitchen, *Solid State Commun.* **6**, 501 (1968).
205a. E. Haller, L. S. Cederbaum, and W. Domcke, *Mol. Phys.* **41**, 1291 (1980).
206. A. Honma, *Sci. Light (Tokyo)* **18**, 33 (1969).
207. A. Matsushima and A. Fukuda, *Phys. Rev.* **B14**, 3664 (1976).
208. D. R. Pooler and M. C. M. O'Brien, *J. Phys. C* **10**, 3769 (1977).
209. J. Duran, *Semiconductors and Isolators*, **3**, 329 (1978).
210. J. Duran, Y. Merle D'Aubigne, and R. Romestain, *J. Phys. C* **5**, 1115 (1972).
211. K. Nasu and T. Kojima, *Prog. Theor. Phys.* **51**, 26 (1974); *Z. Naturforsch.* **309** (1975).
212. M. J. Shultz and R. Silbey, *J. Chem. Phys.* **65**, 4375 (1976).
213. M. Bacci, B. D. Bhattacharyya, A. Ranfagni, and C. Viliani, *Phys. Lett.* **55A**, 489 (1976).
214. S. Muramatsu and K. Nasu, *Phys. Status Solidi B* **68**, 761 (1975).
215. M. Rueff and E. Sigmund, *Phys. Status Solidi B* **80**, 215 (1977).
216. N. N. Kristoffel, *Tr. Inst. Fiz. Astron., Akad. Nauk. Est. SSR* **12**, 20 (1960).
217. B. G. Vekhter and B. S. Tsukerblat, *Fiz. Tverd. Tela* **10**, 1574 (1968).
218. B. S. Tsukerblat, B. G. Vekhter, I. B. Bersuker, and A. V. Ablov, *Zh. Strukt. Khim.* **11**, 102 (1970).
218a. A. Ranfagni and R. Englman, *J. Lumin.* **18/19**, 353 (1979).
219. C. H. Henry, S. E. Schnatterly, and C. P. Slichter, *Phys. Rev.* **137**, 583 (1965).
220. Yu. E. Perlin and B. S. Tsukerblat, in: *Spektroskopia Kristallov*, p. 61, Nauka, Moscow (1975).
221. Ĺ. M. Kushkculei, Yu. E. Perlin, B. S. Tsukerblat, and G. P. Engelgardt, *JETP* **70**, 2226 (1976).
222. Yu. E. Perlin, B. S. Tsukerblat, and T. Singh Dod, in: *Spektroskopiya Kristallov*, p. 115, Nauka, Leningrad (1978); *Fiz. Tverd. Tela* **19**, 1569 (1977).
222a. J. S. Griffith, *Mol. Phys.* **28**, 18 (1956).
223. I. B. Bersuker, *Proc. 2nd Conf. Coord. Chem.*, p. 24, Smolenice–Bratislava (1969).
224. I. B. Bersuker, in: *Novel in Coordination Chemistry, Section Lectures, XIII ICCC, Cracow-Zakopane, 1970*, p. 9, PWN, Wroclaw (1974).
225. I. B. Bersuker and B. G. Vekhter, *Proc. 2nd Symp. Crystal Chem.*, p. 132, Smolenice–Bratislava (1973).
226. I. B. Bersuker, *Teor. Eksp. Khim.* **5**, 293 (1969).
227. I. B. Bersuker, I. Ya. Ogurtsov, and Yu. V. Shaparev, *Teor. Eksp. Khim.* **9**, 451 (1973).
227a. V. L. Ostrovski, I. Ya. Ogurtsov and I. B. Bersuker, *Mol. Phys.* **48**, 13 (1983); *Opt. Spektrosk.* **53**, 536 (1982); *Khim. Fizika* **5**, 579 (1983).
227b. I. Ya. Ogurtsov, V. L. Ostrovski and I. B. Bersuker, *Optica i Spectrosc.* **54**, 442 (1983).
228. R. N. Dixon, *Mol. Phys.* **9**, 357 (1965).
229. P. C. H. Jordan and H. C. Longuet-Higgins, *Mol. Phys.* **5**, 121 (1962).
230. D. A. Ramsey, *Discuss. Faraday Soc.* **35**, 90 (1963).
231. W. L. Jorgensen and L. Salem, *The Organic Chemists' Book of Orbitals*, Academic Press, New York (1973).
231a. S. D. Peyerimhoff, R. J. Buenker, and J. L. Whitten, *J. Chem. Phys.* **46**, 1707 (1967).
232. G. Herzberg and J. W. C. Jahns, *Proc. R. Soc. London, Ser. A* **295**, 107 (1966).
233. R. N. Porter, R. M. Stevens, and M. Karplus, *J. Chem. Phys.* **49**, 5163 (1968).

234. E. Haselbach, *Chem. Phys. Lett.* **7**, 427 (1970).
235. C. Rowland, *Chem. Phys. Lett.* **9**, 169 (1971).
236. H. Bush, M. B. Robin, N. A. Kuebler, C. Baker, and D. W. Turner, *J. Chem. Phys.* **51**, 52 (1969).
237. J. W. Rabalais, I. Bergmark, L. O. Werme, L. Karlson, and K. Diegbahn, *Phys. Ser.* **3**, 13 (1971).
238. J. Arens and L. C. Allen, *J. Chem. Phys.* **5**, 73 (1970).
239. R. N. Dixon, *Mol. Phys.* **20**, 113 (1971).
240. C. A. Coulson and H. L. Strauss, *Proc. R. Soc. London, Ser. A* **269**, 443 (1962).
241. D. F. Smith, *J. Chem. Phys.* **21**, 609 (1953).
242. H. Basch, J. W. Moskowitz, C. Hollister, and D. Hankin, *J. Chem. Phys.* **55**, 1922 (1971).
243. F. A. Gianturco, C. Quidotti, and U. Lamanna, *J. Chem. Phys.* **57**, 840 (1972).
244. L. S. Bartell, *J. Chem. Phys.*, **46**, 4530 (1967); L. S. Bartell and R. M. Gavin, Jr., *J. Chem. Phys.* **48**, 2466 (1968).
245. K. S. Pitzer and L. S. Bernstein, *J. Chem. Phys.* **63**, 3849 (1975).
246. L. S. Bartell, *J. Chem. Phys.* **73**, 375 (1981).
247. I. B. Bersuker, S. S. Stavrov, and B. G. Vekhter, *Biofizika* **24**, 413 (1979); *VI Vsesoyuznoe Soveshchanie po Fiz. i Mat. Metodam v Koord. Khimii, Tezisi Dokladov*, p. 139, Shtiintsa, Kishinev (1977).
248. I. B. Bersuker and S. S. Stavrov, *Chem. Phys.* **54**, 331 (1981); **69**, 165 (1982).
249. M. Gouterman, in: *The Porphyrins*, (D. Dolphin, ed.) Vol. III, p. 1, Academic Press (1978).
250. A. H. Schaffer, M. Gouterman, and E. R. Davidson, *Theor. Chim. Acta* **30**, 9 (1973).
251. W. R. Scheidt, *Acc. Chem. Res.* **10**, 339 (1977).
252. J. E. Kirner, W. Daw, and W. R. Scheidt, *Inorg. Chem.* **15**, 1685 (1976).
253. M. F. Perutz, *Br. Med. Bull.* **32**, 195 (1976).
254. M. Weissbluth, *Hemoglobin. Cooperativity and Electronic Properties*, Chapman Hall, London (1974).
255. W. A. Eaton, L. K. Hanson, P. J. Stephens, J. C. Sutherland, and J. B. R. Dunn, *J. Am. Chem. Soc.* **100**, 4991 (1978).
256. A. V. Ablov, M. M. Botoshanskii, Yu. A. Simonov, T. I. Malinovskii, A. M. Goldman, and O. A. Bologa, *Dokl. Akad. Nauk. SSSR* **206**, 863 (1972).
257. M. M. Botoshanskii, Yu. A. Simonov, and T. I. Malinovskii, *Izv. Akad. Nauk Mold. SSR* **3**, 39 (1973); M. M. Botoshanskii, Yu. A. Simonov, T. I. Malinovskii, and M. A. Simonov, *Kristallografiya* **20**, 63 (1975).
258. F. A. Cotton and J. G. Norman, *J. Am. Chem. Soc.* **93**, 80 (1971).
259. A. C. Villa, A. G. Manfredotti, and C. Cuastini, *Cryst. Struct. Commun.* **2**, 131 (1973).
260. S. S. Budnikov, A. J. Shkurpello, and Yu. A. Simonov, *VII Vsesoyuznoe Soveshchanie po Fiz. i Mat. Metodam v Koord. Khimii*, p. 101, Tezisi Dokladov, Stiintsa, Kishinev (1980).
261. J. H. Enemark and R. D. Feltham, *Coord. Chem. Rev.* **13**, 339 (1974).
262. S. M. Peng and J. A. Ibers, *J. Am. Chem. Soc.* **98**, 8032 (1976); B. B. Wayland, J. W. Minkiewicz, and M. E. Abd-Elmageed, *J. Am. Chem. Soc.* **86**, 2795 (1974).
263. W. R. Scheidt, R. Hatano, G. A. Rupprecht, and F. L. Piciulo, *Inorg. Chem.* **18**, 292 (1979).
264. G. B. Jameson, G. A. Rodley, W. T. Robinson, R. R. Gagne, C. A. Reed, and J. P. Collman, *Inorg. Chem.* **17**, 850 (1978); J. P. Collman, R. R. Gagne, C. A. Reed, T. R. Ralbert, G. Lang, and W. T. Robinson, *J. Am. Chem. Soc.* **97**, 1427 (1975).
265. A. Schweiger, E. Jorn, and Hs. H. Gunthard, in: *Magnetic Resonance and Related Phenomena*, p. 525, Springer, Berlin (1979).
266. J. K. Cashion and D. K. Herschbach, *J. Chem. Phys.* **41**, 2199 (1964).
267. W. H. Gerber and E. Schumacher, *J. Chem. Phys.* **69**, 1692 (1978).

268. P. C. H. Jordan and H. C. Longuet-Higgins, *Mol. Phys.* **5**, 121 (1962).
269. W. L. Clinton and B. J. Rice, *Chem. Phys.* **29**, 445 (1958).
270. W. L. Clinton and B. J. Rice, *Chem. Phys.* **30**, 542 (1959).
271. C. R. Brundle, M. B. Robin, and H. Basch, *J. Chem. Phys.* **53**, 2196 (1970).
272. B. P. Pullen, T. A. Karlson, W. E. Moddeman, G. K. Schweitser, W. E. Bull, and F. Grimm, *J. Chem. Phys.* **53**, 768 (1970).
273. J. E. Collin and J. Delwiche, *Can. J. Chem.* **45**, 1875 (1967).
274. T. Bergmark, J. W. Rabalais, L. O. Werme, L. Karlson, and K. Siegbahn, in: *Proc. Int. Conf. on Electron Spectroscopy*, p. 413, Amsterdam (1972).
275. J. L. Ragle, I. A. Stenhouse, D. C. Frost, and C. A. McDowell, *J. Chem. Phys.* **53**, 178 (1970).
276. H. Köppel, W. Domcke, L. S. Cederbaum, and W. von Niessen, *J. Chem. Phys.* **69**, 4252 (1978).
277. A. Richards, R. J. Buenker, P. J. Bruma, and S. D. Peyerimhoff, *Mol. Phys.* **33**, 1345 (1977).
278. J. W. Rabalais and Ali Katrib, *Mol. Phys.* **27**, 923, (1974).
279. O. Baker and D. W. Turner, *J. Chem. Soc.* **9D**, 480 (1969).
280. F. Brogli, J. K. Crandall, E. Heilberonner, E. Kloster-Jensen, and S. A. Sojka, *J. Electron Spectrosc. Relat. Phenom.* **2**, 455 (1973).
281. L. J. Leng and G. L. Nyberg, *J. Electron Spectrosc. Relat. Phenom.* **11**, 293 (1977).
282. M. J. S. Dewar, G. J. Funken, and T. B. Jones, *J. Chem. Soc.* **1976**, 764.
283. M. J. S. Dewar and G. J. Gleicher, *J. Am. Chem. Soc.* **87**, 3255 (1965).
284. L. S. Snyder, *J. Chem. Phys.* **33**, 619 (1960).
285. W. T. Borchen and E. R. Davidson, *J. Am. Chem. Soc.* **101**, 3771 (1979).
286. C. A. Coulson and A. Golebiewski, *Mol. Phys.* **5**, 71 (1962).
287. G. R. Liebling and H. M. McConnell, *J. Chem. Phys.* **42**, 3931 (1965).
288. L. Asbrink, E. Lindholm, and O. J. Edgvist, *Chem. Phys. Lett.* **15**, 609 (1970).
289. P. N. Sen, B. Chakrabarti, and S. K. Bose, *Tetrahedron*, **23**, 4177 (1967).
290. W. Schmidt, *J. Electron Spectrosc. Relat. Phenom.* **6**, 163 (1975).
291. L. S. Cederbaum and W. Domcke, *J. Chem. Phys.* **33**, 319 (1978); H. Köppel, L. S. Cederbaum, W. Domcke, and S. S. Shaik, *Angew. Chem.* (submitted).
292. R. E. Koning, H. Landvoort, and P. J. Zandstra, *J. Chem. Phys.* **28**, 343 (1978).
293. L. J. I'Haya, M. Nakayama, and T. Iwabuchi, *Int. J. Quantum Chem.* **5**, 227 (1971).
294. M. Koyanagi, R. J. Lwarich, and L. Goodman, *J. Chem. Phys.* **56**, 3044 (1972).
295. M. Koyanagi and L. Goodman, *Chem. Phys. Lett.* **21**, 1 (1973).
296. N. Kanamaru, M. E. Long, and E. C. Lim, *Chem. Phys. Lett.* **26**, 1 (1974).
297. H. Köppel, *Mol. Phys.* **35**, 1283 (1978).
298. E. V. Chisler, *Fiz. Tverd. Tela* **11**, 1272 (1969).
299. A. Carrington, H. C. Longuet-Higgins, and P. F. Todd, *Mol. Phys.* **9**, 211 (1965).
300. E. Kanezaki, N. Nishi, M. Kinoshita, and R. Niimori, *Chem. Phys. Lett.* **24**, 253 (1974).
301. J. A. Stikeleather, *Chem. Phys. Lett.* **24**, 253 (1974).
302. R. F. W. Bader and Kun Po Huang, *J. Chem. Phys.* **43**, 3760 (1965).
303. A. Forman and L. E. Orgel, *Mol. Phys.* **2**, 362 (1959).
304. A. Avdeef and J. P. Fackler, Jr., *Inorg. Chem.* **14**, 2002 (1975).
304a. B. D. Bhattacharryya, *Phys. Status Solidi B* **43**, 495 (1971).
305. J. M. Ammeter and D. C. Schlosnagle, *Chimia* **27**, 372 (1973); *J. Chem. Phys.* **59**, 4784 (1973).
306. S. Mizuhashi, *J. Theor. Biol.* **66**, 13 (1977).
307. D. H. den Boer, P. C. den Boer, and H. C. Longuet-Higgins, *Mol. Phys.* **5**, 387 (1962).
308. A. Alemenningen, E. Gard, A. Haaland, and J. Brunvoll, *J. Organometal. Chem.* **107**, 273 (1976).
309. D. R. Armstrong, R. Fortune, and P. A. Perkins, *J. Organometal. Chem.* **111**, 197 (1976).

310. C. J. Ballhausen and H. A. Johansen, *Mol. Phys.* **10**, 183 (1966); L. L. Lohr, Jr., *Inorg. Chem.* **6**, 1890 (1967).
311. J. P. Fackler, Jr. and I. D. Chawla, *Inorg. Chem.* **3**, 1130 (1964); D. G. Holah and J. P. Fackler, Jr., *Inorg. Chem.* **4**, 1112, 1721 (1965); **5**, 479 (1965).
311a. M. Sano and H. Yamatera, *Chem. Lett.* **1980**, 1495.
312. Masao Kimura, V. Schomaker, O. W. Smith, and B. Weinstock, *J. Chem. Phys.* **48**, 4001 (1968).
313. L. Astheimer, J. Hauch, H. J. Schenk, and K. Schwochau, *J. Chem. Phys.* **63**, 1988 (1975).
314. D. S. Martin and C. A. Lenhardt, Jr., *Inorg. Chem.* **3**, 1368 (1964).
315. D. J. Stujkens, in: *Proc. 2nd Int. Conf. Raman Spectroscopy* (abstracts), Oxford, England, Sept. 13–17, 1970.
316. R. Dingle, *J. Chem. Phys.* **50**, 545 (1969).
317. C. R. Brundle, H. A. Kuebler, M. B. Robin, and H. Basch, *Inorg. Chem.* **11**, 20, (1972).
318. H. Selig, H. H. Claasen, and J. H. Holloway, in: *Proc. 2nd Int. Conf. Raman Spectroscopy*, Oxford, England, Sept. 13–17, 1970.
319. C. P. Prabhackaran and C. C. Patel, *J. Inorg. Nucl. Chem.* **34**, 2371 (1972).
320. P. Spacu, M. Teodorescu, and C. I. Lepadatu, *Z. Phys. Chem., Frankfurt am Main* **88**, 285, (1974).
321. C. Glidewell, *J. Organometal. Chem.* **102**, 339 (1975).
322. H. Koppel, L. S. Cederbaum, W. Domcke, and W. von Niessen, *Chem. Phys.* **37**, 303 (1979).
323. R. N. Dixon, *Phil. Trans. R. Soc. London, Ser. A* **252**, 165 (1960).
324. T. Paramesiwaran and J. A. Köningstein, *J. Mol. Spectrosc.* **66**, 350 (1977).
325. F. A. Blankenship and R. L. Belford, *J. Chem. Phys.* **36**, 633 (1962).
326. C. D. Flint and P. Greenough, *J. Chem. Soc., Faraday Trans. 2* **68**, 897 (1972).
327. Q. W. Mines and R. K. Thomas, *Proc. R. Soc. London, Ser. A* **336**, 355 (1974).
328. D. G. Schmidling, *J. Mol. Struct.* **24**, 1 (1975).
329. J. Pradilla-Sozzano and J. P. Fackler, Jr., *Inorg. Chem.* **12**, 1182 (1973).
330. N. Rumin, C. Vincent, and D. Walsh, *Phys. Rev. B.* **7**, 1811 (1973).
331. K. Sasaki and Y. Obata, *J. Phys. (Paris), Colloq.* **1**, 739 (1971).
332. H. Pollak, R. Quartier, W. Bruynell, and P. Walter, *J. Phys. (Paris), Colloq.* **6**, 589 (1976).
333. B. E. Linder, E. Bunnenberg, L. Seamans, and A. Moscowitz, *J. Chem. Phys.* **60**, 1943 (1974).
334. G. L. Bottger and C. V. Damsgard, *Spectrochim. Acta, Part A* **28**, 1631 (1972).
335. Masafumi Harada and Ikuij Tsijikawa, *J. Phys. Soc. Jpn.* **40**, 513 (1976); **41**, 1264 (1976).
336. G. S. Shephard and D. A. Thornton, *Helv. Chim. Acta* **54**, 2212 (1971).
337. L. D. Cheung, Nai-Teng Yu, and R. H. Felton, *Chem. Phys. Lett.* **55**, 527 (1978).
338. M. S. Gordon and J. W. Caldwell, *J. Chem. Phys.* **70**, 5503 (1979).
339. E. R. Davidson and W. T. Borchen, *J. Chem. Phys.* **67**, 2191 (1977).
340. H. S. Child and A. C. Roach, *Mol. Phys.* **9**, 281 (1965).
341. P. J. Wojtowitz, *Phys. Rev.* **116**, 32 (1959).
342. J. Kanamori, *J. Appl. Phys.* **31**, 14 (1960).
343. R. J. Elliot, R. T. Harley, W. Hayes, and S. R. P. Smith, *Proc. R. Soc. London, Ser. A* **328**, 217 (1972).
344. H. Thomas, in: *Electron–Phonon Interactions and Phase Transitions* (T. Riste, ed.), p. 245, Plenum, New York (1977).
345. R. L. Melcher, in: *Physical Acoustics, Principles and Methods* (Warren P. Mason and R. N. Thurston, eds.) Academic Press, New York–London, Vol. XII, p. 1 (1976).
346. B. G. Vekhter and M. D. Kaplan, in: *Spektroskopiya Kristallov*, p. 149, Nauka, Leningrad (1978).

347. D. Reinen and C. Friebel, *Struct. Bonding (Berlin)* **37**, 1 (1979).
348. P. J. Becker, M. J. M. Leask, and R. N. Tyte, *J. Phys. C* **5**, 2027 (1972).
349. P. J. Becker, M. J. M. Leask, K. J. Maxwell, and R. N. Tyte, *Phys. Lett. A* **41**, 205 (1972).
350. A. H. Cooke, S. J. Swithenby, and M. R. Wells, *Solid State Commun.* **10**, 265 (1972).
351. B. G. Vekhter and M. D. Kaplan, *Fiz. Tverd. Tela* **15**, 2013 (1973).
352. J. E. Battison, A. Kasten, M. J. M. Leask, J. B. Lowry, and K. J. Maxwell, *J. Phys. C* **9**, 1345 (1976).
353. R. L. Melcher and D. A. Scott, *Phys. Rev. Lett.* **28**, 607 (1972).
354. G. A. Gehring, A. P. Malozemoff, W. Staube, and R. N. Tyte, *J. Phys. Chem. Solids* **33**, 1487 (1972).
354a. B. G. Vekhter and M. D. Kaplan, *JETP* **78**, 1781 (1980).
355. L. E. Orgel, *Trans. Faraday Soc.* **26**, 138 (1958); *J. Chem. Soc.* **1958**, 3815, 4186.
356. R. Englman and B. Halperin, *Phys. Rev. B* **2**, 75 (1970).
357. G. Schroder and H. Thomas, *Z. Phys. B* **25**, 369 (1976).
358. B. Halperin and R. Englman, *Phys. Rev. B* **43**, 1698 (1971).
359. S. Kashida, *J. Phys. Soc. Jpn.* **45**, 414 (1978).
360. B. V. Harrowfield and R. Weber, *Phys. Lett. A* **38**, 27 (1972).
361. B. V. Harrowfield, A. J. Dempster, T. E. Freeman, and J. R. Riebrow, *J. Phys. C* **6**, 2058 (1973).
362. C. Fribel, Z. Anorg. *Allg. Chem.* **417**, 197 (1975).
363. B. V. Harrowfield, *Solid State Commun.* **19**, 983 (1976).
364. D. Mullen, G. Heyer, and D. Reinen, *Solid State Commun.* **17**, 1249 (1975).
365. D. Reinen, *Solid State Commun.* **21**, 137 (1977).
366. M. D. Joesten, S. Takagi, and P. C. Lenhert, *Inorg. Chem.* **16**, 2680 (1977).
367. Y. Noda, M. Mori, and Y. Yamada, *Solid State Commun.* **19**, 1071 (1976); **23**, 247 (1977); *J. Phys. Soc. Jpn.* **45**, 954 (1978).
368. Y. Yamada, in: *Electron–Phonon Interaction and Phase Transitions* (T. Riste, ed.), p. 370, Plenum Press, New York and London (1977).
369. D. W. Clack and D. Reinen, *Solid State Commun*, **34**, 395 (1980).
370. L. J. De Jongh and A. R. Miedema, *Experiments on Simple Magnetic Model Systems*, Taylor & Francis, London (1974).
371. D. I. Khomskii and K. I. Kugel, *Solid State Commun.* **13**, 763 (1973).
372. K. I. Kugel and D. I. Khomskii, *JETP* **79**, 987 (1980).
373. C. Fribel and D. Reinen, *Z. Anorg. Allg. Chem.* **407**, 193 (1974).
374. R. Halgele and D. Babel, *Z. Anorg. Allg. Chem.* **409**, 11 (1974).
375. D. Reinen, P. Kohler, and W. Massa, in: *Proc. XIX ICCC*, Vol. 1, p. 153, Prague, September (1978).
376. P. Köhler, W. Massa, D. Reinen, B. Hoffman, and R. Hoppe, *Z. Anorg. Allg. Chem.* **446**, 131 (1978).
377. D. Reinen and S. Krause, *Solid State Commun.* **29**, 691 (1979).
378. C. J. O'Connor, E. Sinn, and R. L. Carlin, *Inorg. Chem.* **16**, 3314 (1977).
379. W. J. Crama, W. J. A. Maaskant, and G. C. Verschoor, *Acta Crystallogr., Sect. B* **34**, 1973 (1978).
380. M. J. Fair, A. K. Gregson, P. Day, and M. T. Hutchings, in: *Proc. Int. Conf. on Magnetism 76*, Part 2, p. 657 (1977).
381. H. A. Goodwin, *Coord. Chem. Rev.* **18**, 293 (1976).
382. A. V. Ablov, I. B. Bersuker, B. G. Vekhter, and B. S. Tsukerblat, in: *Fiz. i Mat. Metody v Koord. Khimii*, p. 151, Shitiintsa, Kishinev (1974).
383. T. Kambara, *J. Chem. Phys.* **70**, 4199 (1979).
384. G. A. Smolenskii, V. A. Bokov, V. A. Isupov, N. N. Krainik, R. E. Pasinkov, and M. S. Shur, *Segnetoelektriky i Antisegnetoelektriky*, Nauka, Leningrad (1971).

385. I. B. Bersuker and B. G. Vekhter, *Fiz. Tverd. Tela* **9**, 2652 (1967); *Izv. Akad. Nauk SSSR, Ser. Fiz.* **33**, 199 (1969).
386. I. B. Bersuker, B. G. Vekhter, and A. A. Muzalevskii, *J. Phys. C* **33**, 139 (1972); *Phys. Status Solidi B* **45**, K25 (1971); *Ferroelectrics* **6**, 197 (1974).
387. N. Kristoffel and P. Konsin, *Phys. Status Solidi B* **21**, K39 (1967); N. N. Kristoffel, in: *Titanat Bariya*, p. 11, Nauka, Moscow (1973).
388. Ya. G. Girshberg and V. I. Tamarchenko, *Fiz. Tverd. Tela* **18**, 1066, 3340 (1976).
389. R. Comes, M. Lambert, and A. Guinier, *Solid State Commun.* **6**, 715 (1968); M. Lambert and R. Comes, *Phys. Lett.* **7**, 305 (1969); M. P. Fontana and M. Lambert, *Phys. Lett.* **10**, 1 (1972); R. Comes, R. Currat, F. Denoyer, M. Lambert, and A. M. Quittet, *Ferroelectrics* **12**, 3 (1976).
390. I. B. Bersuker, B. G. Vekhter, and V. P. Zenchenko, *III EMF*, Abstracts, Section K., p. K3; Zurich, Switzerland (1975); *Ferroelectrics* **12**, 373 (1976).
391. B. G. Vekhter, V. P. Zenchenko, and I. B. Bersuker, *Fiz. Tverd. Tela* **18**, 2325 (1976); *Ferroelectrics* **20**, 163 (1978).
392. S. Takaoka and K. Murase, *Phys. Rev. B* **20**, 2823 (1979).
393. Ismailzade I. H., et al., *Ferroelectrics* (to be published).
394. J. Gazo, *Pure Appl. Chem.* **38**, 279 (1974).
395. J. Gazo, I. B. Bersuker, J. Garaj, M. Kabesova, J. Kohout, H. Langfelderova, M. Melnik, M. Serator, and F. Valach, *Coord. Chem. Rev.* **19**, 253 (1976).
396. J. P. Fackler, Jr. and A. Avdeev, *Inorg. Chem.* **13**, 1964 (1974).
397. F. Hanic and J. Michalov, *Acta Crystallogr.* **13**, 299 (1960).
398. C. K. Prout, R. A. Armstrong, J. R. Carruthers, J. C. Forrest, P. Murray-Rust, and J. C. F. Rossotti, *J. Chem. Soc. A* **1968**, 2791.
399. L. H. Dalh, cited in T. S. Piper and R. L. Belford, *Mol. Phys.* **5**, 169 (1962).
400. D. Hall, A. J. McKinnon, and T. N. Waters, *J. Chem. Soc.* **1965**, 425.
401. M. Bonamico, G. Dessy, V. Fares, and L. Scaramuzza, *J. Chem. Soc., Dalton Trans.* **1972**, 2477.
402. P. C. Healy and A. H. White, *J. Chem. Soc., Dalton Trans.* **1972**, 1913.
403. H. Koizumi, H. Osati, and T. Watanabe, *J. Phys. Soc. Jpn.* **18**, 117, (1963).
404. M. Laight, J. Tordiman, J. C. Gnitel, and G. Boss, *Acta Crystallogr., Sect. B* **28**, 2721 (1972).
405. C. K. Prout, J. R. Carruthers, and J. C. F. Rossotti, *J. Chem. Soc. A* **1971**, 3336.
406. M. S. Hussain, M. D. Joesten, and L. P. G. Lenhert, *Inorg. Chem.* **9**, 162 (1970).
407. C. K. Prout, J. R. Carruthers, and F. J. C. Rossotti, *J. Chem. Soc. A* **1971**, 3350.
408. H. Jaggi, and H. R. Oswald, *Acta Crystallogr.* **14**, 1041 (1961).
409. M. Martinez-Ripoll and S. Martinez-Carrera, *Acta Crystallogr., Sect. B* **27**, 677 (1970).
410. B. Rama Rao, *Acta Crystallogr.* **14**, 321 (1961).
411. M. Langt, J. C. Gnitel, J. Tordiman, and C. Bassi, *Acta Crystallogr., Sect. B* **28**, 201 (1972).
412. S. Siegel and H. R. Hockstra, *Acta Crystallogr. Sect. B* **24**, 967 (1968).
413. M. B. Gingi, C. Cuastini, A. Mussatti, and M. Nardelli, *Acta Crystallogr., Sect. B* **25**, 1833 (1969).
414. C. K. Prout, J. C. Carruthers, and F. J. C. Rossotti, *J. Chem. Soc. A* **1971**, 3342.
415. S. Asbrink and L. J. Norrby, *Acta Crystallogr., Sect. B* **26**, 8 (1970).
416. R. F. Zahrobsky and W. H. Baur, *Acta Crystallogr., Sect. B* **24**, 508 (1968).
417. C. Sabelli and D. F. Zanazzi, *Acta Crystallogr., Sect. B* **24**, 1214 (1968).
418. H. G. Bachman and J. Zemann, *Acta Crystallogr.* **14**, 747 (1971).
419. G. E. Bacon N. A. Curry, *Proc. R. Soc. London, Ser. A* **266**, 95 (1962).
420. B. Rama Rao, *Acta Crystallogr.* **14**, 738 (1961).
421. L. Flügel-Kahler, *Acta Crystallogr.* **16**, 1009 (1963).

422. B. E. Roberson and C. Calvo, *Acta Crystallogr.* **22**, 665 (1967).
423. R. Süsse, *Acta Crystallogr.* **22**, 146 (1967).
424. H. C. Heide and K. Boll-Dornberger, *Acta Crystallogr.* **8**, 425 (1955).
425. C. K. Prout, J. R. Carruthers, and F. J. C. Rossotti, *J. Chem. Soc. A* **1971**, 554.
426. L. Kihlborg and E. Gebert, *Acta Crystallogr., Sect. B* **26**, 1020 (1970).
427. S. Ghose, *Acta Crystallogr.* **15**, 1105 (1962).
428. S. Ghose, *Acta Crystallogr.* **16**, 124 (1963).
429. M. D. Glick, G. L. Downs, and L. F. Dahl, *Inorg. Chem.* **3**, 1712 (1964).
430. I. Hjerten and B. Nyberg, *Acta Chem. Scand.* **27**, 345 (1973).
431. R. V. G. Sundara Rau, K. Sundaramma, and R. Sirasankara, *Z. Kristallogr.* **110**, 231 (1958).
432. M. J. Kay, I. Almodovar, and S. F. Kaplan, *Acta Crystallogr., Sect. B* **24**, 1312 (1968).
433. B.-M. Anthl, B. K. S. Lundberg, and N. Ingri, *Acta Chem. Scand.* **26**, 3984 (1972).
434. D. A. Langs and C. R. Hare, *Chem. Commun.* **1967**, 853.
435. D. E. Fenton, M. R. Truter, and B. L. Vickery, *Chem. Commun.* **1971**, 93.
436. H. R. Oswald, *Helv. Chim. Acta* **52**, 2369 (1969).
437. J. A. Bertland and D. A. Carpenter, *Inorg. Chem.* **5**, 514 (1966).
438. H. Heritsch, *Z. Kristallogr.* **99**, 466 (1938).
439. W. Nowacki and R. Scheidegger, *Helv. Chim. Acta* **35**, 375 (1952).
440. K. Brandt, *Ark. Kemi. Min. Geol. A* **17**, 13 (1948).
441. M. A. Kisvomitre, *J. Chem. Phys.* **37**, 1408 (1962).
442. M. D. Joesten, M. S. Hussain, P. G. Lenhert, and J. H. Venables, *J. Am. Chem. Soc.* **90**, 1623 (1968).
443. B. J. Temple, unpublished results (cited in Ref. [395]).
444. P. T. Miller, P. G. Lenhert, and M. D. Joesten, *Inorg. Chem.* **12**, 218 (1973).
445. H. Heritsch, *Z. Kristallogr.* **102**, 1 (1939).
446. B. Kamenar, *Acta Crystallogr., Sect. B* **25**, 800 (1969).
447. P. T. Miller, P. G. Lenhert, and M. D. Joesten, *Inorg. Chem.* **11**, 2221 (1972).
448. N. V. Mani and S. Ramaseshan, *Z. Kristallogr.* **115**, 97 (1961).
449. F. Hanic, *Czech. J. Phys. B* **10**, 169 (1960).
450. G. B. Brown and R. Chidambaran, *Acta Crystallogr. Sect.* B **25**, 676 (1969).
451. N. V. Mani and S. Rasasetshak, *Z. Kristallogr., Kristallgeom., Kristallphys., Kristallchem.* **115**, 97 (1961).
452. C. J. Brown, *J. Chem. Soc. A* **1968**, 2488.
453. M. Matthew and N. R. Kunchur, *Acta Crystallogr., Sect. B* **26**, 2054 (1970).
454. C. D. Stout, M. Sundaralingam, and G. Hung-Yin Lin, *Acta Crystallogr. Sect. B* **28**, 2136 (1972).
455. M. Bukovska and M. A. Porai-Koshits, *Zh. Strukt. Khim.* **7**, 712 (1961).
456. I. Agrell, *Acta Chem. Scand.* **20**, 1281 (1966).
457. B. J. Hathaway and F. S. Stephens, *J. Chem. Soc. A* **1970**, 285.
458. O. P. Anderson, *J. Chem. Soc., Dalton Trans.* **1973**, 1237.
459. T. Okamoto, K. Matsumoto, and H. Kuroya, *Bull. Chem. Soc. Jpn.* **43**, 1915 (1970).
460. J. Korvenranta and A. Pajunen, *Suomen Kemistilenti B* **43**, 119 (1979).
461. J. Scoulondi, *Acta Crystallogr.* **6**, 541 (1953).
462. F. S. Stephensen, *J. Chem. Soc. A* **1969**, 2233.
463. N. W. Isaacs and C. H. L. Kennard, *J. Chem. Soc. A* **1969**, 386.
464. D. L. Gullen and E. G. Lingafelter, *Inorg. Chem.* **9**, 386 (1970).
465. F. S. Stephens, *J. Chem. Soc. A* **1969**, 883.
466. H. G. von Schnering, *Z. Anorg. Allg. Chem.* **353**, 13 (1967).
467. D. Babel, *Z. Anorg. Allg. Chem.* **336**, 200 (1965).
468. C. Billy and H. M. Haendler, *J. Am. Chem. Soc.* **79**, 1049 (1957).

469. K. Knox, *Acta Crystallogr.* **30**, 991 (1959).
470. A. F. Wells, *Structural Inorganic Chemistry*, 3rd ed., p. 877, Oxford University Press, Glasgow (1962).
471. N. C. Stephensen and D. P. Mellor, *Aust. J. Sci. Res., Ser. A* **3**, 581 (1950).
472. L. Helmholz, *J. Am. Chem. Soc.* **69**, 886 (1947); A. W. Schneter, R. A. Jacobson, and R. E. Rundle, *Inorg. Chem.* **5**, 277 (1966); A. F. Wells, *J. Chem. Soc.* **1947**, 1662.
473. A. F. Wells, *J. Chem. Soc.* **1947**, 1670.
474. A. J. Edwards, *J. Chem. Soc. A* **1971**, 2653.
475. D. R. Sears and J. L. Hoard, *J. Chem. Phys.* **50**, 1066 (1969).
476. A. Okazaki and Y. Suemune, *J. Phys. Soc.* **16**, 176 (1961).
477. M. A. Hepworth and K. H. Jack, *Acta Crystallogr., Sect. B* **28**, 201 (1972).
478. S. Schneider and R. Hoppe, *Z. Anorg. Allg. Chem.* **376**, 268 (1970).
479. J. P. Fackler, Jr., A. Avdeef, and J. Costamagna, in: *Proc. XIV Int. Conf. Coord. Chem., Toronto*, p. 589 (1972).
480. A. Avdeef, J. A. Costamagna, and J. P. Fackler, Jr., *Inorg. Chem.* **13**, 1854 (1974).
481. K. H. Jack and R. Maitland, *Proc. Chem. Soc. London* **1957**, 232.
482. J. W. Tracy, N. W. Gregory, E. C. Lingafelter, J. D. Dunitz, H.-C. Mez, R. E. Rundle, C. Scheringer, H. L. Yakel, Jr., and M. K. Wilkinson, *Acta Crystallogr.* **14**, 927 (1961).
483. J. W. Tracy, N. W. Gregory, J. M. Stewart, and E. C. Lingafelter, *Acta Crystallogr.* **15**, 460 (1962).
484. I. Bertini, P. Dapporto, D. Gatteschi, and A. Scozzafava, *Proc. XIX ICCC, Prague*, Vol. III, p. 14, September (1978).
485. R. Allmann, W. Henke, and D. Reinen, *Inorg. Chem.* **17**, 378 (1978).
486. H. Yamatera, *Acta Chem. Scand., Ser. A* **33**, 107 (1979).
487. J. P. Fackler, Jr., and H. W. Chen, *Proc. XIX ICCC, Prague*, Vol. I, p. 148, September (1978).
488. J. H. Ammeter, *Nouveau J. Chem.* **4**, 631 (1980).
489. J. H. Ammeter, *Nouveau J. Chem.* **4**, 565 (1980).
490. A. A. Levin, A. P. Kliagina, and S. P. Dolin, *Zh. Neorg. Khimii* **24**, 2307 (1979); A. A. Levin and A. P. Kliagina, *Dokl. Akad. Nauk SSSR* **250**, 638 (1980); P. N. Diachkov, *Dokl. Akad. Nauk SSSR* **254**, 920, 1421 (1980).
491. M. E. Diatkina and M. A. Porai-Koshits, *Dokl. Akad. Nauk SSSR* **125**, 1030 (1959).
492. I. B. Bersuker, *Zh. Fiz. Khim.* **35**, 471 (1961).
493. I. B. Bersuker, *Zh. Strukt. Khim.* **16**, 935 (1975).
494. J. Gazo, K. Seratorova, and M. Serator, *Khem. Zvesti* **13**, 5 (1959).
495. B. Papankova, M. Serator, J. Stracelsky, and J. Gazo, in: *Proc. 8th Conf. Coord. Chem.*, Smolenice-Bratislava, p. 321 (1980).
496. T. Obert and I. B. Bersuker, in: *Proc. XIX ICCC, Prague*, Vol. II, p. 94 (1978).
497. I. B. Bersuker, in: *I Vsesoyuzone Soveshchanie po Neorg. Kristallokhimii*, Tezisi Dokladov, p. 4, Zvenigorod (1977).
498. I. B. Bersuker, *Phase Transitions* **2**, 53 (1981).
499. H. Eyring, *J. Chem. Phys.* **3**, 107 (1935); E. Wigner, *Trans. Faraday Soc.* **34**, 29 (1938); H. Pelzer and E. Wigner, *Z. Phys. Chem. B* **15**, 445 (1932).
500. H. Eyring, J. Walter, and G. E. Kimboll, *Quantum Chemistry*, Wiley, New York (1944).
501. E. E. Nikitin, *Teoriya Elementarnikh Protsesov v Gazakh*, Khimiya, Moscow (1970).
502. M. J. S. Dewar, *J. Am. Chem. Soc.* **74**, 3341, 3357 (1952); *The Molecular Orbital Theory of Organic Chemistry*, McGraw-Hill, New York (1969).
503. H. Fujimoto and K. Fukui, in: *Chemical Reactivity and Reaction Paths* (G. Clopman, ed.), Ch. III, Wiley, New York (1974).
504. G. Clopman, in: *Chemical Reactivity and Reaction Paths* (G. Clopman, ed.), Chs. I and IV, Wiley, New York (1974).

505. M. Simonetta, in: *Chemical Reactivity and Reaction Paths* (G. Clopman, ed.), Ch. II, Wiley New York (1974).
506. R. F. W. Bader, *Mol. Phys.* **3**, 137 (1960); *Can J. Chem.* **40**, 1164 (1962).
507. A. V. Eletskii, *Usp. Fiz. Nauk* **125**, 279 (1979).
508. T. F. George and J. Ross, *J. Chem. Phys.* **55**, 3851 (1971).
509. F. D. Mango and J. Schachtschneider, *J. Am. Chem. Soc.* **89**, 2484 (1967); F. D. Mango, *Adv. Catal.* **20**, 291 (1969).
510. R. Ugo, in: *Catalysis, Proc. 5th Int. Congress, Boston, 1972* (J. W. Hightower, ed.), p. 19, North-Holland, Amsterdam (1973).
510a. R. Rericha, *Collect. Czech. Chem. Commun.* **40**, 2577 (1975); **42**, 3530 (1977).
511. L. Pauling, *J. Chem. Soc.* **1948**, 1461; in: *Proc. Symp. Coord. Chem.*, Copenhagen, p. 25 (1953).
512. Yu. Ya. Kharitonov and O. V. Bazileva, *Zh. Neorg. Khim.* **23**, 867 (1978).
512a. A. P. Svitin, S. S. Budnikov, I. B. Bersuker and D. V. Korol'kov, *Teor. i Exp. Chim.* **18**, 694 (1982).
513. M. A. Christiansen and E. A. McCullough, *J. Chem. Phys.* **67**, 1877 (1977).
514. K. S. Krasnov, V. S. Timoshinin, T. G. Danilova, and S. V. Mandozhko, *Molekularnie Konstanty Neorganicheskikh Soedinenii*, Khimiya, Leningrad (1968).
515. A. A. Radsig and B. M. Smirnov, *Spravochnik po Atomnoi i Molekularnoi Fizike*, Atomizdat, Moscow (1980).
516. V. I. Yakerson, L. I. Lafer, and A. M. Rubinshtein, in: *Problemy Kinetiky i Kataliza*, Vol. 16, p. 49, Nauka, Moscow (1975).
516a. I. B. Bersuker, S. S. Budnikov, and A. S. Dimoglo, *Kinetika i Kataliz* **23**, 1454 (1982).
517. V. N. Kondratiev, *Konstanty Skorostei Gazofaznikh Reaktsii*, Nauka, Moscow (1970).
518. *Elektronnie Yavlenya v Adsorptsii i Katalize* (*Sbornik*), p. 247, Nauka, Moscow (1968).
519. *Termodinamicheskie Konstanty Individualnikh Veshchestv, Spravochnik*, p. 352, Izd. AN SSSR, Moscow (1962).
520. S. G. Gagarin, *Koord. Khim.* **6**, 215 (1980).
521. D. Sellman, *Angew. Chem.* **86**, 692 (1974).
522. R. F. Fenske and K. L. LeCock, *Inorg. Chem.* **11**, 437 (1972).
523. I. B. Bersuker, *Teor. Eksp. Khim.* **1**, 5 (1965).
524. G. Cauley, D. Cross, and P. Knowles, *J. Chem. Soc., Chem. Commun.* **20**, 837 (1976).
525. M. J. S. Dewar, S. Kirschner, H. W. Kollmar, and L. E. Wade, *J. Am. Chem. Soc.* **96**, 5242 (1974).
525a. M. S. Gordon and J. W. Caldwell, *J. Chem. Phys.* **70**, 5503 (1979).
526. M. D. Glick, J. M. Kusraj, and J. F. Endicott, *J. Am. Chem. Soc.* **95**, 5097 (1973).
527. B. G. Vekhter and M. L. Rafalovich, *Chem. Phys.* **21**, 21 (1977).
528. D. B. Moss, Lin Chin-tung and D. B. Rorabacher, *J. Amer. Chem. Soc.* **95**, 5179 (1973).

Author Index

The figures given in brackets are the numbers of the full references in the Reference section (pp. 291–305).

Subject Index

www.ingramcontent.com/pod-product-compliance
Ingram Content Group UK Ltd.
Pitfield, Milton Keynes, MK11 3LW, UK
UKHW041859190726
13854UKWH00002B/974